T0190229

Advances in
Chromatography
Volume 57

Advances in Chromatography Volume 57

Edited by
Nelu Grinberg and Peter W. Carr

CRC Press
Taylor & Francis Group
Boca Raton London New York

CRC Press is an imprint of the
Taylor & Francis Group, an **informa** business

ISSN 0065-2415

1st edition published 2021
by CRC Press
6000 Broken Sound Parkway NW, Suite 300, Boca Raton, FL 33487-2742

and by CRC Press
2 Park Square, Milton Park, Abingdon, Oxon, OX14 4RN

First issued in paperback 2022

Visit the Taylor & Francis Web site at
http://www.taylorandfrancis.com

and the CRC Press Web site at
http://www.crcpress.com

ISBN: 978-1-03-239994-2 (pbk)
ISBN: 978-0-367-45612-2 (hbk)
ISBN: 978-1-003-02438-5 (ebk)

DOI: 10.1201/9781003024385

Typeset in Times
by Deanta Global Publishing Services, Chennai, India

Contents

Preface

Lloyd Snyder – A Man for Eternity
(1931–2018)

Lloyd Snyder passed away on September 19, 2018. It is exceedingly difficult to write about a man who contributed so much to the field of liquid chromatography and helped revolutionize separation science through his incisive research, magisterial books, insightful articles, and superb short courses.

Lloyd Snyder was born in 1931 in Sacramento, California. As is mentioned in his autobiography, he felt a great attraction to chemistry from an early age. His fascination with the way atoms combine to form complex molecules led him to enter the University of California at Berkley as a chemistry major. His strong chemistry background, obtained during his undergraduate and graduate studies, both at UCal Berkeley, instilled in him a firm foundation for his deep understanding of chromatographic phenomena. From the separation of small molecules to the chromatography of biomacromolecules, Lloyd Snyder covered virtually every aspect of chromatographic science in his papers and books. The numerous awards he received over the years testify to the quality and worldwide impact of his scientific contributions.

Through his numerous and pioneering short courses on liquid chromatography, he left a huge pedagogical footprint on the whole field for several generations of chromatographers and the literally thousands of students who participated in his courses.

This volume of the *Advances in Chromatography* series is an homage for what he represented to the chromatographic community. Many scientists have contributed their memories of time spent together with Dr. Snyder during international and national meetings and throughout their careers. We are thankful to all the scientists who have collaborated on this volume dedicated to one of the pillars of separation science. For the chromatographic community, Lloyd Snyder's passing represents a big loss which is difficult, if not impossible, to fill. He will remain for us "A Man for Eternity."

Nelu Grinberg, PhD, Co-Editor
Peter W. Carr, PhD, Co-Editor
Advances in Chromatography Volume 57

In Memoriam

Lloyd Snyder (1931–2018)

Lloyd and I were children of the Great Depression, both of our fathers having been unemployed for some time. We also were the oldest children in our families, so perhaps we grew up with a serious attitude about making our way in the world.

Though his family moved many times, Lloyd was very fortunate to live in Sacramento where the schools and city offered many opportunities. Nowadays we would think he was given a great deal of freedom without fear for his safety.

When he was about 13 he built his first home laboratory in the basement, using the money he made making deliveries for a pharmacy. In those days many chemicals could be ordered by mail. One project was to make nitroglycerine, using directions from an 1898 encyclopedia found in his grandmother's house. One day he dropped a container of a mixture of nitroglycerine and acid; it suddenly changed color from clear to dark brown. He quickly turned and was running from the lab bench when it exploded, spraying acid and broken glass everywhere. Several test tubes were shattered, but he was not injured. Since he did not report this to his parents, his mother was surprised to find his shirt with only the neck and sleeves left. There were several similar adventurous experiences.

After graduating from high school with a "scholarship" of $50, he was able to enroll at the University of California (Berkeley) by living for two years with a second cousin in Oakland. Recruited for the crew team, he took part in a race with the University of Washington, which UCal-Berkeley won. About that time, he was invited to a chemistry fraternity, Alpha Chi Sigma; he then lived at the fraternity house for the next one and a half years, serving as housemaster (free room and board) for a year.

The fraternity and my sorority often had "exchanges," and we met on a blind date in our sophomore year. During the summer, after working at summer jobs all day, he would drive from Sacramento to Santa Rosa where I lived and spend the evening dancing at the Russian River resorts, often nearly falling asleep or pulling over for a nap on the way home.

By the end of our junior year we both had met all the requirements to graduate in one more semester at which time, we planned to get married. (He was just 20 so his mother had to give permission!) We were married in the Santa Rosa Presbyterian Church and spent our honeymoon settling into an apartment. He then entered the chemistry doctoral program, and I had one semester of grad school/student teaching. After that I taught fourth grade in Berkeley for a year until our first child, Julie, was born in September 1953.

In his autobiography, Lloyd wrote: "My typical day started in the lab around 8:00 in the morning, maybe broken by a pick-up basketball game with other grad students or a swim at the gym, then home for dinner and back to the lab until 10:00 at night. My intense concentration on my work drew occasional comments from those around me. I do remember several times feeling hungry about 3:00 in the afternoon, then looked around to see that my lunch bag was still unopened. Since I usually got up before Barbara and skipped breakfast, this meant it had been about 20 hours since my last meal!"

I remember typing Lloyd's dissertation – several layers with carbon paper on a manual typewriter – something about the kinetics of methyl ethyl ketone and benzaldehyde. In the Spring of 1954, his doctoral dissertation was approved, and he then became a research chemist at Shell Oil in Houston (Texas), where we got used to humidity, cockroaches, and cute little lizards that inhabited the breezeway.

Our son Tom was born the next year and, though we liked the people and Lloyd liked his job, he accepted an offer at Technicolor in Hollywood and we moved back to California. Shortly after we bought our first home in Van Nuys, but when the company was unable to keep the salary agreement, Lloyd found his next job at Union Oil in Brea (California). That meant a long commute until we finally were able to sell the house and move to Fullerton. Our son James was born about a week after we moved, and David was born three years later. We lived amid orange orchards; much of the time you could smell the orange blossoms.

Those were busy years with all the kids' activities. Every summer we went tent camping in California National Parks and as far off as Yellowstone. Camping was all new to me. We bought a used wood "camp kitchen" (a big cupboard) that fit in our first station wagon; in those days it was bear-proof. One time while camping near a beach somewhere along the Pacific Coast the fog rolled in very quickly and our son Jim, then about nine, could not be found. I went to get a ranger, but by the time we returned to the campsite Jim was back with Lloyd. We tried camping after moving to New York but got rained out every time. We also visited Mexico. I was a Cub Scout leader, Sunday School teacher, etc. Lloyd began reading the philosophers and great literature he had never been able to study in the College of Chemistry program at Berkeley, and we participated in the Great Books program, school-related projects, and duplicate bridge (Lloyd also played during lunch at work). We traveled to Houston for his first award (in petroleum chemistry).

When Dave was in kindergarten, I started grad school. The new Cal State-Fullerton campus was within walking distance from our home. Those were interesting times politically. One day classes at the college were closed because of a raid by the Students for a Democratic Society (SDS), which had just smashed a grand piano! I started taking one class at a time, then two, and after five and a half years (66 units) was credentialed as a school psychologist. Many of my classes were late-day (for working teachers) so, though I prepared the dinner, Julie served it when Lloyd got home and then cleaned up.

Lloyd's portable home office was the dining room table. Among other things, he wrote papers there in the evening and then cleared it for breakfast next day. He had a small desk in the living room and an electric typewriter. His powers of concentration were an important asset at this time.

When the work at Union Oil seemed to approach its limit, he began exploring other jobs and even interviewed for some university jobs. He later said it would have been unfortunate if they had been offered to him. Moving to New York was a huge step for us all when Lloyd started working at Technicon in Tarrytown (New York). My California credentials were useless in New York, and the kids were all in new schools. It was not easy to adapt to the weather, and I had trouble just finding the additional classes New York required. Neighbors (who often came from New York City) seemed suspicious of people from California. Gradually we found our way and became active in a small historic church. My first job in New York was at Wiltwyck School for Boys. When working with a student, I had to lock the door to prevent intruders. (Eventually I got better jobs – even presented a research project at a national convention.)

I remember there were ups and downs at Technicon – with the occasional possible loss of the job. However, Lloyd and I became friends with people from church and with many of his colleagues, some international, and with some of their families on the east coast, including his future business partner, John Dolan. During those busy and productive New York years we enjoyed visiting historic places on the east coast not far from where we lived. I went along with Lloyd to many of the scientific meetings in Europe, and some of his colleagues visited us in our home. We also traveled during vacations with the kids, and later by ourselves, to the Caribbean, Mexico, and Canada.

Lloyd and John started LC Resources as a consulting business in 1982 with John soon moving to his hometown in Oregon where they started a lab. In 1984 Lloyd was invited to the University of Lanzhou, China, where he lectured with a translator for three weeks. I joined him in Beijing for three more weeks in the big cities where we were treated very respectfully by scientists and others. We saw few cars there in those days.

We moved to Orinda, California in 1985 and Tom Jupille joined the growing company. Lloyd's workplace was two adjacent bedrooms with the dividing wall removed. It was hard to convince people that a chemist could work out of his home without violating the zoning rules. I worked for a big school district where I met with a large variety of students and was president of my psychologists' association for a time. I was now able to attend most of the big chemistry meetings in Europe. We visited Poland three times with help from Edward Soczwinski, and I eventually found my maternal grandparents' roots. I retired in 1998 so I could spend more time using the Internet to explore both of our family histories, ending up with six volumes on the various lines. (Years later I escorted my sister and cousins to meet living relatives.)

Lloyd and I moved to a retirement community of 10,000 people in Walnut Creek (California). Lloyd was still active in theoretical collaborations with many working chemists until his health declined.

The amazing part of Lloyd's life and long career is that he also accomplished so much outside his career, reflecting his high energy, idealism, altruistic values, and enthusiasm for whatever he might undertake. Also remarkable were his humility – based on his consideration of all colleagues as equally worthy human beings – and that he had no need for recognition or material possessions beyond those of a modestly comfortable life.

Despite missing liberal arts classes in college, Lloyd was a voracious and fast reader. He made good use of the public libraries; in recent years he kept a reading list with notes on books he had read. He read to our children every night for as long as they were willing and helped them with their hobbies, science projects, and other enterprises. He liked to take them out in the country to explore wildlife and any interesting artifacts they might find. We had many different pets such as tarantulas, white rats, and an alligator.

He also offered his free time to a number of worthy causes. He was a scoutmaster in the early days and later became a Stephen Minister through our church. His role here was, after extensive training, to be an active listener for one man during difficult times of life, with ongoing supervision and continuing education. Though allowed only one Care Receiver a time, through the years a number of men benefited from his care. Once during a visit, he came home to get a mop and cleaning supplies and then returned to the person's home. (This was not part of the job.)

For about ten years he volunteered once a week at Bay Area Rescue Mission in Richmond, teaching basic skills for men in recovery in hopes of their qualifying for a GED certificate (high school equivalency).

Later he trained as a counselor for Senior Peer Counseling, a county program. Here you were allowed as many clients as you wanted – people in need of guidance in making decisions or finding helpful community resources. He did this until his last year. Once he helped a lady clear out her house which was impassable due to hoarding.

And, as others have pointed out, Lloyd was most generous professionally. He appreciated all his peers and became close friends with many. Because his only goal was the search for truth, he shared material and never let personal obstacles stand in the way. Aware of the depth of his theoretical accomplishments, he had no need for praise or rewards – the work was its own reward.

Barbara Snyder

REMEMBRANCES OF LLOYD SNYDER

It is a great pleasure to provide remembrances of my friend and colleague, Lloyd Snyder, who passed away just one year ago. In fact, I had planned to have lunch with him in September, 2018, as I was in San Francisco to speak at the twentieth anniversary of the CASSS-CE Pharm meeting. Lloyd, unfortunately, was in the hospital and passed away one week later.

I first connected with Lloyd in the late 1960s when he was active in modeling chromatographic retention in liquid–solid chromatography for the separation of hydrocarbons, a topic of relevance to his employer, Union Oil in Brea, CA. I remember that in 1970 we both went to a gas chromatography conference (HPLC was in the early stages) in Dublin, Ireland. Prior to that conference, we visited Prof. Istvan Halasz in his home, at that time in Frankfurt, Germany. We had extensive discussions on where LC was headed and how to make it practical. Csaba Horvath, whom I had first met in 1964 when he was a post-doc at Harvard Medical, was also a good friend at Yale University. It is interesting to note that he received his Ph.D. from Prof. Halasz.

In 1971, Lloyd, Csaba, and I decided there was a real need for an up-to-date book on separation science. We wrote many of the chapters ourselves and were fortunate to secure others to write in their areas of expertise. Lloyd always finished his work first, and Csaba and I struggled to keep up. Nevertheless, the book was published in 1973 (1). For the next 25 years, it became the standard graduate text for separation science and was widely cited in many research articles. I know Lloyd was very proud of the book.

The collaboration and friendship of Lloyd, Csaba and myself remained for all the years during which HPLC grew to become the major separation tool it is today. One story which the chromatographic community may not know deals with how the word "isocratic" became widely used in the field to mean constant mobile phase composition throughout the separation, in contrast to gradient elution. In the early 1970s, the three of us were driving together back to Boston after a Gordon Conference, and Csaba suggested we adopt the isocratic term. We didn't announce it, but simply began using it in our talks at scientific conferences. Of course, others picked it up.

Some misspelled the word, e.g. isochratic, but the term became rapidly accepted by the general community.

Both Lloyd and I were very interested in the fundamentals of mobile phase selection in LC. We collaborated in the 1970s on classifying mobile phases in terms of the Hildebrand solubility parameters, based on various interaction parameters, e.g. non-polar, dipole, hydrogen bond, etc. (2). Through this classification scheme, one could estimate the effect of mobile phase mixtures on retention and selectivity. While more sophisticated approaches became available at a later time, our method was a significant step forward for LC in the early stages of its development.

Lloyd and I stayed close when he moved in the late 1970s from Union Oil (CA) to Technicon (NY), a clinical analysis instrumentation firm. In his early days at Technicon, we worked together with my Northeastern University colleague, Roger Giese, to develop a high throughput LC method for analysis of therapeutic drugs in serum (3). The method involved injecting a second or third sample before the first sample eluted from the column. Multiple injections, which we called boxcar chromatography, led to analysis of tens of samples per hour (4), a quite high throughput at that time. Today, clinical liquid chromatography is a major field, with the expectation of significant growth as new biomarkers are discovered and validated.

At that time, John Dolan was a post doc in my lab. He joined Lloyd's group at Technicon near the beginning of our collaboration. That led to a close partnership between Lloyd and John over the next 35 years, especially in the chromatographic training area. John, of course, played an important role over a generation, helping chromatographers with his monthly articles for LC-GC on issues involved in separations and analysis.

Lloyd, with Jack Kirkland, had a major impact on the growth of LC through their multi-edition book on the field and their many, many ACS-sponsored short courses. There are too many contributions from Lloyd to list and describe. However, I was always impressed with his ability to translate complicated phenomena into simple, practical approaches to help workers understand and optimize separation and analysis. I especially believe he was the major contributor to optimization of first gradient elution and then a combination of isocratic and gradient elution for the separation of complex mixtures. These advances have had a major impact, for example on peptide mapping, a key analytical tool in the characterization of proteins, especially biopharmaceuticals. Where would biotechnology be today without peptide mapping?

Lloyd's impact on the field of chromatography, especially LC, has been immense. He was always an honest debater – if you had a better argument, he would accept your point of view. His enthusiasm, depth of knowledge, and, especially, kindness stand out in my memory. He was a giant in the field of chromatography. He is sorely missed.

Barry L. Karger
Director Emeritus, Barnett Institute
James L. Waters Chair and Distinguished Professor Emeritus
Northeastern University, Boston, MA

REFERENCES

1. Karger, B.L., Snyder, L.R. and Horvath, C., *An Introduction to Separation Science*, Wiley Interscience, New York, 1973.
2. Karger, B.L., Snyder, L.R. and Eon, C., "Expanded Solubility Parameter Treatment for Classification and Use of Chromatographic Solvents and Adsorbents," *Anal. Chem.*, 50, 2126 (1978).
3. Snyder, L.R., Karger, B.L. and Giese, R.W., "Clinical Liquid Chromatography," *Contemporary Topics in Analytical and Clinical Chemistry*, Vol. 2, D.M. Hercules, G.M. Hieftje, L.R. Snyder and M.A. Evenson, Eds., Plenum Publishing Corp., New York, Chapter 5 (1978).
4. Karger, B.L., Giese, R.W. and Snyder, L.R., "Automated Sample Cleanup in HPLC Using Column-Switching Techniques," *Trends in Anal. Chem.*, 2(5), 106 (1983).

LLOYD SNYDER FOCUSES ON THE HUMAN SIDE OF SCIENTISTS

Science is a uniquely human endeavor. But, what motivates humans to become scientists? About 50 years ago, high-performance liquid chromatography (HPLC) and related techniques in separation science burst over the horizon, attracting interest from many gifted scientists. In the mid-1960s, a small cohort (~seven) devoted their careers to understanding and advancing separation science, predominately liquid chromatography. Over the next decade, the cohort of HPLC experts grew to about 50.

The scientific literature presents an impersonal, factual history of the technical evolution in HPLC the five decades from about 1965 to 2015.

Lloyd Snyder, Ph.D., felt that the personal trials, tribulations and eventual successes of scientists are an important story that compliments the technical and commercial history. After all, these leading scientists each invested decades of their lives in advancing separation science. They made a human choice. What motivated them to make their choice? What hurdles did they face and clear? These are the human challenges that we all face in our lives.

Dr. Snyder recruited Frank Svec (University of California at Berkley, CA) and Robert Stevenson, Ph.D., of American Laboratory to collaborate in fleshing out this topic. We all shared the vision that recording the human side of science is also important.

Being members of CASSS (formerly the California Separation Science Society, Emeryville, CA), we started with the recipients of the CASSS Award for Outstanding Achievements in Separation Science as subjects of our study, which is a collection of anthologies. We selected a cohort of living awardees of the CASSS Award for Separation Science since 1995. The complete list of awardees is available at http://www.casss.org/?561

Under Lloyd's leadership, CASSS has also curated and posted (http://www.casss.org/?BIOINTRO) a collection of anthologies from leading specialists in separation science.

Lloyd Snyder starts his anthology (*An "Accidental" Career in Science*) with a quote from George Orwell: *Autobiography is only to be trusted when it reveals something disgraceful. A man who gives a good account of himself is probably lying...*

By nine years of age, Lloyd Snyder had decided to be a chemist after his grand-mother gave him a chemistry set. Lloyd starts off with a short discussion of and an adolescent's interest in explosives facilitated by visits to the public library. "…So I started reading chemistry books, and at that age I could comprehend the basics without too much difficulty; I was also absorbing lots of chemical facts and history, the sort of stuff that has (regrettably) largely been removed from high school and college courses."

This led to:

- An experiment to produce nitroglycerine that went very wrong
- The discovery that HCl plus $KClO_4$ does not produce Cl_2 (my first conclusion), but the highly explosive Cl_2O_7
- Numerous recreational "experiments" that involved the combination of sodium and water
- Mistaking kerosene for water when transferring white phosphorus from one container to another

All of which he survived. Other accidents followed. So with confessions out of the way he addresses his interest in chromatography.

Lloyd's career in chromatography started at Shell Development Co. in Houston, Texas, when he was assigned to a team to build a high-temperature gas chromatograph for the characterization of feeds and products of petroleum processing. Lloyd played a small role in assembling the instrument but as they collected data he noticed a relationship between retention time and analyte structure. Thus started his interest in understanding and applying chromatographic retention behavior to structure activity/relationships. This interest was still active in the summer of 2018, when he passed. The details will be expanded on by others. Or, you can read Lloyd's own words on the CASSS website.

Lloyd lived one hill west of me in Orinda (CA). In 2017 he downsized to Rossmoor in Walnut Creek, which is one hill east. After the move, he said that he wished he had done it years ago. I recall several chance meetings at hardware stores, markets and, of course, conferences. I was also honored to serve as a commercial mentor at DryLab, which LC Resources marketed as a computer-aided method of development.

Lloyd also served as a Deacon at Moraga Valley Presbyterian Church, where he was also active as a Stephen Minister for about 20 years.

Although separation science will continue to advance technically with new tools and problems to solve, future scientists will face qualitatively similar problems in their careers. Technical careers are tough. So much is governed by serendipity. Networking and dedication are also essential. In the sciences, each day can present a new challenge and learning opportunity. Lloyd was interested in helping us all along life's path. Even after his passing, I often flash on "What Lloyd would say about…"

Robert Stevenson

If you are reading this, I am going to assume that you almost certainly knew of Lloyd Snyder. This means you're probably familiar with at least some of Lloyd's many achievements: his justly famous solvent selectivity triangle; his work, along with others', developing the linear solvent strength relationship; and the development of the hydrophobic-subtraction model (HSM). If you are from an older generation of chromatographers, as I am, perhaps you've read his treatises on the physiochemical basis of normal phase separations. These writings, long out of print, are the essence of trenchant, succinct technical prose. And at the time – the late 1960s – they did much to clear away the common misconceptions that had cluttered and constrained the field up to that point.

None of this is to say that Lloyd labored alone in these pursuits. Quite the opposite. He sought out and thrived on long-term collaborations. And anyone collaborating with Lloyd was likely to find it a deeply rewarding experience, for he was a generous colleague, quick to recognize his fellow researchers. It was this aspect of Lloyd's personality that made him perfectly suited for serving on the editorial board of the *Journal of Chromatography*. In his hands, a submission to the journal was sure to receive a timely, even-handed review. The reward for Lloyd was that, as editor, he was furnished with a complete set of the journal's back issues, which he kept on several large cabinets in his home office. These were well used. Lloyd had a tremendous talent for recalling, with great fidelity, any article he had read. When Lloyd went silent on the other end of a phone conversation, you knew he had swiveled his office chair around in order to scan his shelves. Once he'd found the relevant article and he was back on the line, it invariably confirmed what he'd remembered about the piece.

Lloyd's career path was not an unbroken series of successes. As he described it in an autobiographical sketch, his path was peripatetic. Interspersed with his achievements were episodes of professional self-doubt and questioning. His early positions in the oil industry quickly confirmed that he was unsettled by some of the demands of the corporate world. He considered but then discarded the idea of pursuing an academic position, anticipating that he would soon tire of the endless committee work, curricula demands, and grant writing. Lloyd eventually landed at Technicon, a small east coast company. This job did not play to his strengths. As he once told me, somewhat ruefully, his position at Technicon demanded more of his people-managing skills than his technical skills. This mismatch, together with regret over leaving his home state of California, left him increasingly dissatisfied. Years later, he would smile when recounting this chapter of his life, but it was not hard to see it pained him at some level.

Lloyd's career turned a corner when, together with John Dolan and Tom Jupille, he founded a small company, LC Resources. I joined LC Resources in 1991, when I, too, was struggling with career choices. I had just completed a temporary position as a college lecturer and needed a more permanent position. My hiring by LC Resources was predicated, in part, on my experience with biochromatography. In reality, I had very little. My first project at LC Resources sought to optimize the ion exchange separation of a protein. At the project's completion, Lloyd asked me what I had observed about the role of mobile phase pH on the protein's resolution. I recall the lengthy silence that ensued when I told him I hadn't bothered to investigate this property. To my wonderment, he chose not to terminate me, as prudence might have dictated. Appreciating that I was in over my head, he chose instead to provide me with numerous articles and textbooks. Reading these quickly enabled me to become much better versed in all forms of chromatography.

As time went on at LC Resources, I interacted less and less with Lloyd. This evolution reflected not only the progression in my own skills, but also that many projects were frankly more mundane, and as a consequence didn't require Lloyd's expertise. I left LC Resources in 2001 to become an assistant professor at a small midwestern university. At this point I assumed, erroneously, that my departure would mean I would have little future contact with him. Aware of the pressures facing young faculty to establish a research program, Lloyd invited me to join the multiyear effort in developing the HSM. His support came in many tangible forms: funds for purchasing supplies, detailed experimental plans, and the initial drafting of manuscripts. This support went a long way towards helping me obtain tenure. What I really treasured, however, was our frequent phone calls. Our chats were always amiable, low-keyed, and at times humorous. As our work on the HSM together wound down, I asked him if he thought HSM would be his most enduring contribution to chromatography. He laughed. "No," he replied, "I have given up trying to predict which ideas will survive and live on." He hazarded a guess that few future chromatographers would consult or even know about the model. This exchange captures his self-effacing modesty as well as the long view he took on the field. Despite his immense stature, he was the least arrogant person I knew.

Lloyd will be missed both for his innovations in the field of chromatography and for his approach to the scientific endeavor. His innate curiosity was the driving force behind his approach to science, and he asked of others only that they share this passion. He conducted himself with a calmness and directness that were reflected in both his writing and speaking styles. He never tried to buffalo others with his reputation or knowledge. Lloyd may have been skeptical about his lasting contributions; I am not.

D. Marchand

A TRIBUTE TO LLOYD SNYDER AND HIS LEADERSHIP ROLE IN HPLC DEVELOPMENT

After completing a Ph.D in Analytical Chemistry at the Pennsylvania State University with Prof. Joseph Jordan and a postdoctoral position at Purdue University with Prof. Lockhart (Buck) Rogers, I joined DuPont in 1967 at the Experimental Station in Wilmington, Delaware. My first book purchase in preparation for a chromatography career was *Dynamics of Chromatography, Principles and Theory* by J. Calvin Giddings (1965). I was fortunate to have selected a book that I still refer to often after more than 50 years of practicing separation techniques. It is clearly written and still relevant to modern HPLC. My second book purchase was *Principles of Adsorption Chromatography, The Separation of Nonionic Organic Compounds* by Lloyd R. Snyder (1968). The Snyder book covers important concepts such as column band spreading, solvent elution strength and nonionic molecular interactions. Chromatographers will quickly notice that both Giddings and Snyder wrote many books and papers before LC instruments became widely available *so there are no chromatograms in either book – the word chromatogram does not appear.*

Soon after I joined DuPont, Jack Kirkland started preparation of a book called *Modern Practice of Liquid Chromatography* (1971) that went beyond the theory of packed column operation to include instrument design. Jack asked me to write a chapter on apparatus for high-speed LC. Compared to GC systems, dealing with high pressure was the primary obstacle in designing and operating HPLC systems. Lloyd contributed two chapters to the Kirkland book: one on mobile phase properties where he differentiated between liquid (bonded phase) partition and adsorption, and another on adsorption chromatography. Snyder also discussed the principles of reversed-phase partition for the first time that I recall seeing in a book. Chromatograms with time scales were used throughout *Modern Practice of Liquid Chromatography*. It was exciting to play a small part in that project with Jack, Lloyd and others so early in my career.

It is difficult to pay tribute to Lloyd Snyder without mentioning Jack Kirkland because they collaborated so well for so long. The popular Snyder–Kirkland ACS Short Course, *Modern Liquid Chromatography*, took shape during 1970–71 while Lloyd was working at Technicon in Tarrytown, New York. It began a long period of collaboration and friendship between Lloyd Snyder and Jack Kirkland that exposed users to rapid developments in HPLC and propelled the technique forward. Course notes were published in about 1974 as a first edition. In 1979, Snyder and Kirkland followed up with *Introduction to Modern Liquid Chromatography, Second Edition*, an amazing book that became known as "the LC bible" because of its size and comprehensive nature. Lloyd Snyder founded LC Resources with John Dolan in 1984 and returned to the west coast. John and Lloyd had worked together at Technicon. In 2010, Snyder and Kirkland updated the LC bible to a third edition and added John Dolan as a major coauthor. During this period, Snyder and Kirkland published another popular book in 1988 called *Practical HPLC Method Development* with Joe Glajch of DuPont as a coauthor. An expanded version of the book was published as a second edition in 1997.

Lloyd published hundreds of articles and nine books and had an enormous impact on the development and growth of HPLC. He was quite humble about his scientific accomplishments and very generous about sharing his knowledge. Lloyd often included multiple authors in his steady stream of publications. Although extremely productive, he was not one-dimensional and had many interests. One of my chances to see a different side of Lloyd came at the Montreal 2002 HPLC Meeting. I had just retired from both Keystone Scientific and Penn State University and was feeling overloaded with HPLC. My wife Ginger and I had decided to sneak away from the meeting and take a guided tour of Quebec City to learn more about its rich Canadian history. As we boarded the bus, I was feeling guilty about attending the social program at a scientific meeting when we ran into Lloyd and Barbara Snyder, who invited us to sit near them. Lloyd was an avid reader and student of history. He set a good example for me by taking the day off to enjoy something unrelated to science. I remember HPLC 2002 for that more than anything else.

Certain types of individuals always seem to pioneer the growth and development of new chemical techniques. Lloyd will be remembered in HPLC as one of those scientists who led the charge. He was available at the perfect time but still leaves a void that will be strongly felt.

Richard A. Henry

I first met Lloyd Snyder in the late 1970s when I was working at DuPont Central R&D and Lloyd had started an association with the DuPont Instrument Products Division. Of course, Lloyd had already worked for many years with our colleague Jack Kirkland, teaching HPLC courses for the American Chemical Society and writing the classic book *Introduction to Modern Liquid Chromatography*. One of our early collaborations with Lloyd was based on his solvent selectivity triangle, which Jack and I used as the basis for our theory and practice of HPLC method development, especially for reversed-phase separations. This was followed shortly by work on normal-phase method development and Lloyd patiently educated us on the theory of adsorption chromatography. I still have a copy of his 1968 book, *Principles of Adsorption Chromatography*, which Lloyd claimed was an academic success. By that he meant that it was pretty well written but did not sell many copies! This was certainly not true of his other books which sold as many as 22,000 copies.

In addition to our early work together on published papers, Lloyd, Jack, and I wrote two editions of the book *Practical HPLC Method Development*, where I came to know firsthand what a wonderful and meticulous writer we had on our team with Lloyd. I also started teaching a method development course with Lloyd and Jack (later joined by John Dolan) and I quickly was impressed with not only his knowledge of the subject, but his ability to explain to all students in a very patient manner.

One of Lloyd's great strengths was his knowledge and ability to learn many areas of chromatography and many applications as well. Lloyd was trained as a physical organic chemist under Don Noyce but like many of us in the field, his first work was in gas chromatography (at Shell Oil). However, he quickly transitioned into adsorption chromatography with a new position at Union Oil and this started his lifelong study of liquid chromatography. His work on the solvent selectivity triangle applied equally well to normal-phase and reversed-phase LC and he then transitioned to doing most of the rest of his work on reversed-phase. This included the theory and practice of gradient elution and work with other types of samples such as peptides and eventually even other biomolecules. The culmination of much of this work resulted in the development of DryLab with John Dolan, which revolutionized and greatly shortened the method development process. Not to be content with this achievement, Lloyd spent the last 15–20 years of his career working on theory again, this time with the hydrophobic-subtraction model, working with Jack Kirkland, Pete Carr, John Dolan, John Dorsey, Uwe Neue, and Dan Marchand in a truly multi-lab collaborative effort.

Finally, near the end of his career, Lloyd collaborated once again with Jack Kirkland and John Dolan to produce the third edition of *Introduction to Modern Liquid Chromatography*, which was a fitting culmination of a long and productive career.

I will always remember Lloyd as a great friend and mentor, but also as a very patient and dedicated teacher and a prolific author of books and papers. In addition to his great technical knowledge, he was also a very humble person with a great sense of humor and humanity.

Joseph Glajch

As a very young assistant professor of analytical chemistry in Hungary in the early 1970s, I became aware of Lloyd's work through his books *Principles of Adsorption Chromatography* (1968) and *Introduction to Modern Liquid Chromatography* (1974, with Barry Karger and Csaba Horvath), and his seminal papers in *Analytical Chemistry, Journal of Chromatography*, and *Journal of Chromatographic Science*. Later, I had the good fortune of being able to translate his *Introduction to Modern Liquid Chromatography* (with Jack Kirkland) into Hungarian (1979, see photo), which became the *de facto* bible of HPLC for our generation. Despite a general lack of funds in Hungary in those years for acquisition of western-made scientific equipment, these, along with the efforts of the *Chromatographic Working Group of the Hungarian Chemical Society* and the contributions of the Hamilton Foundation's *Scientific Exchange Agreement* (administered by Georges Guiochon), created an exuberant epoch of growth in the separation sciences (for a comprehensive history see *Szepesy László: A kromatográfia és rokon elválasztási módszerek története és fejlesztése Magyarországon*, Müszaki Publishing House, Budapest, 2007, ISBN: 978-963-06-1854-0).

I met Lloyd (and Jack Kirkland) in person only in 1986, after I moved to Texas A&M University (TAMU). Being new on the US scene, I was quite surprised not only that a chance encounter turned into a lengthy, animated conversation, but also by his detailed knowledge of our previous work on ion-pair liquid chromatography. At the end of the conversation he asked me if I would be interested in serving as a reviewer and later a member of the editorial advisory board for *Journal of Chromatography A*, an invitation I was quite happy to accept, and which led to my reading the literature more broadly and deeply in order to cope with the ever more frequent review requests, which soon grew to weekly assignments and which continued until his retirement as editor in 2000. Lloyd was a very good – and very practical – mentor and I benefitted from his guidance greatly during my years of service as one of the Symposium Volume editors of *Journal of Chromatography A*.

My separation science graduate classes also benefitted from Lloyd's (and John Dolan's, and, much later on, Imre Molnar's) generosity: they donated copies of subsequent versions of DryLab that we installed on a separate walk-up computer in my group for use by all members of the annual graduate separations science classes. Since class members (numbering over several hundred by the time of my retirement) came from different colleges of TAMU and had "real life" separation problems in their research, Lloyd's gesture provided them with invaluable immediate help and a life-long skill that they otherwise would not have had.

After Lloyd's retirement from the editorship our previously weekly communications became less frequent but he maintained an interest in our research – despite the fact that we became focused on electrophoresis, with HPLC (both RP and HILIC) serving "only" as the indispensable daily tool that made our synthetic efforts successful (single isomer strong electrolyte cyclodextrins, isoelectric buffers, isoelectric membranes, rapidly focusing isoelectric point markers, and pyrene-based fluorophores having tunable charge and spectral properties). Nevertheless, I was fortunate enough to be able to enjoy Barbara's and Lloyd's dinner hospitality – and vibrant conversation – in their home on my drive up from San Francisco to our summer place in Gold Beach, Oregon.

With the successive passing of Joseph Huber (2000), Csaba Horváth (2004), Georges Guiochon (2014), and now Lloyd (2018), it seems that the twenty first century brought us the gradual loss of many of the founders of HPLC. Thus, I believe it behooves us – who knew them personally – to transfer to the upcoming generations – who won't have had this opportunity – not only their lasting contributions to our field, but also their unique take on it (just think of the wealth of new information Lloyd could tease out from anybody's published, correctly measured data set, or Csaba's ability to capture the essence of complicated principles or processes in masterfully crafted words, such as *pellicular, isocratic, isopicnic*). For a starter, please, do read Lloyd's humbling autobiography (*An "Accidental" Career in Science* https://cdn.ymaws.com/www.casss.org/resource/resmgr/Biographies/Snyder LRBiography.pdf). It is too bad the other "greats" didn't get to write theirs. May they all rest in peace and never be forgotten!

Gyula Vigh

MEMORIES ON LLOYD SNYDER

In 1981 I was having a hard time in my life, as I started my private research institute in Berlin in 1981 and gave courses about "solvophobic theory," which I developed together with Csaba Horváth at Yale, explaining how a reversed phase would work. As I was regularly talking to Csaba on the phone, he suggested, one day in 1984, that I should get in touch with Lloyd Snyder, who would be planning to leave Technicon and who also wanted to establish a private company, like myself. Lloyd was already a great teacher of HPLC, giving courses together with Jack Kirkland at the American Chemical Society, and was in the process of moving from the east coast to his native place in the area of San Francisco. Lloyd was very glad about Csaba's effort to connect us and soon we started communicating and designing HPLC courses. I invited him to Berlin and we taught short courses together at pharma companies such as Schering. As at that time the first "IBM-compatible" computer appeared on the market, myself, Lloyd, John Dolan – the founders of LC Resources – saw a chance to develop some kind of a software for HPLC. Later Tom Jupille joined the team also.

We started in 1986 with several software variants, which we exhibited at the PittCon every year, e.g. a database ("Marian the Librarian") that helped to collect and document chromatography literature, or the "HPLC Doctor" for troubleshooting in HPLC. This latter software was John Dolan's favorite, as he could put all the know-how he had learned in Barry Karger's lab into it, and also sell videotapes along with the software that showed visually how to exchange the column or the injection loop, clean the detector cell, cut capillary tubing, etc. We also had a Prep DryLab version, which was developed by Lloyd. My favorite project was DryLab, which we planned for chromatographers in industrial laboratories. The name came from Lloyd, from his way of describing how important it is to test experiments in a dry fashion before doing them "wet." We developed different building units, such as DryLab 1, 2, 3, and 4, which were elements of the "isocratic" version (DryLab I), which calculated the effects of band spreading, capacity factors, etc., helped by all the data Lloyd had from DuPont, measured with Jack Kirkland and a few graduate students. The calculation of mass transfer resistance inside the pores was a major issue, which depended on the ligand length, density, and pore diameter, and this factor contributed to peak broadening too. Also the approximate pressure could be calculated from the length and ID of the column and from the particle size of the packing at a given flow rate of mixed eluents in RP systems. The quality of a method was characterized by the capacity factor range, which was supposed to stay in isocratic work between $1 < k < 10$ and which depended on the %B-value of the organic modifier. The goal was to find conditions for baseline resolution with a critical resolution value of the two closest peaks being $> 1,5$. This was an important measure of the method quality and Lloyd was producing excellent publications, which supported the marketing of DryLab. I suggested early on using chromatograms to see the different influences better in a visual way. Although Lloyd mentioned that he didn't want any "slick stuff," later he agreed and John started to program visual ways using an asterisk at the beginning to show chromatograms.

Later we recognized the need for a gradient version, where Lloyd showed his great talent for thinking as a chromatographer and a mathematician at the same time. DryLab started from a "one factor at the time" (OFAT) type of software, which helped us to understand the influence of pH, %B, and temperature. On the other hand, we could calculate the influence of the column length, column diameter, particle size, flow rate, and the extra column volume at the same time. So far DryLab was already, as early as 1990, a multifactorial modeling tool for HPLC and was a driving force of the Quality by Design (QbD) paradigm change introduced by the FDA in 2002 and with the ICH Q8 R2. The new request was a description not just of how the method had to be used, but also of how the method was developed, including the use of tolerance limits.

We were a team for over 20 years and Lloyd was our leader in many ways. He was guiding the group to scientifically solid solutions. He was an extremely productive scientist. He wrote, between 1958 and 2013, more than 218 publications, ca. 100 review articles, seven books. He had ten patents.

My job was to control the predictions of DryLab in reality, and as the "devil is hiding in the details" I was finding the bugs and letting the programmers correct them. Lloyd was reliable and loyal and was always available for discussions, also on personal issues. In this way we were a successful team and created something great together. I will always be thankful to Lloyd for going through this part of my life with me. DryLab, from our teamwork, is the only software on the planet giving a scientific background of HPLC, and it will continue for years to come to support chromatographers in industry, academia, and regulatory sciences, supporting a better understanding of chromatographic fundamentals in a method and enabling the faster development of new drugs for the patients in need.

The founders of DryLab at the 43rd Pittsburgh Conference (March 1992): (from left) John Dolan, Lloyd Snyder, Tom Jupille, and Imre Molnár (with Margaret Watkins).

One of the first short courses in Berlin in 1985 on Principles of HPLC with Lloyd Snyder, Imre Molnár, and participants from Schering.

John Dolan and Imre Molnár at the Pittsburgh Conference in 1998.

Lloyd Snyder (middle) and Barry Karger, recipient of the Waters HPLC price (right), with Joe Glajch (left) at the Pittsburgh Conference in 2000.

Thank you Lloyd for the years with us, for all your friendship and good spirit in making us successful.

Rest in peace.

Imre Molnár
Molnár-Institute
Berlin, Germany

MY MEMORIES OF LLOYD SNYDER

In 1969, as a fresh postgraduate student who just had selected liquid chromatography as the topic of his dissertation, I found a treasure in our university library – the book *Principles of Adsorption Chromatography* by Lloyd Snyder, published just a year before. The clarity and comprehensiveness of Lloyd's text fascinated me and drew my attention to the investigation of the principles of separation based on the distribution between the stationary and the mobile phase in HPLC, which has remained the main topic of my lifetime research. Soon after the publication of our first series of articles on the gradient-elution models, Lloyd contacted me (I was too shy to address him), and from then onwards we remained in contact, exchanging opinions on the drafts of the "gradient" articles. Lloyd sent me his positive opinion on our book *Gradient Elution in Liquid Chromatography*, published in 1985 in the Elsevier J. Chrom. Library series. The book *High-Performance Gradient Elution – The Practical Application of the Linear-Solvent-Strength Model*, published by Lloyd at Wiley in 2007, summarizes Lloyd's systematic work in this field. In the introduction, Lloyd kindly quoted my contribution: "During the past 35 years, another scientist, Pavel Jandera from the University of Pardubice, has similarly devoted much of his career to the study and elucidation of the principles and practice of gradient elution. The present book owes much to his many contributions in this area…" In his LSS model, Lloyd excellently explained close relations between the retention phenomena under isocratic and gradient elution conditions, to elucidate the difficult topics and to make gradient elution as easy to understand as possible.

Lloyd's scientific interests were much broader than in only gradient elution. Just to mention two very important achievements: Lloyd and his co-workers introduced the hydrophobic subtraction model (HSM) for characterization and classification of the reversed-phase columns; and (together with John Dolan and Imre Molnár) Lloyd developed the user-friendly and commercially very successful "DryLab" HPLC method optimization software (he generously presented me a copy).

Lloyd applied his extraordinary didactic skills in the numerous HPLC courses he organized all over the world. I admired his ability to immediately distinguish the main aspects of a problem from the less important ones. Lloyd compiled his encyclopedic knowledge of all theoretical and practical aspects of chromatography in the bible of the HPLC technique, the book *Introduction to Modern Liquid Chromatography*, where the reader finds all the important information he or she may need.

For many years, Lloyd was an editor of the *Journal of Chromatography*. He sent me for review theoretical, often "difficult" manuscripts, sometimes giving his short opinion. He was always helping and inspiring when I asked him for his opinion on my own manuscript drafts. I met him at several HPLC symposia in the US and Europe. The discussions with him were always a pleasure and a source of inspiration for me. I remember meeting Lloyd and his charming wife Barbara at a pleasant afternoon cocktail organized by Erich Heftman in his house during the HPLC 1996 symposium in San Francisco. Unforgettable was a micro-symposium organized by Elsevier in Ellecom (the Netherlands) on the occasion of Lloyd's seventieth birthday in 2001. (The journal *Analytical Scientist* reproduced a photo from this event in the September 2018 issue.)

I will never forget Lloyd Snyder, an excellent scientist, one of the founders of HPLC, but above all a wonderful person, a kind and gentle friend.

Pavel Jandera

LLOYD R. SNYDER – A MEMORIAL

When I joined the DuPont Company in 1965 as a newly minted BS chemist, one of the greatest benefits I received was having Jack Kirkland as my mentor. Through him, I was introduced to most of the early developers of modern high-performance liquid chromatography. One of the HPLC pioneers that Jack introduced me to was Lloyd Snyder. Jack and Lloyd had an especially close relationship that included collaborations on each other's research efforts and publications, and that eventually led them to co-teach ACS courses for many years to thousands of aspiring separation scientists on the basics of modern liquid chromatographic theory and techniques. Together, and with other authors such as John Dolan and Joe Glajch, they wrote seminal textbooks that served for decades as the go-to books for practicing liquid chromatographers.

Looking back on my 40+ years of experiences with Lloyd fills me with an impression of a kind, gentle man who was always teaching others. Besides the ACS courses and textbooks, and his many publications, including his early work on the elution strengths of solvents for liquid–solid chromatography which helped me immensely during my research for my master's degree, Lloyd worked for the DuPont Company for several years as a consultant during which he mentored several pre-doctoral students from local universities in the Wilmington, Delaware area. These students all co-authored peer-reviewed publications with Lloyd and went on to receive their PhDs from their respective universities and to enjoy successful scientific careers. A lasting tribute to Lloyd's ability to mentor and teach others.

Another memory of Lloyd is that he was a terrible jokester. He would tell us what he believed to be funny stories, at the end of which he would laugh and laugh but most of us would simply smile. The joke would usually just be corny or he may have just told it wrong. It didn't matter. The important thing is he was trying to bring a light moment into our lives and not always be serious, showing his human side.

Lloyd's ability to interpret the results of research experiments and draw conclusions from them is legendary. I have never encountered anyone else who could take raw data and come up with a graph or an equation to fit the data the way Lloyd could. Amazingly, the end results have withstood the test of time as well, which is even more important. Consequently, Lloyd's contributions to the science of HPLC and to the development of liquid chromatographic theory are among the most important in the field.

Lloyd Snyder, the man, will be missed as we move forward in our day-to-day lives but his DryLab for HPLC method development, the hydrophobic subtraction model for comparing HPLC columns from different manufacturers, and his many publications, PhD researchers, colleagues, and friends will continue to push forward the frontiers of science. We will forever be grateful for having known him.

Joseph DeStefano

I first encountered Lloyd Snyder at the HPLC 1995 Symposium held in Innsbruck, Austria, organized by Wolfgang Lindner. This was the first conference I had attended as a presenter rather than a mere participant – my contribution was a poster on silanol effects in reversed-phase separations. In one of the quieter moments I noticed Lloyd and his long-time friend Jack Kirkland discussing my work at some length, but rather than approach these distinguished researchers, being wary of difficult questions, I kept in the background on this occasion. This was a mistake, because both Lloyd and Jack were most generous to early career researchers and were incredibly helpful to me in my subsequent studies.

Nevertheless, encouraged by this interest, I sent Lloyd a fax message a while later asking him his opinion on some other studies I had performed on the overloading of basic compounds. I hardly thought that this busy and eminent scientist would even bother to reply. I was astonished to receive, the very next day, a detailed and helpful critique of my work, which resulted in its publication in the *Journal of Chromatography*. This experience established a pattern for my discussions with Lloyd, which continued first by fax and subsequently by email over a period of more than 20 years. Before this time (and without the benefit of more modern electronic communications), I had been rather isolated as a researcher in the U.K. and had few opportunities for interaction with world-class scientists. Without a strong research background and funding at that time, travel to international conferences was difficult, and the benefit of communication with Lloyd was tremendous for me. I remember sometimes arriving at my desk in Bristol in the morning and spending most of the day composing a subject for discussion with Lloyd, who, living in Oakland, California, was nine hours behind U.K. time. I would send the fax message to Lloyd just before leaving in the evening. Invariably, coming in the next day there would be a detailed reply from Lloyd, discussing the results and debating conclusions that were possible. Sometimes the reply would come on the very same day, suggesting that Lloyd had got up around 7:00 a.m. and replied to me immediately.

It was a great privilege for me to be invited to attend the celebration for Lloyd's retirement from the editorship of the *Journal of Chromatography* organized by Elsevier in Ellecom, the Netherlands, in 2001. This was a great opportunity to meet all the leading lights of separation science, including Csaba Horvath, Georges Guiochon, Peter Carr, John Dolan, Nobuo Tanaka, Hans Poppe, Phyllis Brown, Jack Kirkland, and Peter Schoenmakers. A couple of years later, Lloyd and his wife Barbara visited London, where the three of us spent a most enjoyable weekend in scientific discussions and in visits to the theatre, St Paul's cathedral, and a classical music concert at Wigmore Hall.

In the last 15 years or so, Lloyd had become rather disillusioned with traveling to international conferences, declining even to attend the 2006 HPLC conference in neighboring San Francisco. Nevertheless, I was able to meet up with Lloyd for a day, in which we visited the impressive Muir Woods Redwood forest and his lovely house in Oakland, and of course continued our discussions on chromatography.

Lloyd was essentially a rather private person, and a modest genius. I once likened him to a "simultaneous chess grandmaster" who alone played different games at the same time with multiple opponents. His games included adsorption chromatography, the solvent selectivity triangle, prediction of retention in RP, basic compounds and silanol effects, theory and practice of gradient elution, preparative chromatography, and many others. I had only the ability to play against him (or more correctly, "with him" – Lloyd was never competitive) in one or two of these topics. Lloyd's generosity towards fellow scientists, and, in his semi-retirement, towards others (for instance in his voluntary educational work with offenders in the local prison) were remarkable. He was a great man.

David McCalley

Lloyd Snyder was truly one of the pre-eminent figures in the development of modern liquid chromatography. Over the 40 years I personally knew him, I counted Lloyd – a great mentor and exceedingly generous human being – among my best friends.

As is the case with so many others who have written about him, I first came to know Lloyd through his books, starting with *Adsorption Chromatography*, which developed a simple but powerful model of liquid–solid chromatography. Indeed, the development of theoretically sound yet easily applied – and therefore useful – models was the driving force behind much of Lloyd's approach to science. I suspect that he was strongly influenced by the great physical chemist Joel Hildebrand, who was his freshman chemistry instructor at U Cal Berkeley. Nearly 50 years later, I still teach my graduate course using sections of the above book, as well as *Principles of Separation Science*, which Lloyd wrote with Barry Karger and Csaba Horvath and which is, in many ways, still the best, rather comprehensive textbook dealing with the broad fundamentals of our field.

Again, as with so many others who knew him, it was Lloyd who initiated our first personal contact; he reached out to invite me, then a newcomer to the field, to contribute to a symposium he was organizing at a meeting of the International Conference on Liquid Chromatography (ICLC). Over the intervening years, we collaborated on some 25 papers, most recently an extremely important series that Lloyd initiated on the hydrophobic subtraction model (HSM), which is used to characterize stationary phase selectivity in reversed phase chromatography. Along the way we worked on an extension of his famous and extraordinarily highly cited paper on the *"solvent triangle"* by incorporating the newly developed, solvatochromically-based empirical solvent strength scales of Mort Kamlet and Bob Taft, plus additional studies of various aspects of reversed phase separations. Indeed, it is difficult to find an area of liquid phase separations to which Lloyd has not made major contributions. He was certainly a master of both normal and reversed phase chromatography at both the analytical and preparative scales, of column dynamics, of method development methods and chromatographic retention modeling, and, of course, gradient elution chromatography, on which he did extensive, pioneering work.

I stated above that Lloyd was very generous. Indeed, I vividly recall a conversation in which he said (I paraphrase just a bit), "On a paper with five co-authors, everyone gets one-third of the credit." That is the spirit in which Lloyd worked. Anyone with whom Lloyd worked, especially when he was one of the US editors of the *Journal of Chromatography*, knows that he reached out repeatedly to established scientists to invite them to assist less-experienced newbies. In every sense Lloyd was a great editor, teacher, and mentor.

Since his passing, I cannot count the many times I have wanted to call or email Lloyd about a scientific puzzle, recommend a book on history (which was a second common interest), or simply share some thoughts. I, along with the entire chromatographic community, miss Lloyd's presence tremendously.

Peter W. Carr

Editors

Dr. Nelu Grinberg retired in 2016 as a Distinguished Research Fellow in the Chemical Development Department at Boehringer Ingelheim Pharmaceuticals in Ridgefield, CT. Prior to this, he worked for 16 years in the Analytical Department at Merck Research Laboratories in Rahway, NJ, where he was a Senior Research Fellow. He has authored and coauthored over 160 publications, including articles and book chapters. He is currently Editor-in-Chief of the *Journal of Liquid Chromatography and Related Techniques*, editor of the book series *Chromatographic Science Series*, and co-editor of the *Advances in Chromatography* series. He is also the President of the Connecticut Separation Science Council and a Koltoff fellow of the Hebrew University of Jerusalem. Dr. Grinberg obtained his Ph.D. in Chemistry from the Technical University of Iasi in Romania. He conducted postdoctoral research with Prof. Barry Karger at Northeastern University in Boston, Massachusetts, and with Prof. Emanuel Gil-Av at The Weizmann Institute of Science in Rehovot, Israel.

Dr. Peter W. Carr earned his B.S. in Chemistry (1965) from the Polytechnic Institute of Brooklyn where he worked with Prof. Louis Meites, and a Ph.D. in Analytical Chemistry at Pennsylvania State University (1969) under the guidance of Prof. Joseph Jordan. Following appointments at Brookhaven National Laboratory, Stanford University Medical School and the University of Georgia (Athens), in 1977, Dr. Carr joined the faculty at the University of Minnesota where he became Professor of Chemistry in 1981. He has been a consultant to Leeds and Northrup Co., Hewlett-Packard and the 3M Company, and was the founder and first President of ZirChrom Separations Inc. In 1986, he became an Associate Director of the Institute for Advanced Studies in Biological Process Technology at the University of Minnesota. He has been President of the Symposium on Analytical Chemistry in the Environment (1976), founder and first President of the Minnesota Chromatography Forum, and Chairman of the Subdivision of Chromatography and Separation Science of the Analytical Chemistry Division of the American Chemical Society (1988–1989). Prof. Carr has served on the Editorial Advisory Boards of *Analytical Chemistry*, *Talanta*, *The Microchemical Journal*, *LC/GC*, *Journal of Chromatography*, *Chromatographia* and *Separation Science and Technology*.

List of Contributors

Rigoberto Blanco
Universidad de Costa Rica
School of Chemistry

and

Caja Costarricense de Seguro Social
San José, Costa Rica

Carlos Calleja-Amador
Centre for Research, and
    Microbiological and Chemical
    Services CEQIATEC
Environmental Protection Research
    Center CIPA
Instituto Tecnológico de Costa Rica
Escuela de Química
Cartago, Costa Rica

Rui P. V. Faria
Department of Chemical Engineering
Faculty of Engineering
University of Porto
Porto, Portugal

Masahiro Furuno
Graduate School of Engineering
Osaka University
Suita, Japan

Jonathan C. Gonçalves
Department of Chemical Engineering
Faculty of Engineering
University of Porto
Porto, Portugal

Tomáš Hájek
Department of Analytical Chemistry
Faculty of Chemical Technology
University of Pardubice
Studentská, Czech Republic

Anthony R. Horner
Department of Chemistry
University of Pittsburgh
Pittsburgh, Pennsylvania

Tohru Ikegami
Graduate School of Science and
    Technology
Kyoto Institute of Technology
Kyoto, Japan

Pavel Jandera
Department of Analytical
    Chemistry
Faculty of Chemical Technology
University of Pardubice
Studentská, Czech Republic

Eisuke Kanao
Graduate School of Engineering
Kyoto University
Kyoto, Japan

Kazuhiro Kimata
Nacalai Tesque, Inc.,
Kyoto, Japan

Kodai Kozuki
Graduate School of Science and
    Technology
Kyoto Institute of Technology
Kyoto, Japan

Takuya Kubo
Graduate School of Engineering
Kyoto University
Kyoto, Japan

J. F. Ogilvie
Centre for Experimental and
    Constructive Mathematics
Department of Mathematics
Simon Fraser University
Burnaby, Canada

and

Universidad de Costa Rica
Escuela de Química
San Pedro, Costa Rica

Koji Otsuka
Graduate School of Engineering
Kyoto University
Kyoto, Japan

Michael T. Rerick
Department of Chemistry
University of Pittsburgh
Pittsburgh, Pennsylvania

Alírio E. Rodrigues
Department of Chemical Engineering,
    Faculty of Engineering
University of Porto
Porto, Portugal

Erin P. Shields
Department of Chemistry
University of Pittsburgh
Pittsburgh, Pennsylvania

Nobuo Tanaka
Graduate School of Science and
    Technology
Kyoto Institute of Technology
Kyoto, Japan

and

Graduate School of Engineering
Osaka University
Suita, Japan

Stephen G. Weber
Department of Chemistry
University of Pittsburgh
Pittsburgh, Pennsylvania

1 Mobile Phase Effects in Reversed-Phase and Hydrophilic Interaction Liquid Chromatography

Pavel Jandera and Tomáš Hájek

CONTENTS

1.1 INTRODUCTION

The ultimate target of high-performance liquid chromatography (HPLC), like of other analytical separation techniques, is the resolution and determination of sample components. Retention, separation selectivity and chromatographic efficiency control the resolution and separation quality in liquid chromatography (LC). The development of efficient chromatographic columns has experienced remarkable success in the last two decades. Some of the modern types of efficient stationary phases, including the sub-2 μm particle, core-shell, monolithic and micropillar-array columns, challenge the correct definition of the volumes of the stationary and mobile phases, between which the solute distributes during the migration along the column. Adopting a phase volume convention can mitigate the problems caused by the possible mixing of the adsorption and partition mechanism and by the selective preferential adsorption of mobile phase components in the stationary phase.

The retention and resolution on even the most efficient LC column results from the interactions between the solute, the stationary phase and the mobile phase, the cocktail of which characterizes different LC separation modes. Non-polar interactions are the predominant (but not the only) forces controlling the separation mechanism in reversed-phase (RP) systems employing low-polarity columns and polar aqueous-organic mobile phases. In contemporary HPLC, most frequently used are the reversed-phase separation systems. However, many polar compounds elute too early in RP LC. Hydrophilic interaction liquid chromatography (HILIC) is becoming increasingly popular, as it often provides significant improvement in the retention and separation efficiency for the separation of polar and weakly ionic compounds [1]. HILIC is essentially a normal-phase (NP) mode employing a polar column, like classical adsorption chromatography with a mixed organic solvent mobile phase. However, HILIC employs an aqueous-organic mobile phase with a high concentration of the organic solvent; it is therefore also known as aqueous normal-phase (ANP) liquid chromatography. There are a plethora of polar columns suitable for HILIC separations, including silica gel, bare or with various bonded polar ligands, and polar organic polymers [2], showing different properties such as chromatographic selectivity and water adsorption [3]. The adsorbed water forms a part of a HILIC stationary phase, and therefore the appropriate convention defining the volumes of the stationary and mobile phases is especially important for the quantitative description of retention.

In both reversed-phase and HILIC systems, the mobile phase is a very active – but often underestimated – player affecting the retention, separation selectivity and ultimately the sample resolution in HPLC. The present work focuses on the role of the mobile phase in RP LC and in HILIC. It compares several two- and three-parameter models describing the effects of the mobile phase on the separation, dating from the early days of HPLC, with the ABM three-parameter model, which does not presume either adsorption or partition retention mechanisms [4]. In gradient elution, increasing (in RP LC) or decreasing (in HILIC) the concentration of the organic solvent in water accelerates the elution of strongly retained compounds and improves the resolution of complex samples. The retention models allow a prediction of the retention data in gradient LC from the isocratic experiments [5].

Several theoretical models correlate the contributions of various interactions characterized by structural parameters to the retention. The linear solvation energy relationships (LSER) model applies in HILIC, like in RP LC [6]. The mixed-mode columns improve separation of ionic (ionizable) compounds combining ion exchange with either RP or HILIC retention mechanisms. A single polar column often shows a dual HILIC/RP retention mechanism, depending on the mobile phase. In the organic solvent-rich mobile phase, polar interactions control the retention (HILIC or ANP mode), whereas in more aqueous mobile phases the column shows essentially reversed-phase behavior with hydrophobic interactions playing a major role. Alternating RP and HILIC runs may provide different (even orthogonal) separation selectivity on a single column and complementary information on the sample composition [7].

1.2 STATIONARY PHASES IN LIQUID CHROMATOGRAPHY

1.2.1 SAMPLE DISTRIBUTION IN LIQUID CHROMATOGRAPHY

In high-performance liquid chromatography, the stationary phase is usually a bed of fine solid particles with narrow size distribution, densely packed in a metal, glass or plastic tube – a chromatographic column. The particles may be either fully or only partially porous, such as core-shell columns with a layer of the stationary phase chemically bonded to a support material. On the contrary, monolithic columns do not contain particles; instead, a continuous chromatographic bed fills the full inner column volume. The mobile phase (eluent) is a liquid, usually a mixture of two or more solvents (often containing suitable additives) forced through the column by applying elevated pressure in HPLC. The sample compounds move at different velocities along the column, together with – but more slowly than – the mobile phase. The elution process ideally leads to the eventual sample separation. The separated compounds appear at different times at the outlet from the column as the elution waves (peaks) monitored by a detector attached to the outlet of the column. The elution (retention) time, t_R, of the peak maximum is a characteristic property of each sample compound, depending on the distribution constant between the stationary and the mobile phases in the chromatographic column. Hence, the t_R, or the retention volume V_R, is a useful tool for solute identification.

HPLC has become one of the most powerful tools for the separation and determination of even very complex samples containing non-polar, moderately or strongly polar, and ionic compounds, either simple species or high-molecular synthetic polymers or biopolymers. These features are especially useful in pharmaceutical, biomedical and clinical analyses.

In an ideal chromatographic process, the equilibrium distribution of the sample compounds between the stationary and the mobile phases establish at any time in any part of the chromatographic bed. The changes in the partial molar Gibbs free energy, ΔG, of the solute transfer from the mobile to the stationary phase control the thermodynamics of the chromatographic process [8]. For strongly diluted samples:

$$\Delta G = -RT \cdot \ln K_D = -RT \cdot \ln \frac{c_s}{c_m} \qquad (1.1)$$

R is the gas constant, T is the temperature (in Kelvins) and K_D is the distribution (partition) constant, which gives the equilibrium ratio of the concentrations of the solute in the stationary, c_s, and in the mobile, c_m, phases, respectively. The velocity of a solute moving along the column is controlled by the ratio of the time spent by the solute in the stationary phase, t_s, to the time spent in the mobile phase, t_m. This ratio, the retention factor k, is equal to the ratio of the masses of the solute in the stationary, m_s, and in the mobile, m_m, phases, respectively, in the column it is directly proportional to the distribution constant of the solute, K_D, between the stationary and the mobile phase:

$$k = \frac{t_s}{t_m} = \frac{t_R - t_m}{t_m} = \frac{V_R - V_m}{V_m} = \frac{m_s}{m_m} = \frac{c_s}{c_m} \cdot \frac{V_s}{V_m} = K_D \cdot \frac{V_s}{V_m} = K_D \cdot \Phi \qquad (1.2)$$

The proportionality constant Φ in Equation (1.2) is the phase ratio, i.e., the ratio of the volumes of the stationary, V_s, and of the mobile, V_m, phases in the column. From Equation (1.2) it follows that [9]:

$$t_R = t_m \left(1+k\right) = \frac{L}{u}\left(1+k\right); \quad V_R = V_m\left(1+k\right) = t_m \cdot F\left(1+k\right) \qquad (1.3;\ 1.4)$$

t_m and V_m are also known as the column hold-up time and hold-up volume, respectively. The terms t_0, and V_0 are sometimes used instead of t_m and V_m. t_m (t_0) is equal to the ratio of the column length, L, and the linear velocity of the mobile phase along the column, u. F is the flow rate of the mobile phase, a simple conversion factor between the retention times and retention volumes. The retention factor, k, depends on the nature of the solute, on the character of the stationary and the mobile phases, and on temperature, but is independent of the flow rate of the mobile phase, the dimensions of the column (provided that the density of packing is uniform, i.e., a constant phase ratio along the column). Hence, k is a fundamental parameter in the method development and optimization of HPLC separations. Theoretically, k is suitable for measuring thermodynamic quantities by chromatography, such as the Gibbs free energy, enthalpy or entropy. Unfortunately, the retention factors determined from the experimental retention data – Equation (1.2) – do not provide reliable information on the presumed mechanism of retention, as several different mechanisms may contribute to the actual k.

1.2.2 STATIONARY AND MOBILE PHASE VOLUMES IN COLUMN LIQUID CHROMATOGRAPHY

The retention factors depend on the ratio of the volumes of the stationary and mobile phases in the column, i.e., on the column phase ratio, $\Phi = V_s/V_m$. In spite of continuing research, we still lack a reliable universal method for the determination of the phase volumes in liquid chromatography.

A simple, but questionable, method for the estimation of the column hold-up time is the time of the appearance of the first base line disturbance peak on the detector baseline. The thermodynamic void volume of the column can be obtained by integrating the plot of the retention times of the perturbation peaks from 0% to 100% of the organic modifier [10]. This approach is time-consuming, however, and the peak is not always very apparent or may not even appear, due to the impurities excluded, before the real column hold-up time.

Most frequently, the void volume is set equal to the elution volume of an "inert" compound, which does not interact with the solid phase (after the correction for extra-column volumes). The selection of a suitable marker compound is generally pragmatic and may not always yield the exact V_m values. Small polar molecules such as uracil or thiourea are often used since the column markers in reversed-phase chromatography do not guarantee 100% accurate thermodynamic data in all particular separation systems. Benzene or toluene are more suitable column hold-up volume markers in acetonitrile-rich mobile phases (used in the HILIC systems) [11]. However, the elution times of benzene and toluene on silica gel columns may slightly increase when the mobile phase reaches concentrations of 30% water. This means that the water amount adsorbed close to the polar adsorbent surface depends on the water concentration in the bulk mobile phase [12]. Even the retention time of a component of the mobile phase or of a monovalent salt may not provide 100% accurate V_m values [13].

Static methods of the determination of the column hold-up volume, such as the subsequent weighing of the column filled with two solvents of different densities (e.g., tetrachloromethane and methanol) [13] do not rely on the elution volume of an "inert" marker. McCalley and Neue recommended filling the column alternatively with water and methanol by flushing thoroughly with the solvent. After each flushing and capping the column ends, the excess solvents are wiped off and the column is weighed. Combining the weights of the liquid-filled columns and the solvent densities allows one to estimate the volume of the mobile phase in the column. However, this method cannot account for the preferential solvation of the stationary phase. The presence of the ions in the mobile phase may also strongly affect the phase volumes in the column, either inducing the exclusion or promoting the adsorption in the column [14].

Linearization of the logarithmic net retention times of the members of a homologous series was also suggested for the determination of V_m, but is lengthy and presumes the validity of the ideal reversed-phase model [14]. A more recent approach for the calculation of the phase ratio in reversed-phase chromatography from the measurements of retention factors k for two hydrocarbons, for which the octanol/water partition coefficients, log K_{ow}, are known, yields results strongly depending on the mobile phase composition [15].

The determination of the volume of the stationary phase or the interfacial area is even more difficult and usually requires a presumption of a particular retention model. The assignment of the solvent adsorbed on the solid phase surface either to the stationary phase or to the mobile phase volume is still controversial [16]. Further, the exact position of the boundary (dividing plane) between the bulk mobile phase and the liquid occluded on the stationary phase is difficult, if possible at all [17]. Hence, the phase ratio employed in the calculations of the distribution constants in LC depends on the retention mode and – deliberately or unintentionally – adopts a phase definition convention.

Convention 1 (classical). V_S is the column volume inaccessible to a non-retained marker compound, i.e., the empty column inner volume, V_{column}, minus the total pore volume, V_T (the inner pores, V_i, plus the inter-particle pores, V_0): $V_s = V_{column} - V_i - V_0$. This is the standard convention used in the traditional determination of the retention factors in LC. The volume of the mobile phase in reversed-phase LC is usually estimated as the elution volume of uracil or thiourea in 80–100% acetonitrile/water: $V_m = V_i + V_0$. In the normal-phase and HILIC, a small non-polar molecule (benzene, toluene) is the most frequent V_m marker. However, the phase volumes determined using the "inert" markers might be subject to errors.

Convention 2 (core-shell). Some columns contain "inert" parts, which do not take part in the retention process, as they are inaccessible to the sample and to the mobile phase, such as the solid non-porous core in superficially porous columns, V_{core}. Including the inactive part of the particles in the stationary phase yields overestimated retention factors in comparison to the thermodynamic k, considering only the active stationary phase (Equation (1.2)) [18]. Phase volumes are determined according to *convention 1*, after subtracting V_{core} (obtained from the manufacturer's data) not included in the stationary phase: $V_s = V_{column} - V_i - V_0 - V_{core}$.

Convention 3 (inner pore). In the (idealized) size-exclusion chromatography (SEC), the solid particle skeleton, V_{skelet}, is inert and does not participate in the distribution process, like the solid non-porous core. The full volume of the inner pores contains the liquid stationary phase, $V_s = V_i$, which has the same composition as the mobile phase in the inter-particle volume (such as tetrahydrofuran), $V_m = V_0$. The molecules of the solute distribute between V_i and V_0, based on their size, which controls the proportion of the accessible pore volume. The elution volume of a high-molecular weight standard for which the inner pores are inaccessible, (e.g., polystyrene with $M_r \geq 10^6$), estimates the volume of the mobile phase, V_m; the elution volume of a small hydrophobic molecule (benzene, toluene) is equal to the sum of the volumes of the inner pores and of the inter-particle space, $V_i + V_0$.

In the liquid–liquid (partition) LC, the liquid stationary phase volume, V_s, may fill only a part of the inner pore volume, but it differs from the bulk liquid mobile phase. The mobile phase is contained in the inter-particle volume, and in a part of the inner pores. This situation is typical for hydrophilic interaction liquid chromatography (HILIC), where, according to the original model, the sample distributes between the bulk organic solvent-rich mobile phase and water adsorbed on the polar solid surface [1]. The amount of the adsorbed water strongly depends on the composition of the bulk mobile phase, which makes the exact determination of the volume of the stationary phase difficult. *Convention 3*, setting the whole inner pore volume equal to the volume of the stationary phase in the column, $V_s \approx V_i = V_{column} - V_{skelet} - V_0$, overcomes this inconvenience, but does not provide the true volume of the stationary phase. Rather, it can be useful as a constant reference value independent of the mobile phase composition. The solid particle skeleton contributes to the retention mainly by the surface adsorption and we can neglect the contribution of its volume, V_{skelet}, to a first approximation.

1.2.2.1 Fully Porous Packing Particles

For rapid and efficient separations, the migration path of sample compounds to the active interaction sites should be as short as possible to minimize band-broadening by diffusion and to provide narrow peaks and high sample resolution. Since the advent of HPLC, there has been a continuing advancement in the reduction of particle size, d_p, to allow faster separations and higher separation efficiency in the terms of theoretical plate counts, N [19].

The length of the column needed to obtain the required number of theoretical plates, N, in the shortest separation time, depends on the optimum mobile phase flow velocity [20]. As a rule, 3–5 μm porous particles are used in conventional analytical columns and 2–3 μm porous particles in short (3–5 cm) columns for fast simple separations. Cheaper 10 μm or larger diameter particles are used for preparative separations in large columns.

Very fine particles (sub-2 μm in diameter) produce narrow peaks and allow very fast efficient separations at very high pressures (over 1,000 bar). Utilizing these advantages in practice requires minimizing extra-column dispersion. The ultrahigh-performance (UHPLC) technique requires special instrumentation with minimized extra-column flow-through volumes [21], for which various commercial UHPLC systems with pressure capabilities over 1,000 bar have been introduced [22]. Due to a very high operating pressure, UHPLC columns may show a short service life.

The mechanical friction between very small particles and the mobile phase flowing through the bed creates heat, which increases the temperature in the column. The heat generation is directly proportional to the pressure drop across the column and to the flow rate of the mobile phase; hence it increases with finer particles. As the heat generation is more significant in the center of the column than near the column wall through which the heat is dissipated, a radial temperature gradient forms in the column so that the viscosity and the flow characteristics change across the column cross-section. The retention factor and the diffusion coefficient of sample compounds depend on temperature, so that the solute migrates faster along the column at the center than near the walls. These effects cause additional band-broadening, which decreases the column efficiency and the beneficial effect of small particle size on the plate height. Hence, there are ultimate limits under which the diameter of fully porous particles cannot decrease unless deterioration rather than improvement in the column efficiency occurs [19]. With conventional-diameter analytical columns, these limits are close to $d_p \approx$ 1.5–1.7 μm [23]. In the columns of a smaller internal diameter, the radial temperature gradient is less significant due to a faster heat dissipation through the column walls, especially with efficient capillary columns of a diameter of 0.5 mm or less. The heat formation is also less important with non-porous particles due to the readily accessible adsorption centers on the surface. For example, fast separation (<1 min) of small molecules (barbiturates) on a capillary column packed with 1 μm non-porous C_{18} silica particles at 2,400 bar column pressure was reported [24].

1.2.2.2 Core-Shell Columns

The controlled surface porosity (CSP) materials, developed originally in the late sixties, represent a moderate-pressure alternative to UHPLC. The modern CSP (core-shell) particles consist of a thin (sub-1 μm) outer porous layer of an active stationary phase deposited on a solid inert impervious spherical fused-core sphere (1.5 to 5 μm) (Figure 1.1). Core-shell materials provide reduced band-broadening and outstanding efficiency, while preserving sufficient particle size to allow acceptable operation

FIGURE 1.1 Structure of a core-shell particle with a solid core surrounded by a porous active layer.

pressure [25]. A short diffusion path in a thin porous shell layer provides fast mass transfer kinetics and reduced band-broadening due to the shallow pores of the thin active adsorbent layer. The non-porous inner core of the shell particles represents between 25% and 36% of the particle volume, which is not accessible to any solute. For more details on the fabrication and the properties of the core-shell columns see, e.g., the review [26].

Not including the core volume, V_{core}, in the stationary phase, we can define the "thermodynamic" shell retention factor, $k_{shell} = k_{exp} \cdot f_{cor}$, considering only the volume of the active shell stationary phase that really participates in the chromatographic distribution process: $V_{s,shell} = V_C - V_M - V_{core} = V_s$. (*convention 2*). The shell correction factor, f_{cor}, accounts for the ratio of the thickness of the shell layer, d_{shell}, and the mean particle radius, r_{partic}, $\rho = d_{shell}/r_{partic}$ [18]:

$$f_{cor} = \left[1 - \left(1 - \frac{d_{shell}}{r_{partic}} \right)^3 \right] \tag{1.5}$$

Figures 1.2 and 1.3 illustrate the effects of the three stationary phase conventions (classical, core-shell and inner pore) on retention and selectivity in the example of two short core-shell columns (30×3 mm i.d.), 2.7 µm particle diameter, 1.7 µm

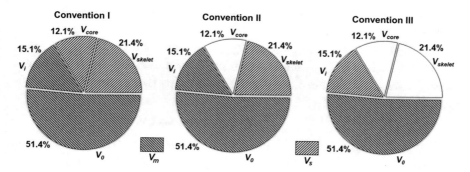

FIGURE 1.2 The three conventions defining the stationary phase. The Ascentis Express C18 (30×3 mm, particle i.d. 2.7 μm, solid core i.d. 1.7 μm) core-shell column (reversed-phase system). V_i – inner pore volume (32 μL), V_0 – interstitial volume (109 μL), V_{core} – volume of core in the particles (26 μL), V_{skelet} – volume of the solid shell skelet (45 μL).

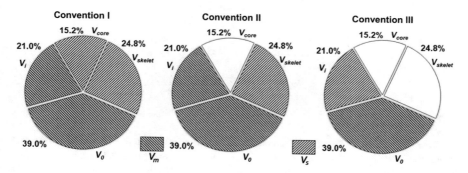

FIGURE 1.3 The three conventions defining the stationary phase. The Ascentis Express HILIC (30×3 mm, particle i.d. 2.7 μm, solid core i.d. 1.7 μm) core core-shell column (aqueous normal-phase system). V_i – inner pore volume (44.5 μL), V_0 – interstitial volume (82.7 μL), V_{core} – volume of core in the particles (32.5 μL), V_{skelet} – volume of the solid shell skelet (52.5 μL).

core diameter and 0.5 μm porous shell layer thickness. Uracil is the V_m marker for the Ascentis Express C18 column in the reversed-phase mode; for a polar Ascentis Express HILIC column, toluene is the V_m marker in the HILIC mode. Table 1.1 illustrates the differences in the stationary and mobile phase column volumes according to the classical, core-shell and inner pore conventions in the example of two core-shell columns in the aqueous acetonitrile mobile phases: A: Ascentis Express C18 (reversed-phase mode, uracil as the column hold-up volume marker); B: Ascentis Express HILIC column (HILIC mode, toluene as the column hold-up volume marker). The retention factors of hesperidine (k_1), hesperetine (k_2) in the RP system and of quercetin (k_3), esculin (k_4) in the HILIC system depend on the phase ratio convention, and are higher with *convention 3* than with *convention 2*; however, their relative retention (the selectivity factor) is independent of the phase-adopted convention.

TABLE 1.1

Distribution of the Column Volume, V_{column}, among the Solid Core, V_{core}, the Shell Solid Skeleton, V_{skelet}, the Inner Pores, V_i, and the Inter-Particle Space, V_0, in the Superficially Porous (core-shell) Columns Using Three Different Conventions

A – Ascentis Express C18

	V_{column} (µl)	V_m (µl)	V_S (µl)	k_1	k_2	$\alpha = k_2/k_1$
convention I	212	$V_0 + V_i$	$V_{core} + V_{skelet}$	2.5	14.6	5.7
(classical)		141 (uracil)	71 (33.5% V_{column})			
convention II	212	$V_0 + V_i$	V_{skelet}	1.6	9.3	5.7
(core-shell)		141 (uracil)	45 (24.3% V_{column})			
convention III	212	V_0	V_i	1.5	8.5	5.7
(inner pores)		109	32 (22.7% V_{column})			

B – Ascentis Express HILIC

	V_{column} (µl)	V_m (µl)	V_S (µl)	k_3	k_4	$\alpha = k_4/k_3$
convention I	212	$V_0 + V_i$	$V_{core} + V_{skelet}$	7.8	11.3	1.44
(classical)		127 (toluene)	85 (40.0% V_{column})			
convention II	212	$V_0 + V_i$	V_{skelet}	4.9	7.0	1.44
(core-shell)		141 (toluene)	52.5 (29.2% V_{column})			
convention III	212	V_0	V_i	6.3	9.1	1.44
(inner pores)		82.7	44.5 (35.0% V_{column})			

A – Ascentis Express C18 (30×3 mm, particle i.d. 2.7 µm), k_1=Hesperidin, k_2=Hesperetin (19% acetonitrile/water – RP mode). **B** – Ascentis Express HILIC (30×3 mm, particle i.d. 2.7 µm), k_3=Quercetine, k_4=Esculine (4% water/acetonitrile – HILIC mode). V_{column} – empty column inner volume; V_m – volume of the mobile phase in the column; V_s – volume of the stationary phase in the column; (A: $V_i = V_{skelet} + V_{e,uracil} - V_{e,PSI\,800\,000}$; B: $V_i = V_{skelet} + V_{e,toluen} - V_{e,PSI\,800\,000}$), skelet; V_{PS} – elution volume of polystyrene, Mr = 1 800 000); $k_{1,3}$ – retention factor of the less retained compounds; $k_{2,4}$ – retention factor of the more retained compounds; α – selectivity factor.

Figure 1.4 shows the effect of the volume fraction of acetonitrile on the retention of two flavonoid compounds over a broader mobile phase composition range (0–30% acetonitrile in water) on an Ascentis C18 column. Figure 1.5 illustrates the analogous effects of the water concentration on the retention of another pair of compounds on an Ascentis HILIC column in the RP system (in 0–10% water in acetonitrile).

Convention 2 (core-shell), not including the core volume, V_{core}, in the stationary phase provides theoretically more correct retention data. The distribution constant, $K_{D,core-shell}$, which decreases proportionally to the core correction factor, f_{cor}, $K_{D,core-shell} = K_D \cdot f_{cor}$. The main advantage of *convention 3* (inner pore) is that the volume of the stationary phase does not depend on the solvation of the solid phase skeleton, but rather necessitates the determination of the inner pore volume using a large molecule

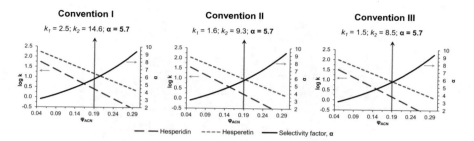

FIGURE 1.4 The effects of the volume fraction of acetonitrile on the retention factors of hesperidine, k_1, and hesperetine, k_2, on the Ascentis Express C18 column (30×3 mm, particle i.d. 2.7 µm, solid core i.d. 1.7 µm), calculated using different phase volume conventions; α – selectivity factor (k_2/k_1). Reversed-phase system.

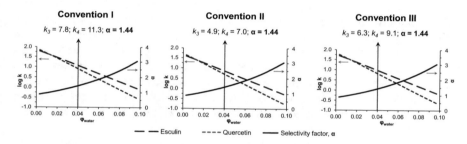

FIGURE 1.5 The effects of the volume fraction of water on the retention factors of esculin, k_3, and quercetin, k_4, on the Ascentis Express HILIC column (30×3 mm, particle i.d. 2.7 µm, solid core i.d. 1.7 µm), calculated using different phase volume conventions; α – selectivity factor (k_4/k_3). HILIC system.

(such as polystyrene with $M_r \geq 10^6$). *Convention 3* (inner pores), considering principally a liquid stationary phase, is potentially useful in HILIC and in dual-mode HILIC/RP separation systems. Hence, various conventions are suitable for method development and optimization, if consequently used, but the specification of the convention used for the calculation of the phase volumes in the column is recommended.

1.2.2.3 Monolithic Columns

Monolithic columns for HPLC do not contain particles, but rather consist of a single piece of continuous separation media (rods) with dual pore morphology: a network of small pores interconnected by large flow-through pores, providing a good bed permeability and a low flow resistance. Approximately three-times faster analyses are possible in comparison to the particulate packed columns of the same length at the same operating pressure [27]. Two types of monolithic columns with different skeletons and pore morphologies are suitable for LC separations: silica gel-based monoliths [28, 29] and organic polymer monoliths [30]; see Figure 1.6A and B.

Co-polymerization of tetramethylsiloxane and methyltrimethoxysiloxane is the usual technique for preparation of silica monolithic columns [31].

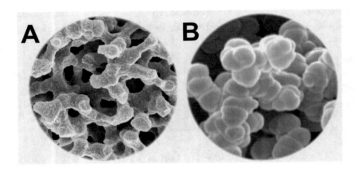

FIGURE 1.6 The internal structures of a silica-based monolith (A) and a polymer-based monolith (B).

Direct *in situ* polymerization in fused silica capillaries is possible for preparing micro-columns with inner diameters of 0.5 mm or less. Synchronizing phase separation and gelation prevents polymer skeleton shrinkage and the formation of void volumes between the silica monolith and the chromatographic column walls during polymerization, which would destroy the column performance [32]. However, it is difficult to achieve this effect during polymerization in wider tubes used as the traditional HPLC formats. To solve this issue, the manufacturers fabricate polymeric monolithic rods, which they firmly stick inside dual-wall organic polymer stainless steel tubes.

More recently, a group of Korean researchers prepared hybrid monolithic–particulate stationary phases from ground silica monoliths, chemically modified and then slurry-packed into stainless steel tubes, like classical column particles [33]. The packed bed has a partially monolithic structure enabling fast slurry-packing and, in spite of using column frits, the permeability of the column based on the ground silica monolithic particles is better than that of the column based on the spherical silica particles at similar column efficiencies.

Silica gel monoliths (Figure 1.6A) have a bimodal pore structure with significant representation of 7–12 nm mesopores (~13%). The specific surface area is in the range of several hundred m²/g and the silica monoliths are suitable for the separations of small molecules, as the mesopore size allows fast diffusion and easy solute penetration of the adsorption sites, which results in an efficiency of separation comparable to 5 μm particles [34]. Monolithic silica rods with chemically bonded alkyl or other functionalities allow fast separations of low-molecular samples, with column efficiencies up to 100,000 theoretical plates/m [29]. Unfortunately, silica monoliths generally show less good performance for separations of macromolecular compounds such as proteins and other biopolymers. Further, limited stability at temperatures higher than 60°C and pH > 8.5, similar to silica particles, might affect their performance for the analysis of polar – especially basic – compounds.

Organic polymer monoliths are usually prepared by *in situ* radical polymerization of monovinyl and cross-linking divinyl monomers in the presence of pore-forming solvents, typically alcohols [35, 36]. The polymerization provides microglobules resembling cauliflower, interconnected by large through-pores (15–100 nm)

(Figure 1.6B). A relatively low specific surface area (10–150 m^2/g) is more suitable for the separation of large biopolymers, retained on the organic polymer monoliths due to the convection rather than to the diffusion [30].

Organic polymer monoliths include polystyrene-, polymethacrylate-, polyacrylamide- and other matrices, which offer a variety of stationary phases by selecting the components of polymerization mixture or by post-polymerization modification. Their advantage is a simple preparation in various column diameters, high temperature and pH stability [37].

Organic polymer monoliths provide relatively low separation efficiency for small molecules, which is due to a restricted mesopore size, causing slow diffusion [38]. Increasing the proportion of mesopores with larger diameter improves the diffusion and efficiency of organic polymer monoliths for small molecules [39]. Optimizing a single-step polymerization by adjusting the polymerization time, temperature and composition of the polymerization mixture, or a post-polymerization modification of the monoliths, can be applied for this purpose [30].

The independent control of the pore size distribution of both large flow-through pores and small mesopores in the organic polymer monoliths during a single-step polymerization is a challenge [11]. Terminating the polymerization reaction at an early stage is a simple way to control both the porosity and the related efficiency of polymer monoliths [40]; however, the repeatability depends on the exact control of the polymerization time.

Another approach for controlling the pore morphology is optimizing the temperature or (and) the composition of the polymerization mixture. The polarity and the size of both the functional monomer and of the cross-linker, and their respective concentrations in the polymerization mixture, affect the early stages of the polymerization, where the phase separation and formation of the first cross-linked polymeric nuclei occur. The efficiency of the poly(methacrylate) monolithic columns for low-molecular compounds improves with increasing the chain length of the poly(methylene) dimethacrylate cross-linkers, decreasing the proportion of the large through-pores (>50 nm) and increasing the proportion of smaller mesopores (1–50 nm) [41].

The proportion of the large through-pores decreases and the separation efficiency for small polar molecules significantly improves when substituting non-polar poly(methylene dimethacrylates) with more polar poly(oxyethylene) dimethacrylate cross-linkers with surface-active properties, potentially affecting the gel solvation of the monoliths. Columns prepared with tetraoxyethylene dimethacrylate cross-linkers show high permeability, excellent reproducibility and long-term stability at elevated temperatures, with high efficiencies of up to 70,000 theoretical plates/m for the separations of both biopolymers and small polar molecules in the reversed-phase [41] or HILIC systems [42]. Using living radical polymerization instead of free radical polymerization could possibly improve the band-broadening and separation efficiency, leading to more homogenous monolithic stationary phases [39].

Another approach to improving the efficiency of the separation of small molecules is the two-step post-polymerization modification of polymeric monolithic columns [43]. The approaches employed to this aim include the reaction of the pre-existing surface groups, the addition of polymeric chains, the addition of novel, highly ordered nanomaterials [44] or, finally, photo-grafting, which forms a dense polymeric network with a separate zwitterionic functionality at the hydrophobic scaffold, shifting the proportion of flow-through pores in favor of inner pores <50 nm due to the additional cross-linker in the zwitterionic layer (Figure 1.7) [45].

Hyper-cross-linking surface modification in the second step is a useful tool for adjusting the density of monolithic micro-columns. Generally, the polymerization mixtures based on monomers such as styrene or vinylbenzyl chloride, and divinylbenzene as a cross-linker, provide monolithic stationary phases suitable for a hyper-cross-linking modification. Because of its higher reactivity, divinylbenzene participates dominantly at the beginning of the polymerization reaction, but its concentration in the polymerization mixture rapidly decreases as the polymerization proceeds. At the end of the first-step polymerization reaction, the polymer chains at the surface of the monolith are only lightly cross-linked with a layer rich in monovinyl monomers with chloromethyl functionalities. Then, the Friedel–Crafts alkylation reaction modifies the surface of the polymer monolith, increasing the cross-linking density of the monolithic stationary phase [46].

1.2.2.4 Micro-Pillar Array Columns

Micro-pillar arrays are a powerful alternative for classical packed bed columns and monoliths, especially for separations of complex samples requiring a high number of theoretical plates, such as in the analysis of proteomes and peptidomes [47]. Freestanding micro-pillars on a thin silicon wafer are produced by a lithographic etching process (Figure 1.8). The accurately positioned arrangement of micro-pillars forms a perfectly ordered separation bed with reduced sample dispersion, as all molecules follow identical paths through the column. In comparison to a column densely packed with particles of the same size, the flow resistance of pillar columns drops by approximately 32%. High permeability allows for the operating of long columns at moderate pressures and for the generating of highly efficient peptide maps, comparing favorably to the particulate or monolithic nano-columns. To arrange a micro-pillar column several meters long on the surface of a silicon wafer of only several

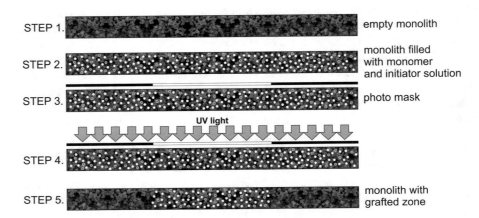

FIGURE 1.7 Schematic of single step photo-grafting of a zwitterionic zone upon a monolithic hydrophobic scaffold column.

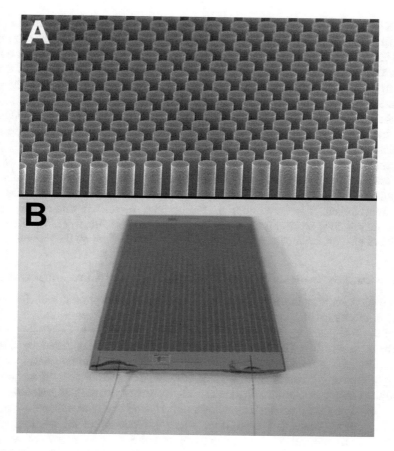

FIGURE 1.8 A micro-pillar stationary phase array. A – micro-pillar arrangement detail; B – a silicon wafer carrying a folded 150 cm-long folded pillar array column.

centimeters, the channel design must include a large number of turns to connect the different bed segments of a limited length (e.g., up to 5 cm) to minimize variations in packing dimensions across the column length. The connecting turns slightly affect the efficiency of the micro-pillar array arrangement.

A 3-m-long column with micro-pillars, 5 μm in diameter, 20 μm high (positioned at a distance of 2.5 μm from each other), covered with a 300 nm surface porous layer and chemically modified with C18 ligands allowed achievement of the efficiency of one million theoretical plates in the relatively short time of 20 minutes at 350 bar [48].

1.3 PHASE SYSTEMS AND SEPARATION MODES IN HPLC

The differences in the LC separation of non-ionic compounds depend mainly on the differences in their polarities. In normal-phase separation systems (NP), the stationary phase is more polar than the mobile phase, and the retention increases proportionally to the solute polarity and decreases with increasing mobile phase polarity, opposite to what happens in reversed-phase (RP) chromatography. Many ionizable compounds are difficult to separate in the RP systems because of too low a retention. Ionic additives to the mobile phase often suppress the ionization of weakly basic or weakly acidic compounds, increasing their retention and enabling the separation. Ion exchange chromatography (IEC), a traditional technique for the separation of small organic ionic compounds, nowadays serves mainly for the analysis of small inorganic ions or of ionic biopolymers, because of a relatively low separation efficiency for small organic molecules.

In the real world, pure separation modes rarely control the separations of both ionic and non-ionic compounds in the NP and RP systems. Most often, polar and non-polar (and possibly ion exchange or steric) interactions participate in the separation mechanism. It is very important to note that the relative contribution of the individual retention mechanisms depends not only on the chemistry of the stationary phase, but also on the mobile phase, which has a strong effect on the retention.

1.3.1 ORGANIC SOLVENT NORMAL-PHASE (ADSORPTION) CHROMATOGRAPHY

Normal-phase chromatography (NPC), known also as adsorption liquid chromatography, is the oldest liquid chromatographic mode, having been already employed more than 100 years ago by M.S. Tswett. The stationary phase in NPC is a polar adsorbent, most often silica gel, either bare or chemically modified by bonding amino, diol, etc. groups, more polar in comparison to the mobile phase, which is usually a mixture of two or more organic solvents of different polarities. Alumina, zirconium dioxide or titanium dioxide are relatively rarely used. The adsorption sites occupy fixed positions on the surface of a polar adsorbent. The molecules with the polar functional groups in the positions fitting the location of the adsorption centers show a stronger retention in comparison to the molecules with another geometry, which is the reason of excellent isomer separation selectivity of the NPC. The polarity and the elution strength, i.e. the ability to enhance the elution, generally increases in the following order of the most common solvents: hexane ≈ heptane ≈ octane < methylene chloride < methyl-t-butyl ether < ethyl acetate < dioxane < acetonitrile ≈ tetrahydrofuran < 1- or 2-propanol < methanol < water. The preferential adsorption of polar solvents, especially water, often causes long equilibration times when changing the separation conditions. Further, the poor solubility of polar samples in non-polar solvents seriously limits the application possibilities of NPC separations in purely organic mobile phases.

1.3.2 REVERSED-PHASE CHROMATOGRAPHY

Reversed-phase chromatography (RPC), even though introduced later than NPC, is nowadays widely used in more than 90% of HPLC applications. The RPC is the first choice among the LC separation modes, because even minute structural differences in the non-polar hydrocarbon part of the molecules often allow for successful separation of a great variety of samples, containing non-polar, polar and even ionic compounds. In contrast to NPC, the majority of stationary phases in RPC are hydrophobic, whereas the mobile phase is usually an aqueous solution of one or more polar organic solvents (usually methanol or acetonitrile, but other solvents may be also used, if compatible with the detection). The solute polarity decreases its retention, which also drops as the concentration of the organic solvent in the aqueous-organic mobile phase increases. In RPC, the ionic forms of weak acids or bases elute very early; ionic additives to the mobile phase (buffers, neutral salts, weak acids or ionic surfactants) usually enhance the retention and improve the separation.

The stationary phases for RPC are usually prepared by covalently bonding organo-silane reagents on the silanol (Si-OH) groups of the from spherical silica gel particles. A plethora of columns with bonded siloxane Si-O-Si-R groups is available for RPC. The bonded ligand R is usually an alkyl, most often C_8 or C_{18}, but many other bonded phases are available. Theoretically, the alkylsilica stationary phases are similar to liquid alkanes immobilized on a solid support, but the bonded alkyl chains differ from the free molecules of liquid hydrocarbons in terms of their limited mobility. The organic solvents used as the components of the mobile phases preferentially solvate the stationary phase and modify its properties [49, 50]. Finally, the attractive specific polar interactions with the stationary phase may more or less contribute to the retention.

For steric reasons, at least 50% of the original silanol groups of the silica gel remain unreacted after chemical modification. The residual silanols may give rise to unwanted interactions with polar (especially basic) solutes, which results in their poor and irreproducible separation, band-tailing or distorted peak shape. Many RP stationary phases are stable only up to pH 8–9 and at temperatures up to 60°C. To improve the chemical and temperature stability of the RP columns, "endcapping" reactions with a small molecule of trimethylchlorosilane or hexamethyldisilazane bond some residual silanol groups. This approach improves the stability of the stationary phases. Diisopropyl- or diisobutyl-alkylchlorosilanes utilize the branched side alkyls to shield the Si–O–Si bonds. Some fully endcapped alkyl bonded phases are claimed to resist the mobile phases up to pH 11. The stationary phases with an embedded amide or carbamate group in the bonded chain close to the silica gel surface provide polar interactions with the residual silanols on the surface of the chemically bonded stationary phase, which decrease the silanol activity toward the sample compounds. Moreover, the embedded groups improve the "wettability" of the stationary phase surface in water-rich mobile phases to avoid the collapse of the non-polar alkyl chains.

Materials made up of ethylene-bridged hybrid porous silica particles exhibit high stability even at the eluent temperatures higher than 100°C in an enlarged pH range. The superior mechanical stability allows their use in the UHPLC up to 1,000 bars [51].

Type C silica, i.e., the "hydride silica," has up to 95% of the original silanol surface groups replaced by the non-polar silicon hydride Si–H groups during the hydrosilation process (Figure 1.9), which changes the fundamental properties of the original silica gel type B. The surface of the silica hydride materials populated with the Si–H groups does not significantly attract water, which results in improved reproducibility of retention [52]. Hydrosilated silica gel chemically modified by low-polarity bonded groups, such as bidentate undecyl silica or cholesterol, provides improved selectivity for the separations of weakly polar compounds [53].

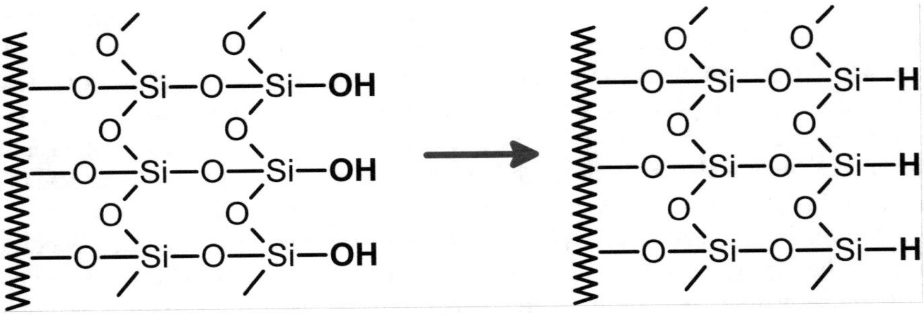

FIGURE 1.9   Hydrosilation of silica (type B) surface to a silica hydride stationary phase.

The chemistry of the bonded ligand, the amount and type of residual silanols, "endcapping" and pore size, among other factors, affect the chromatographic properties. Generally, the retention in RP LC increases with increasing content of carbon atoms in the chemically bonded phase, i.e., with increasing the length of the bonded alkyl chains. The individual columns can exhibit unique characteristics due to specific polar and ionic properties responsible for secondary intermolecular interaction mechanisms, providing advantages for specific separation cases. A recently published review [54] provides a comprehensive survey of the silica bonded stationary phases (not only) for RP LC and presents useful hints for their selection.

The exact retention mechanism in reversed-phase chromatography is still controversial [50]. The retention of a solute may result from the partitioning between the layer of the solvated bonded non-polar groups of the stationary phase, from the adsorption or from their combination [55]. In a first approximation, the interactions in the non-polar stationary phase are less significant than the polar interactions in the mobile phase, which are the main factor controlling the retention.

According to the solvophobic partitioning model, the transition of a solute molecule from the bulk mobile phase to the surface of the stationary phase results from a decrease in the contact area of the solute with the mobile phase. Replacement of weaker interactions between a moderately polar solute and a strongly polar mobile phase with mutual interactions between strongly polar molecules of the mobile phase in the space originally occupied by a solute molecule results in an overall energy decrease in the system. The solvophobic interactions are the driving force of the retention in the absence of strong (polar) interactions of the solute with the stationary phase [56].

A modified partitioning model proposed a three-step process involving the creation of an appropriate sized cavity in the stationary phase, the transfer of the solute from the mobile phase to the stationary phase and the closing of the solute-sized cavity in the mobile phase. The solute penetrates in between the stationary-phase chains [57].

The interphase model attributes the retention to the active region on the surface of stationary phase, but does not allow quantitative predictions. A thin diffuse interphase region between the bonded alkyl chains and the bulk mobile phase is heterogeneous and its thickness depends on the character of the bonded ligands, the composition of the mobile phase and other factors [58].

1.3.3 HYDROPHILIC INTERACTION LIQUID CHROMATOGRAPHY (HILIC)

Very hydrophilic samples such as carbohydrates or small polar compounds usually elute close to the column hold-up volume in RPC, so that their separation from one another and from polar matrix interferences may be difficult to accomplish, even in highly aqueous mobile phases [59]. On the other hand, some polar compounds elute late, if at all, in organic solvent normal-phase systems. The separation on polar stationary phases often improves after adding water to the mobile phase. Alpert introduced the name "hydrophilic interaction liquid chromatography" (HILIC) for the separation mode employing a "normal phase column" in combination with a "reversed-phase mobile phase," containing less than 50% water [1]. The term "hydrophilic" refers to the affinity to water. The HILIC technique provides appropriate retention and resolution for many polar compounds, often with better separation efficiency in comparison to RP chromatography. The diffusion coefficients of ionized basic compounds in less viscous organic-rich mobile phases under HILIC conditions are approximately twice those under RP conditions, leading to improved separation efficiency (lower height equivalent of a theoretical plate, H) [60]. Further, the peak symmetry of basic compounds in the HILIC mobile phases (acetonitrile – ammonium acetate or formate buffers) often improves in comparison to reversed-phase HPLC [61]. HILIC separation of ionic/ionizable compounds needs ionic additives to aqueous-organic mobile phases [62].

Finally yet importantly, the popularity of HILIC is due to the excellent suitability of the technique for coupling to mass spectrometry (LC/MS) [63].

The quality of HILIC separations strongly depends on both the stationary phase and on the composition of the mobile phase [64]. Bare silica gel is frequently used as the stationary phase in HILIC applications. The low viscosity of acetonitrile-rich (75–95%) mobile phases allows operating columns packed with sub-2 µm particles at moderate pressures, providing fast (< 1 min) separations of polar drugs with conventional HPLC instrumentation. 1.7 µm ethylene-bridged hybrid organic-silica (BEH) particles show an improved chemical resistance at pH < 5 with respect to the bare silica in HILIC [65].

Silica gel chemically modified with polar functionalities such as diol-, amino-, amide-, cyclodextrin, ion exchange, zwitterionic, poly(2-hydroxyethyl aspartamide) or poly(succinimide) show improved retention and separation selectivity for various sample types in aqueous-organic mobile phases [66, 67]. Sufoalkylbetaine stationary phases are especially suitable for HILIC separations [68]. The active zwitterionic layer grafted onto a wide-pore silica gel or on a polymer support contains both strongly acidic sulphonic acid groups and strongly basic quaternary ammonium groups, separated from each other by a short alkyl spacer. The two oppositely charged groups are present in a 1:1 molar ratio, so that there is only a very low net negative surface charge of the bonded layer, attributed to the larger distance of the sulphonic groups from the silica gel surface [69].

The HILIC retention mechanism is obviously complex; the solid adsorbent is not just an inert support for the adsorbed water layer which, as a liquid stationary phase, is in the partition equilibria with the bulk aqueous-organic mobile phase. Obviously, adsorption of polar compounds on the solid phase may participate in the HILIC retention mechanism, due to hydrogen-bonding, ionic and other interactions with bonded polar functional groups, or residual silanols on the silica-based polar bonded phases. Ion exchange or ion repulsion may be also involved in the distribution of partially ionized analytes [70].

The water adsorbed on a polar column is probably the most important factor controlling the retention [71]. The ion exchange interactions obviously contribute more or less to the retention of ionizable compounds, often giving rise to a mixed retention mechanism [2]. Obviously, adsorption and partition actually coexist in HILIC and the contribution of each mechanism depends on the solute, the hydration and the charge of the stationary phase polar functional groups and on the eluting conditions.

Silica hydride (silica gel type C) shows low attraction for water; hence, Pesek and Matyska [52] claim that the predominant role of the adsorption mechanism in the HILIC is played by hydrosilated silica, which they prefer to call "aqueous normal phase chromatography" (ANPC).

In non-aqueous HILIC chromatography (NA-HILIC), water is replaced by a polar organic solvent, an organic "protic modifier" such as ethylene diol, methanol or ethanol (with the elution strength decreasing in this order). NA-HILIC covers the gap between the non-aqueous NP and the aqueous-organic HILIC of polar samples and is helpful in the analysis of some poorly soluble oligomers or weakly polar compounds, which may precipitate in water-containing HILIC mobile phases. On the bare silica, diol, thioglycerol and oxidized thioglycerol polar bonded phases, the retention and separation selectivity of nucleic bases, nucleosides and deoxynucleotides depends on the type of protic solvent added to acetonitrile [72]. Methanol is the first choice protic solvent for NA-HILIC. A change in the concentration of methanol in the polar organic mobile phases affects the retention more significantly than a change in the concentration of water in the HILIC mode. The retention of phenolic acids on the hydrosilated silica bonded columns is very weak in mobile phases containing more than 20% methanol in acetonitrile. Resolution and selectivity are generally better in the aqueous systems. At increased temperatures, the retention factors and peak widths decrease in both ANP and NANP (non-aqueous normal phase), showing linear ln k versus 1/T plots, due to a single retention mechanism over the temperature range from 25°C up to the column stability limit; however, low temperatures improve the resolution. The differences between the aqueous and non-aqueous modes are possibly due to the absence of the adsorbed water layer in NANP [73].

1.4   MOBILE PHASE EFFECTS ON RETENTION

A single solvent only rarely provides suitable separation selectivity and retention in LC. The mobile phase in the classical adsorption chromatography is a mixture of two (or more) organic solvents and a non-polar one, e.g., n-hexane and 2-propanol. Selecting appropriate concentration ratios of the organic solvents in a two- or a multi-component mobile phase is an efficient tool for adjusting the retention in the NPC.

The mobile phase in RP LC contains water and one or more polar organic solvents. The most useful are, in order of decreasing polarities, acetonitrile, methanol, dioxan, tetrahydrofuran and propanol. By the choice of the type of organic solvent, selective dipole–dipole, proton-donor or proton-acceptor, polar interactions with the analytes can be enhanced or suppressed, to adjust chromatographic separation. Binary mobile phases often make the adequate separation of various samples possible. Ternary or, less often, quaternary mobile phases offer the advantage of fine-tuning the optimum selectivity of more difficult separations [74].

The mobile phase in HILIC systems contains 5–40% water in a polar organic solvent. Protic solvents (lower alcohols) are more similar to water in providing proton donor-acceptor interactions. Consequently, they show lower retention and selectivity in HILIC in comparison to the aprotic acetonitrile, which is the most frequently used organic solvent in HILIC [72].

1.4.1 Solvent Adsorption in the Stationary Phase

As a rule, HPLC employs mixed mobile phases, comprising solvents showing different affinities to the stationary phases. The composition of the mobile phase at the solid phase surface is different from its bulk composition due to the selective solvation of the stationary phase by the molecules of the solvent with a higher affinity to the stationary phase [75, 49]. The solvation changes the properties of the stationary phase and affects the distribution equilibrium of the analytes.

Polar adsorbents preferentially take up polar solvents, especially water, from mixed mobile phases. In adsorption NPC, fully organic mobile phases may contain different levels of trace water concentrations. A compact water layer forms on the adsorbent surface. When adjusting the proportion of the organic solvents in the binary mobile phases, the trace concentration of water changes, too, sometimes with unexpected effects on the separation.

Water does not adsorb significantly on the non-polar phases with chemically bonded alkyl or other non-polar ligands used in reversed-phase chromatography. These phases usually show preferential adsorption of the organic components from the mixed aqueous-organic mobile phases [76]. The adsorption of the organic solvents affects the hydrophobicity and polarity of the stationary phase surface. [50] The surface coverage of the endcapped C18-silica materials affects the excess adsorption of commonly used organic solvents from water [77].

The polar stationary phases show large variability in the water amount adsorbed from the aqueous-organic mobile phases. Various, fully porous, core-shell silica gel, hybrid organic-silica and hydrosilated silica, bare and silica bonded polar stationary phases with bonded cholesteryl, phenyl, nitrile, pentafluorophenylpropyl, diol, zwitterionic sulfobetaine, phosphorylcholine and other ligands take up water from aqueous-organic mobile phases. The adsorbed water layer participates in the retention mechanism. Water is miscible at any proportion with acetonitrile, acetone, methanol or other polar organic solvents used in the HILIC mode. Consequently, the water adsorbed from aqueous-organic mobile phases forms a diffuse layer lacking fixed boundaries. Hydrogen bonds of water to silanol or other polar groups almost immobilize water molecules close to the surface of the bare silica or to the bonded zwitterionic sulphobetaine groups. The concentration of water progressively decreases from the polar solid surface toward the bulk organic-rich mobile phase outside (possibly, even partly inside) the pores of the stationary phase [2].

Computer molecular dynamics simulation studies indicate that the relative proportion of the amount of water contained in the pores of silica-based phases to the water concentration in the bulk mobile phase increases at low total water concentrations in the column [78]. This may be the reason why the LC separations are often irreproducible or fail in mobile phases containing less than 2% water in acetonitrile. The water molecules close to the silica surface strongly adhere to the silanol groups by the hydrogen bonds. Three types of water molecules coexist inside the 6–10 nm pores: free water molecules, "freezable" bound water and bound water that does not freeze at the regular water freezing temperature [79].

Because of the diffuse character of the adsorbed water layer, it is not possible to determine its thickness exactly, but we can measure the amount of adsorbed water. In the acetonitrile-water mobile phases the elution time of toluene slightly decreases as the water content increases, even though the non-polar toluene cannot penetrate inside the water layer [80, 81]. This behavior indicates an increasingly thick water-rich liquid layer on the stationary phase and allows for an estimation of the volume of the water-rich liquid layer as the difference between the toluene elution volume in a mobile phase containing a specific volume fraction of acetonitrile and its elution volume in the pure acetonitrile [82].

The amount of water taken up by the column depends on the distribution isotherm of water between the stationary and mobile phases. The static method of the isotherm measurement employs several vials containing small amounts of the dry column packing material in solvent mixtures containing varying water concentrations in an organic solvent. After equilibration, the supernatant solution in each vial is subject to direct analysis, best by Karl–Fisher titration, to acquire the data for the construction of the excess water isotherm [71].

The water isotherm data can also be acquired using dynamic methods not needing the bulk column packing material, which may not be readily available. The most frequently used dynamic methods measure the non-adsorbed amount of a component of the mobile phase continuously introduced onto the column, monitoring the column effluent using a non-specific (refractive index or low-wavelength UV) detector [83].

Direct methods of water determination are also suitable for a dynamic frontal analysis technique. The analyzed aqueous-organic solvent mixture continuously flows through the column and water in the collected small effluent fractions are determined by the Karl–Fischer titration. The difference between the water amount fed onto the column and the residual water amount determined by the Karl–Fischer titration in the effluent fractions corresponds to the adsorbed amount of water in the individual fractions [3].

The Langmuir model Equation (1.6) describes the water isotherms satisfactorily [84]:

$$q_i = \frac{a \cdot c_m}{1 + c_m \left. a \middle/ q_s \right.} \tag{1.6}$$

q_i is the concentration of adsorbed water, c_m is the concentration of water in the bulk mobile phase and a is the distribution constant of water in the pores of the stationary phase at very low c_m. q_s in Equation (1.6) gives the maximum (saturation) column adsorption capacity for water. Figure 1.10 compares three water isotherms on a zwitterionic ZIC-HILIC, a polymeric TSKGel Amide and a hydrosilated Cogent Silica C column [3].

There are large differences in water uptake between the individual stationary phases (Figure 1.11). Less than 9% v/v water in the mobile phase is sufficient for the full water saturation of porous, core-shell and hybrid ethylene-bridged silica gel columns. On the other hand, the water adsorption isotherms on the ZIC-HILIC and TSK gel amide 80 columns are shallow (Figure 1.10); even in 20% water in acetonitrile the columns are not fully saturated with water. The columns with bonded polar ligands (hydroxyl, diol, nitrile, pentafluorophenylpropyl, diol, zwitterionic sulphobetaine and phosphorylcholine) show stronger water adsorption in comparison to bare silica. At full column saturation, the excess adsorbed water, V_{ex}, fills up to 45% of the pore volume of silica-based columns. This corresponds to 6–9 monomolecular water layer equivalents for the ZIC-cHILIC and ZIC-HILIC zwitterionic stationary phases, to 3–5 monomolecular water layer equivalents for the stationary phases with bonded hydroxyl groups (Luna HILIC, YMC Triart diol and Ascentis Express OH5) and to 1–2 water layer equivalents for the XBridge HILIC and Atlantis HILIC columns (Figure 1.11). Due to low affinity to water, the silica hydride-based stationary phases adsorb less than one monomolecular water layer equivalent (full horizontal line), corresponding to 0.2–0.4% water in the 2.6–5.5% of the inner pore volume [3].

A hydrosilated Cogent Silica C column shows a steep isotherm with a low plateau water concentration (water uptake of 2.6% inner pore volume, i.e., 0.45 water monolayer on the adsorbent surface – Figure 1.10).

The presence of electrolytes in the mobile phase usually increases the adsorption of water and consequently the retention of non-polar compounds [12].

A sub-monomolecular layer of adsorbed water does not provide enough space for a sample partition. Hence, competition between the surface-adsorbed water and polar solutes controls the retention [85]. A low number of the adsorbed monomolecular equivalents may distinguish the aqueous normal-phase (ANP) from the traditional HILIC systems [3].

1.4.2 THE MOBILE PHASE-RETENTION MODELS

Some of the retention models describing the theoretical effects of the mobile phase on retention date from the early days of HPLC, such as the Snyder–Soczewinski displacement model of retention in normal-phase adsorption chromatography [86]. Retention results from the competition between the molecules of the solute and of the strong solvent for the active sites on the adsorbent surface. In a two-component mobile phase comprised of a stronger (more polar) solvent B and a weak (less polar) solvent A, the solute retention factor, k_{ab}, depends on the specific adsorbent surface, A_s, the adsorbent activity, α', and the solvent strength of the mixed mobile phase, ε_{ab}.

$$\log k_{ab} = \log k_a + \alpha' \cdot A_s \left(\varepsilon_a - \varepsilon_{ab} \right) \qquad (1.7)$$

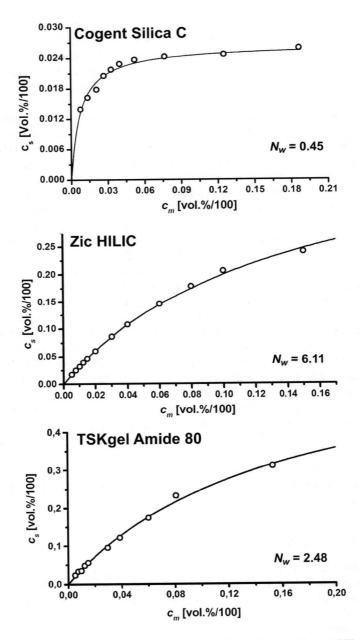

FIGURE 1.10 The Langmuir isotherms of water adsorbed on a Cogent Silica C silica hydride, a zwitterionic sulfobetaine (ZIC-HILIC) and an amide (TSkgel Amide 80) silica bonded stationary phases. c_s – the volume fraction of excess water contained in the pores of the stationary phase; c_m – the volume fraction of water in the mobile phase in equilibrium with the stationary phase; N_w – number of monomolecular water layer equivalent. Based on the data in [2].

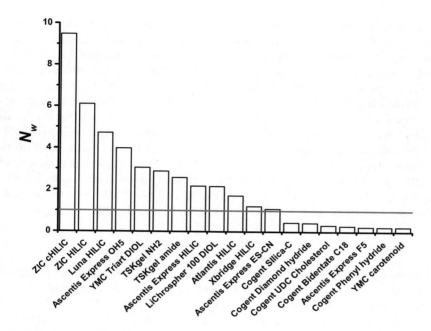

FIGURE 1.11 The equivalent number of the adsorbed monomolecular water layers, N_w, per the inner pore surface unit at the full saturation capacity on different stationary phases. The horizontal line denotes the hypothetical monomolecular water adsorption. Based on the data in [3].

The solvent strength ε_{ab} of a mixed binary phase that results from the solvent strengths of the stronger (more polar) solvent B, ε_b, and a less polar solvent A, ε_a, depends on the molar fraction, x_b, of the strong solvent B in the mobile phase:

$$\varepsilon_{ab} = \varepsilon_a + \frac{\log\left[x_b\left(10^{\alpha' \cdot n_b(\varepsilon_b - \varepsilon_a)} + 1 - x_b\right)\right]}{\alpha' \cdot n_b} \tag{1.8}$$

With some simplification, the displacement adsorption model yields the simple Equation (1.9) describing the retention factor, k, as a function of the volume fraction of the strong solvent B, φ, in the binary mobile phase [87, 88].

$$\log k = a - m \cdot \log \varphi \tag{1.9}$$

$a = \log k_b$, where k_b is the solute retention factor in the pure polar solvent B; the parameter m characterizes the number of molecules of the strong solvent B necessary to displace one adsorbed molecule of the solute.

Snyder introduced the linear solvent strength model (LSS) to describe the effects of the volume fraction of the strong (polar organic) solvent, φ, on the retention in binary aqueous-organic mobile phases in reversed-phase chromatography [89]:

$$\log k = \log a' - m' \cdot \varphi \tag{1.10}$$

$a' = k_w$, i.e., the solute retention factor in the weak solvent A (water) and m' (S in the original Snyder work), characterizes the organic solvent elution strength. Some compounds may show the LSS behavior in the NP or ion exchange (IEX) separation modes. Equation (1.9), originally derived for the adsorption-displacement model of NPC chromatography, provides a better fit for some RP experimental data in comparison to the LSS equation (Equation (1.10)). This is understandable as in the real separation systems a pure separation mechanism is an exception rather than a rule and both partition and adsorption may participate in the retention mechanism [55].

The two-parameter equations, Equation (1.9) and Equation (1.10), often do not satisfactorily describe the experimental data. Several three- or more parameter models often provide an improvement in the data fit. A formal extension of Equation (1.10) to Equation (1.11) by adding a quadratic term $d\varphi^2$ to characterize the curvature of the log k versus φ plots sometimes improves the model fit [90]:

$$\log k = a - m \cdot \varphi + d \cdot \varphi^2 \tag{1.11}$$

However, the physical meaning of the second-order term is not clear enough and even the quadratic model usually does not allow an accurate description of the RP retention over the full range of aqueous-organic mobile phases [91].

To account for possible combination of the adsorption and partition effects, Jin et al. presented a three-parameter mixed-mode retention model, which combines Equation (1.7) and Equation (1.9) to yield Equation (1.12) [92]:

$$\ln k = a + b + \ln \varphi_{H_2O} - c \cdot \varphi_{H_2O} \tag{1.12}$$

Neue and Kuss introduced a more complex empirical model using three parameters (k_0, b, m) [93]:

$$k = k_0 \left(1 + b \cdot \varphi\right)^2 \cdot e^{-m \cdot \varphi / (1 + b \cdot \varphi)} \tag{1.13}$$

k_0 is k in pure water. For a very low parameter b, Equation (1.13) becomes identical to the LSS Equation (1.10).

The comparison of the prediction errors of the two-parameter and more complex retention models in HILIC LC yielded ambiguous results [94, 95].

Usually, the three- or more parameter models fit the data better than the two-parameter models do. However, a good fit alone does not prove the validity of the underlying model. If the amount of experimental data is too low relative to a high number of parameters, "over-fitting" of an equation due to a low number of degrees of freedom may result in a close fit to the biased data subject to experimental errors [96]. For the sake of model robustness, the model-based retention equation should be as simple as possible and include the smallest number of meaningful parameters that provide a good fit for the experimental data.

The two-parameter model equations, Equation (1.9) and Equation (1.10), and some three-parameter equations fail to describe the retention at very low concentrations of the strong solvent, φ, in the mobile phase. The term b, introduced into Equation (1.9), corrects for possible retention in a pure weak solvent, $k_0 = 1/(b)^m$, in the ABM model Equation (1.14) [97]:

$$k' = \left(b + a \cdot \varphi\right)^{-m'} \tag{1.14}$$

When the retention in the pure strong solvent is small enough, the parameter b is negligible and Equation (1.14) becomes formally identical to Equation (1.9).

The ABM model does not strictly distinguish between the adsorption and the partition retention mechanisms. It describes retention in the organic NP chromatography very well, however, where φ stands for the volume fraction of the polar solvent in the non-polar one [98]. The model is suitable also for RP chromatography, where φ denotes the volume fraction of the organic solvent; or in HILIC systems where φ denotes the volume fraction of water or of an aqueous buffer. For example, in the reversed-phase LC on core-shell columns the ABM Equation (1.14) fits the experimental data better in comparison to the two-parameter equations (Figure 1.12). Equation (1.9) shows a slight systematic positive and Equation (1.10) slight systematic negative deviations from the experimental data (points) at lower acetonitrile concentrations in the mobile phase [18].

Rearranging the ABM model Equation (1.14) to Equation (1.15) reveals the relation of the parameters a, b to the retention factors in the pure weak solvent, k_0, and in the pure strong solvent, k_1, in the mobile phase. Hence, estimating the retention factors k_0 and k_1 from the experimental elution data is principally possible [4]:

$$k' = \left(b + a \cdot \varphi\right)^{-m'} = \left[\left(\frac{1}{k_1^{\frac{1}{m'}}} - \frac{1}{k_0^{\frac{1}{m'}}}\right)\varphi + \frac{1}{k_0^{\frac{1}{m'}}}\right]^{-m'} \tag{1.15}$$

$k_0 = \dfrac{1}{b^{m'}}$ is the retention factor in the pure weak solvent A, $k_1 = \left(b + a\right)^{-m'}$ is the retention factor in the pure strong solvent B and φ is the concentration of the strong solvent; m' is proportional to the area of the solid surface occupied by one molecule of the solvent B. The strong solvent can be either water (aqueous buffer) in HILIC, or a polar organic solvent (acetonitrile) in RP LC. Table 1.2 shows a few examples of the experimentally evaluated parameters a, b, m' of the ABM equation for benzene and a few alkylbenzenes, phenolic acids and flavones on two core-shell columns, Kinetex EVO C18 and Ascenti Express Phenyl-Hexyl, in acetonitrile-water mobile phases. The table also shows the estimated retention factors in the pure aqueous buffer, k_0, and in the pure acetonitrile, k_1. The parameter b decreases with increasing analyte lipophilicity and is almost insignificant for pentylbenzene and higher alkylbenzenes, so that Equation (1.15) eventually becomes identical to Equation (1.9). In spite of the relatively low accuracy of the estimated k_0 and k_1 values, which are not accessible to direct determination, the data calculated from the parameters a, b, m' may illustrate some trends in retention in reversed-phase and HILIC

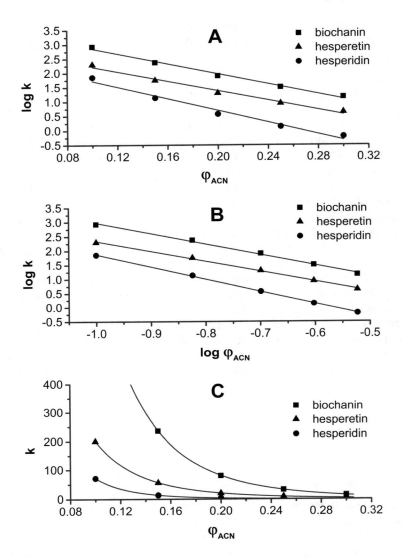

FIGURE 1.12 Effects of the volume fraction of acetonitrile, φ_{ACN}, on the retention factors, k, of biochanin A, hesperidin and hesperetin for LSS (A, Equation (1.10)), NP (B, Equation (1.9)) and ABM (C, Equation (1.14)) retention models on a Kinetex BiPhenyl column.

systems. The retention factors, k_l, in pure acetonitrile are low (0.1–0.4). On the other hand, the term b is important with phenolic acids and flavones. All compounds, except for phenolic acids ($k_0 = 43$–220), show very high retention in pure water, ($k_0 \geq 10^4$). For polar flavones and phenolic acids, the estimated k_l in acetonitrile are very low, in the range 10^{-5}–10^{-3}. The Phenyl-Hexyl siloxane bonded phase shows some preference for flavones in pure acetonitrile with respect to the octadecyl siloxane bonded phase.

TABLE 1.2

Experimental Values of the Parameters *a*, *b* and *m'* of the ABM Retention Model, Equation (1.14), for Alkybenzenes, Phenolic Acids and Flavones on the Kinetex EVO C18 and Ascentis Express Phenyl-Hexyl Columns

Comp.	Equation (1.14)			D^2	Equation (1.15)	
	b	*a*	*m'*		k_0	k_1
Kinetex EVO C18						
Alkylbenzenes (30–70% acetonitrile/water)						
B	0.40±0.02	1.06±0.03	5.34±0.21	1.0000	$1.3 \cdot 10^2$	0.13
EB	0.14±0.02	1.23±0.02	4.80±0.14	1.0000	$1.5 \cdot 10^4$	0.22
PeB	0.01±0.01	1.16±0.00	5.44±0.01	1.0000	$2.1 \cdot 10^{16}$	0.44
Phenolic acids (3–13% acetonitrile/water)						
SYR	0.41±0.07	3.65±0.33	4.31±0.77	0.9998	44.9	$2.4 \cdot 10^{-3}$
HPA	0.59±0.12	2.77±0.71	5.17±1.85	0.9996	15.8	$1.9 \cdot 10^{-3}$
SIN	0.42±0.11	2.86±0.35	6.18±1.75	0.9998	221.8	$6.5 \cdot 10^{-4}$
Flavones (10–30% acetonitrile/water)						
LUT	0.10±0.01	2.83±0.02	5.00±0.11	1.0000	$9.2 \cdot 10^4$	$4.6 \cdot 10^{-3}$
NGI	0.24±0.01	3.48±0.05	6.64±0.15	1.0000	$1.3 \cdot 10^4$	$1.6 \cdot 10^{-4}$
HED	0.10±0.02	4.03±0.08	5.61±0.23	1.0000	$4.0 \cdot 10^5$	$3.5 \cdot 10^{-4}$
Ascentis Express Phenyl-Hexyl						
Alkylbenzenes (30–70% acetonitrile/water)						
B	0.52±0.01	0.87±0.03	6.68±0.23	1.0000	$8.0 \cdot 10^1$	0.11
EB	0.22±0.02	1.20±0.03	5.31±0.21	1.0000	$3.5 \cdot 10^3$	0.16
PeB	0.04±0.03	1.25±0.02	5.65±0.30	1.0000	$1.6 \cdot 10^8$	0.24
Phenolic acids (5–13% acetonitrile/water)						
SYR	0.27±0.02	3.99±0.10	2.87±0.16	1.0000	43.4	$1.6 \cdot 10^{-2}$
HPA	0.31±0.02	4.22±0.13	2.49±0.11	1.0000	17.6	$2.3 \cdot 10^{-2}$
SIN	0.22±0.02	3.12±0.03	3.48±0.13	1.0000	203.6	$1.5 \cdot 10^{-2}$
Flavones (15–30% acetonitrile/water)						
LUT	0.09±0.05	2.70±0.12	4.70±0.44	1.0000	$9.3 \cdot 10^4$	$8.0 \cdot 10^{-3}$
NGI	0.14±0.01	3.51±0.03	5.98±0.06	1.0000	$9.0 \cdot 10^4$	$4.3 \cdot 10^{-4}$
HED	0.11±0.01	3.60±0.05	5.83±0.11	1.0000	$4.4 \cdot 10^5$	$4.9 \cdot 10^{-4}$

k_0 – Retention Factor in Pure Water; k_1 – Retention Factor in Pure Acetonitrile, Equation (1.15); D^2 – Coefficient of Determination.

1.4.3 MODELING GRADIENT-ELUTION LIQUID CHROMATOGRAPHY

In the course of gradient elution, the composition of the mobile phase changes by the mixing of two or more components, according to a pre-set program. A weaker mobile phase used in the initial part of the gradient elution provides adequate retention of weakly retained compounds, while increasing the concentration of the strong eluting component gradually decreases the retention of late-eluting compounds. Eventually, the retention of the too early eluting compounds increases and the strongly retained solutes elute faster in comparison to the isocratic elution. In addition to improved resolution, the gradient elution provides narrower bandwidths, increased peak capacity and shorter separation time with respect to the isocratic separations.

In the reversed-phase gradient systems, the proportion of a polar organic solvent in the aqueous-organic mobile phase increases with time. On the contrary, in the HILIC gradients, the retention of polar compounds decreases by increasing the water proportion in the mobile phase. The theory of gradient-elution chromatography allows one to predict the gradient-elution retention times and bandwidths and to optimize gradients in various reversed-phase, normal-phase and ion exchange systems. This is important for compound identification and for setting the time windows for peak integration, and in fundamental studies of retention [99].

Unlike isocratic elution, the retention factors k_i during gradient elution can be considered constant only within a very small (differential) time interval dt, in which the sample zone migrates a very short distance along the column, corresponding to the differential of the column hold-up time, dt_m. The solute migration during the interval dt corresponds to an increase in the final net retention time, t'_R, by the differential increment $d(t'_R)$:

$$d\left(t'_R\right) = k_i \cdot d\left(t_m\right) \tag{1.16}$$

The solution of Equation (1.16) enables calculations of the gradient retention data in various LC systems. The integration of the equation requires the combining of the equation controlling the gradient program with the appropriate model retention equation (such as Equation (1.9), Equation (1.10) or Equation (1.14)). The regression analysis of the experimental isocratic or independent gradient retention data provides the necessary retention equation parameters [100].

In linear gradient elution, the volume fraction of the strong solvent B, φ increases proportionally to the volume of the mobile phase passed through the column, V, from the initial concentration, A, to the φ_G at the end of the gradient: $\varphi = A + B \cdot V$. The effects of the gradient ramp (B) and gradient range ($\varphi_G - A$) on the elution volumes, $V_{R(g)}$, can be predicted from the parameters of the isocratic model equations, assuming that the column hold-up volume, V_m, does not change significantly in the gradient range [101, 5]. Only a few retention equations (such as Equation (1.9), Equation (1.10), Equation (1.13) and Equation (1.14)) allow a direct analytical solution; other models require numerical integration. The reversed-phase Equation (1.10) yields the gradient Equation (1.17), identical to the solution of the Snyder LSS gradient model [102–105]:

$$V_{R(g)} = \frac{1}{m \cdot B} \log\left[2.31 \cdot m \cdot B \cdot \left(V_m \cdot 10^{(a-m \cdot A)} - V_D\right) + 1\right] + V_m + V_D \tag{1.17}$$

The isocratic NP Equation (1.9) provides the gradient equation, Equation (1.18) [102, 106]:

$$V_{R(g)} = \frac{1}{B}\left[(m+1)B\cdot\left(k_0\cdot V_m + V_D\cdot A^m\right) + A^{(m+1)}\right]^{\frac{1}{m+1}} - \frac{A}{B} + V_m + V_D \qquad (1.18)$$

The three-parameter isocratic Equation (1.14), applied to the gradient elution, yields Equation (1.19) [100]:

$$V_{R(g)} = \frac{1}{a\cdot B}\left[(m'+1)\cdot b\cdot B\cdot\left[V_m - V_D\left(b + a\cdot A\right)^{m'}\right] + \left(b + A\cdot a\right)^{(m'+1)}\right]^{\frac{1}{m'+1}}$$

$$- \frac{b + A\cdot b}{b\cdot B} + V_m + V_D \qquad (1.19)$$

V_m is the column hold-up volume, A and B are the parameters of the linear gradient ($\varphi = A + B\cdot V$). $V = t\cdot F_m$ is the volume of the mobile phase that has flown through the column in the time, t, that has elapsed since the start of gradient elution at the mobile phase flow rate F_m. a, b, k_0, m and m' are the best-fit regression parameters of Equation (1.9), Equation (1.10) and Equation (1.14), respectively. V_D is the instrumental dwell volume (i.e., the volume of the gradient mixer and of the connecting tubing between the mixer and the column inlet) containing the starting mobile phase in which less retained compounds may move some distance along the column under isocratic conditions, before the front of the gradient program.

Like the ABM model, the Neue–Kuss three-parameter model equation, Equation (1.13), also allows for direct analytical calculations of the gradient retention times from the isocratic equation parameters, yielding Equation (1.20) [93]:

$$t_{R(g)} = \frac{1}{B\cdot b}\cdot \frac{\left(1+b\cdot A\right)^2\cdot\ln\left[\text{Bak}_0 e^{-aA/(1+bA)}\cdot\left(t_0 - \frac{t_D}{\left(1+b\cdot A\right)^2\cdot k_0\cdot e^{\frac{aA}{(1+bA)}} + 1}\right) + 1\right]}{1 - \left(\frac{b}{a}\right)\cdot(1+bA)\cdot\ln\left[\text{Bak}_0 e^{-aA/(1+bA)}\cdot\left(t_0 - \frac{t_D}{\left(1+b\cdot A\right)^2\cdot k_0\cdot e^{\frac{aA}{(1+bA)}} + 1}\right) + 1\right]} + t_m - t_D$$

$$(1.20)$$

In Equation (1.17)–Equation (1.20), $t_{R(g)} = V_{R(g)}/F_m$ is the gradient retention time, $t_m = V_m/F_m$ is the column hold-up time and t_D is the gradient delay time due to the instrumental gradient dwell volume between the injection port and the gradient mixer. Equation (1.20) is more complex in comparison with the ABM gradient Equation (1.19), hence is more sensitive to calculation errors.

Equation (1.17) is formally identical with the LSS gradient equation introduced by Snyder et al. [89], widely applied in RP systems. In some RP systems, the gradient elution volumes predicted from Equation (1.18), Equation (1.19) and Equation (1.20) provide comparable or even better agreement with the experimental data, for example, fast (1–2 min) gradients in the second dimension of comprehensive two-dimensional chromatography [107] using short packed [108], core-shell [109] or silica monolithic columns [110]. Figure 1.13 compares the prediction errors of the two-parameter displacement and LSS models and the three-parameter ABM model on short 5 cm C18 and Phenyl-Hexyl core-shell columns for 1–5 min acetonitrile-water gradients run at 2 mL/min and 4.5 ml/min [4]. The errors in the gradient retention volumes of alkylbenzenes and flavones predicted with the best-fit model parameters determined by the regression analysis of the isocratic and gradient input data decrease in the order: LSS > displacement > ABM models. The parameters of the ABM model acquired under gradient conditions provide prediction errors in the gradient elution volumes of alkylbenzenes ($\leq 0.4\%$) and flavones, ($\leq 1.2\%$), in agreement with the results reported earlier [111]. The computation is relatively easy with the generally available statistical software. The strategy of the commercial DryLab gradient optimization software employs the LSS model and the experimental data acquired in two scouting gradient experiments [112, 113].

In the HILIC gradient elution, the concentration of water (or an aqueous buffer) increases. For the prediction of the HILIC gradient retention data, Pirok et al., employing the model parameters acquired in two scouting gradient runs, find the two-parameter displacement NP model, Equation (1.9), more accurate in comparison to the two-parameter LSS model, Equation (1.10), the mixed-mode three-parameter model, Equation (1.12), the quadratic model, Equation (1.11), or the Neue–Kuss model, Equation (1.13). The prediction accuracies depended on the analyte class and on the stationary phase, with better results for a diol column than for an amide column [114].

In a similar comparison study, Tyteca et al. [115] found the mixed model (Equation (1.12)) most suitable for describing and predicting the gradient data for twelve nucleobases and nucleosides, while the quadratic relationship (Equation (1.11)) was the worst. The empirical Neue-model (Equation (1.13)) provided predictions close to those of the mixed model. However, since the mixed model cannot be integrated analytically the authors recommend the empirical Neue-model (Equation (1.20)) for the prediction of the gradient retention in the HILIC.

The gradient retention models assume that the phase ratio remains constant during gradient elution. This is not the case in HILIC, where the gradients of increasing water concentration continuously increase the thickness of the adsorbed water layer. As the adsorbed water layer forms a part of the stationary phase in HILIC, the continuously increasing water adsorption may affect the gradient prediction errors even more than the retention model equation employed. The hydrosilated silica gel columns with less than a monomolecular layer of adsorbed water (already 58–78% saturated at the gradient starting at water concentrations as low as 2%) showed excessively high errors in the predicted gradient retention volumes, which could be suppressed to 5% or less by correcting the actual gradient profile for the increasing amount of adsorbed water [116]. On the other hand, the TSK gel Amide-80 and YMC Triart Diol columns strongly adsorbing water are obviously less sensitive to the water uptake during the gradient elution and provide the low uncorrected HILIC gradient prediction errors of Equation (1.19), 1.2–1.5% for gradients in the range of 96–70% water [117].

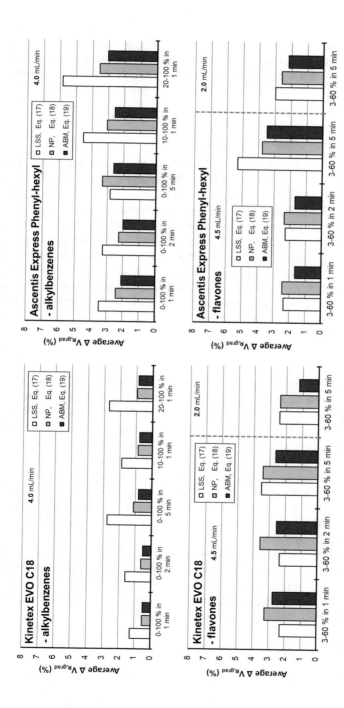

FIGURE 1.13 Average per cent differences between the experimental, $V_{R,g,exp}$ and predicted, $V_{R,g,calc}$ gradient elution volumes based on three retention models (Equation (1.15)–Equation (1.17)) for alkylbenzenes (ethylbenzene–pentylbenzene) and flavones; the Kinetex EVO C18 and Ascentis Express Phenyl-Hexyl columns. Flow rate 4.5 mL/min or 2.0 mL/min (flavones), 4.0 mL/min (alkylbenzenes); linear gradients of 10 mM NH$_4$Ac in acetonitrile. Based on the data in [4].

A numerical approach based on a targeted trial-and-error search allowed the (approximate) prediction of the retention of the first and the last peak in the HILIC gradient chromatogram [95]. The gradient modeling based on the non-linear Neue–Kuss model applied to the coupled column systems with different stationary phase chemistries (an amide and a pentahydroxy functionalized silica gel) with a gradient from 95% to 85% acetonitrile was claimed to provide prediction errors of less than 2% [115].

1.4.4 STRUCTURAL RETENTION CORRELATIONS

The stationary phases used in HPLC contain various chemical moieties, which provide different interactions with the individual sample classes. The phases chemically bonded on silica gel used in reversed-phase and HILIC separations contain residual silanol groups providing selective dipole–dipole, proton-donor, proton-acceptor, π–π electron and electrostatic interactions, in addition to the lipophilic interactions of the bonded moieties (such as the alkyl chains). The linear solvation energy relationships (LSER) model quantitatively describes the structural contributions to retention in liquid chromatography, employing multiple linear correlations between the retention and the molecular structural of a solute [118, 119]. The original model considering only the non-ionic compounds was later enlarged by including the solvation parameters describing the ionic interaction [120, 6]:

$$\log k = c + v \cdot V + s \cdot S + a \cdot A + b \cdot B + d^- \cdot D^- + d^+ \cdot D^+ \tag{1.21}$$

V, S, A, B, D^- and D^+ are the structural descriptors characterizing the sample: the molar volume of the solute, V, the dipole–dipole interactions, S, the hydrogen-bonding acidity, A, and the hydrogen-bonding basicity, B. D^- represents the negative charge carried by anionic species, and D^+ represents the positive charge carried by cationic species. The coefficients v, s, a, b, d^- and d^+ of Equation (1.21) provide a measure of the response of the separation system (the stationary and the mobile phase) to the selective properties of the analytes obtained using multivariate simultaneous least-squares regression of the experimental retention data. The products of the system parameters and the corresponding molecular structural descriptors in Equation (1.21) characterize the contributions of the selective interactions to retention in particular separation systems.

The LSER model, intended for the classification of the column for RP HPLC, provides a good data fit in the HILIC LC (up to 95% acetonitrile) for the silica hydride, bonded polyethylene glycol, diol and zwitterionic sulfobetaine columns. The model is a useful tool in column design for verifying the expected effects on the retention of various ligands immobilized on the silica surface [72].

1.4.5 COMBINED-RETENTION-MECHANISM LC SYSTEMS

Polar and ion-exchange stationary phases contain some hydrocarbon moieties, which can give rise to non-polar (solvophobic) interactions. The classical ion-exchange resins, which contain cation- or anion-exchange groups bonded on organic polymer matrices with lipophilic properties, were used for the (essentially reversed-phase) separations of small non-ionic molecules long before the advent of silica bonded stationary phases [121].

The electrostatic-repulsion hydrophilic-interaction chromatography (ERLIC) on ion exchangers in highly organic mobile phases offers possibilities for the independent adjusting of the HILIC and the ion-exchange selectivities. The anionic solutes retained *via* hydrophilic interactions are subject to the electrostatic repulsion, which decreases the retention of basic peptides [70].

1.4.5.1 Mixed-Mode and Zwitterionic Stationary Phases

The mixed-mode retention mechanism is due to a cocktail of different solute / stationary phase / mobile phase interactions, largely independent of the mobile phase composition, such as solvophobic, polar, attractive or repulsive electrostatic effects [122, 123]. The LSER model (Equation (1.21)) in fact characterizes a mixed mode RP NP behavior, of course, with a dominant RP contribution and the possible participation of the ion-exchange/repulsion effects.

Mixed-mode HILIC/ion-exchange stationary phases with a long-alkyl chain ligand and a hydrophilic polar terminal ion-exchange group bonded on the silica gel support have been designed to improve the separation selectivity for a wide range of polar and non-polar compounds in organic-rich mobile phases by combining the HILIC partition and ion exchange mechanisms [124–127].

The Acclaim mixed-mode WAX-1 and WCX-1 columns, or a weak anion exchanger, the PolyWAX LP column, prepared by modifying silica gel with a cross-linked coating of linear poly(ethyleneimine), Figure 1.14 [128], offer complementary application possibilities to the typical HILIC stationary phases such as TSKGel Amide-80, ZIC-HILIC or polysulfoethyl A [129]. At high concentrations of acetonitrile, the separating of basic and acidic peptides is possible in a single run, either on a strong anion exchange (SAX) or on a weak anion exchange (WAX) column. At a low pH, the HILIC and electrostatic repulsion retention mechanisms superimpose in the ERLIC mode. This enables the HILIC and the ion-exchange selectivities to be adjusted independently in highly organic mobile phases [130].

More recently reported mixed-mode HILIC/ion-exchange columns include, for example, a glutamine silica bonded stationary phase containing an amino alcohol group and two different amide groups, one a polar head and the other embedded into an aromatic phenyl ring [131]. A glutathione mixed-mode HILIC/cation-exchange stationary phase enables separations of peptides varying in both hydrophobicity/hydrophilicity and charge [132], probably by the combined electrostatic repulsion and HILIC mechanisms [70].

Thiol-ene click chemistry, based on the reactions between a thiol and an alkene group, provides useful mixed-mode columns [133], for example, a thiol-Click-COOH column (thioglycolic acid bonded onto vinyl-bonded silica, Figure 1.14G), which provides selective separations of nucleosides, bases and water-soluble vitamins due to the combined HILIC–ion exchange mechanism [134]. Neomycin, a hydrophilic aminoglycoside containing six amino groups, seven hydroxyl groups and six glycosidic oxygen functions, grafted onto silica gel, enables separations of organic acids by mixed-mode hydrophilic/ion-exchange interactions [135].

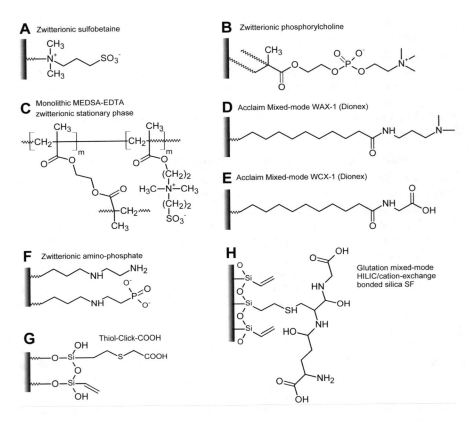

FIGURE 1.14 Structures of some zwitterionic and mixed-mode HILIC–ion-exchange stationary phases.

A trimodal stationary phase prepared by coating porous spherical silica particles with charged organic polymer nano-beads (0.1 μm), covalently modifying the inner-pore area with an organic layer, and the outer surface being modified with strong cation-exchange groups, shows both reversed-phase and weak anion-exchange properties. The spatial separation of the anion exchange and cation-exchange regions provides possibilities for simultaneous separations of acids, bases and neutral compounds. At high acetonitrile concentrations, the material exhibits a mixed HILIC/ion-exchange retention mechanism enabling simultaneous separation of ionized hydrophilic drugs and their counter-ions (e.g., penicillin G potassium salt) [136].

Zwitterionic stationary phases also contain spatially separated positively and negatively charged moieties. The sulfobetaine stationary phases contain the active zwitterionic layer grafted either on a silica gel or on a polymer support [137, 138]. An alkyl chain separates the strongly acidic sulfonic acid group at the end of the ligand from a strongly basic quaternary ammonium group closer to the silica gel support. The two oppositely charged groups are present in a 1:1 molar ratio, so that there is only a very low net negative surface charge of the bonded layer [69]. Polar (hydrogen-bonding and dipole–dipole) interactions are of primary importance, even though weak electrostatic interactions may affect the separation of partially ionized analytes, too. The sulfobetaine-bonded ZIC-HILIC silica (Figure 1.14A) is suitable for HILIC separations of a wide range of small polar metabolomes [139], glucosinolates [140], aminoglycosides [141], peptides, purine and pyrimidine bases and nucleosides [142], or glycopeptides [143, 144]. The HILIC separation of peptides on ZIC-HILIC columns at pH = 3 resembles the separations on strong cation exchangers (SCX), but improves at a higher pH (7–8) [145].

Capillary sulfobetaine organic polymer monolithic columns prepared by copolymerization of methacryloxyethyl-N-(3-sulfopropyl) ammonium betaine (MEDSA) with ethylene dimethacrylate (EDMA) are suitable for separations of nucleic bases and other neutral, basic and acidic polar analytes in aqueous-organic mobile phases containing 60% or more acetonitrile [42] or with 1.2-bis(p-vinylphenyl) ethane (BVPE) [146].

Phosphorylcholine stationary phases (Figure 1.14B) prepared by graft polymerization of 2-methacryloyloxyethyl phosphorylcholine onto the surface of silica gel support differ from the sulfobetaine ZIC-HILIC material not only by the nature of the negatively charged group, but also by the charge arrangement, causing significantly different separation selectivity for peptides, free amino acids and carboxylic acids [68]. HILIC on a capillary zwitterionic ZIC-cHILIC column enabled the separation and identification of more than 100 N-glycopeptides and O-glycopeptides in a single run [147].

An amino-phosphate zwitterionic stationary phase containing a negatively charged phosphate group bonded via an ester spacer and a positively charged quaternary amine group (Figure 1.14F) can separate vitamins, nucleosides, deoxynucleosides, nucleobases and aromatic acids under HILIC conditions [148].

1.4.5.2 Serially Coupled HILIC/RP Columns

The on-line coupling of two columns with different polarities allows the simultaneous separation and determination of polar, moderately polar and non-polar compounds in complex samples. A set of serially coupled columns containing different stationary phases behaves like a new column with a modified selectivity [149]. For example, serially coupled ZIC-HILIC sulfobetaine and amide columns allowed simultaneous separations of polar and non-polar metabolites in a mouse serum sample with a HILIC gradient from 95% to 35% acetonitrile in an aqueous buffer [150]. A single gradient of simultaneously increasing the concentration of acetonitrile and decreasing the salt concentration separated polar and non-polar phenolic compounds in wine on serially coupled C_{18} and zwitterionic columns [151].

Unfortunately, the on-line combinations of HILIC and reversed-phase HPLC are often subject to compatibility problems originating from the differences in the mobile phase elution strengths in the HILIC and in the RP modes. High concentrations of the organic solvents used for the HILIC separations usually provide a weak retention in the RP systems, whereas the mobile phases rich in water used in RP HPLC are usually excessively strong HILIC eluents. This may significantly decrease the retention; un-symmetric or even split peaks may appear, with detrimental effects on the separation [152, 153].

To mitigate the problem, a C_{18} reversed-phase column serially coupled via a T-capillary piece with a (polyhydroxyethyl aspartamide) HILIC column was employed for separation of quaternary ammonium compounds in brain extracts, including acylcarnitines of low polarity. The addition of the make-up flow of 95% acetonitrile to the effluent from the RP column improved the retention of the polar compounds transferred onto the HILIC column [154].

A column-switching setup with a HILIC BEH micro-column serially coupled with a Phenyl-Hexyl reversed-phase micro-column *via* a switching valve allowed separations of compounds widely differing in polarities. A trapping RP column retained a plug of weakly polar compounds eluting early from the tandem columns, until the separation of polar compounds on the HILIC column had been finished. Then the connecting valve was switched to redirect the weakly polar compounds onto the RP column for separation by a gradient of increasing acetonitrile concentration in 0.02% aqueous formic acid [155]. Without a column-switching approach, the mobile phase incompatibility is a major challenge when using a broad range of decreasing acetonitrile concentration on the RP and HILIC serially coupled columns.

1.4.5.3 Dual-Mode Retention Mechanism on a Single Column

Some polar columns containing significant non-polar structural moieties usually provide both polar and non-polar interactions, with the role of the individual separation mechanisms being predominant in limited mobile phase composition ranges. In this aspect, the dual-mode retention mechanism on a single column differs from the common mixed-mode mechanism, less affected by the mobile phase range. The dual-mode HILIC/RP retention mechanism results in the predominating HILIC (ANP) mode at the high organic solvent concentrations (>60%) in the aqueous-organic mobile phases; the RP mode primarily controls the retention in the highly aqueous mobile phase range [117].

The dual-mode HILIC/RP behavior depends on the column type and on the nature of the analytes. With ionic compounds, the dual-mode mechanism may include the RP/IEX or the HILIC/IEX mixed-modes. The separation selectivity in the HILIC mobile phase range largely differs from the RP range. Such a single polar column provides useful information on the sample by combining the data obtained in the acetonitrile-rich (HILIC) and in the highly aqueous (RP) mobile phases (often orthogonal) [156].

Some columns may provide practically useful separations only in either high organic solvent concentrations (HILIC) or highly aqueous mobile phases (RP), depending both on the stationary phase and on the analyte. In a major part of the medium mobile phase composition range, the retention of polar compounds is too low. On aminopropyl, amide, diol and cyanopropyl bonded stationary phases, many acidic and neutral pharmaceutically important compounds show stronger retention under RP conditions than in the HILIC range of the aqueous-organic mobile phase [157].

A Discovery HS PEG column with bonded poly(oxyethylene) groups, a LiChrocart 100 DIOL column with bonded glycerol groups and a Luna HILIC 200A with diol stationary phase bonded via cross-linked oxyethylene bridges all show a dual-mode retention mechanism for phenolic compounds. The Luna HILIC column retains the phenolic acids and flavone compounds more strongly in both the HILIC and the RP modes in comparison to the PEG and DIOL columns [158].

In spite of a very low affinity to water, the silica hydride columns with bidentate C_{18}, cholesterol, undecanoic acid (UDA), perfluorinated phenyl and other non-polar surface-bonded ligands show dual retention mechanisms, HILIC in buffered mobile phases containing more than 50–70% acetonitrile and RP at higher concentrations of water [159]. The unmodified hydrosilated columns do not show significant retention in the reversed-phase mode. The UDA silica hydride stationary phase shows increased separation selectivity for mono-, di- and tri-phosphate nucleotides with respect to the unmodified silica hydride column [160]. At increasing temperature, the retention factors and peak widths decrease both in the aqueous normal phase and in the reversed phase mobile phase range. In agreement with the van 't Hoff model, linear ln k versus $1/T$ plots indicate a predominating single retention mechanism for each mobile phase range (k is the solute retention factor and T is the thermodynamic temperature in Kelvins). The Cogent UDC cholesterol column is stable up to 100°C and provides selective and efficient separations of flavones [161] and phenolic acids [53], both in the ANP and in the RP modes, with the (almost) reversed elution order.

Depending on the mobile phase composition, the zwitterionic silica bonded columns often show a dual HILIC/RP retention mechanism [145, 68]. The retention on a ZIC-HILIC column with sulfobetaine stationary phase bonded on silica gel is relatively weak in the RP mode. On the other hand, the organic polymer zwitterionic columns offer relatively broad mobile phase ranges both for the HILIC and the RP separation modes and provide excellent separations of flavonoids, phenolic acids and other polar analytes in the two retention modes [162, 163]. Various functional monomers and cross-linkers provide useful dual-mode zwitterionic polymethacrylate monolithic columns; longer-chain cross-linking agents improve the inner pore morphology and enhance chromatographic selectivity [41].

In situ (co)polymerization of [2-(methacryloyloxy)ethyl]-dimethyl-(3-sulfopropyl)-ammonium hydroxide (MEDSA) zwitterionic monomer and bisphenol A glycerolate dimetacrylate (BiGDMA), or dioxyethylene dimethacrylate (DiEDMA), cross-linking monomers in fused silica capillaries, yields efficient monolithic micro-columns, providing adequate retention of various polar compounds, such as nucleic bases, nucleosides, sulphonamides, heterocycles, aromatic acids, polyphenolic compounds, etc., both in the HILIC mode and in the RP mode [164].

Some chiral columns show a dual-mode HILIC/RP mechanism, in addition to enantiomeric selectivity. For example, after the separation of aqueous plant extracts into fractions on a beta-cyclodextrin bonded phase in the RP conditions, the same column allows later separation of the individual fractions in the HILIC mode [165].

In the presence of a dual HILIC/RP mechanism, the graphs of the sample retention factors, k, versus the volume fraction of water, φ_{H_2O}, in binary aqueous-organic mobile phases including the HILIC and the RP ranges ($\varphi_{H_2O} > 0.02$), show characteristic U-profiles. The four-parameter Equation (1.22) often can describe these profiles [156]:

$$\log k = a + m_{RP} \cdot \varphi_{H_2O} - m_{HILIC} \cdot \log\left(1 + b \cdot \varphi_{H_2O}\right) \qquad (1.22)$$

The parameter m_{RP} characterizes the effect of the increasing concentration of water in the mobile phase on retention due to the RP mechanism in water-rich mobile phases, whereas the parameter m_{HILIC} is a measure of the water contribution to the decrease of retention in the HILIC range. Equation (1.22) accounts for a low, but finite, retention at very low water concentrations [158]. Figure 1.15A and B show examples of the dual-mode U-shape graphs described by Equation (1.22) for several sulfonamides on the MEDSA-BiGDMA and the MEDSA-DiEDMA monolithic capillary polymethacrylate zwitterionic columns [163]; Figure 1.15C illustrates this behavior for several flavonoids on a commercial Luna HILIC column [117].

Equation (1.22) employs fitting parameters determined using the retention factors, k, determined separately in the RP and HILIC ranges of the organic solvent–water mobile phase. In between the two ranges, around the (hypothetical) minimum on the U-shape graph, φ_{min}, there is a more or less broad intermediate aqueous-organic mobile phase range, where the HILIC/RP retention mode transition occurs (Figure 1.15). The width and the position of the transition mobile phase area, providing very low retention, which generally increases with the solute polarity and depends on the stationary phase; for example, 70–90% acetonitrile for bonded polyethylene glycol columns in comparison to 40–70% acetonitrile for more polar diol and zwitterionic columns. The Luna HILIC column with significant proton-donor and proton-acceptor properties provides the mode transition for sugar-containing glycosides at higher water concentrations with respect to the corresponding aglycones [156].

The HILIC/RP transition range of the mobile phase practically rules out the usage of a dual-mode column for the simultaneous separation of non-polar and polar samples in a single broad range gradient run. The continuously increasing concentration of the organic solvent would elute all polar compounds when the gradient mobile phase composition meets the transition mobile phase range; this also applies for non-polar compounds with the HILIC water gradients starting in a highly organic mobile phase. Hence, the separations of broad-polarity-range samples on dual-mode columns need two independent subsequent sample injections, one in the HILIC range mobile phase and the other in the RP mobile phase range.

The exact application of Equation (1.22) needs different V_m values for the RP and the HILIC range, determined with the marker compounds appropriate for the individual retention modes (e.g., uracil in the RP range and toluene in the HILIC range, respectively).

Table 1.3 lists the best-fit parameters of Equation (1.14) for several sulfonamides on the MEDSA-DiEDMA micro-column determined with $V_m = 8.1$ μL (toluene) in the HILIC range, 90–98% acetonitrile/water and $V_m = 8.8$ μL (uracil) in the RP range, 20–40% acetonitrile/water). The difference of approximately 10% is probably due to different solvation of the zwitterionic polymethacrylate stationary phase in the acetonitrile-rich and in the highly aqueous mobile phases [4].

Defining the volume of the stationary phase, V_s, as the volume of the inner pores, V, (convention 3) avoids this inconvenience; V_m, determined as the retention volume of a large molecule, of polystyrene with $M_r = 1,800,000$, on the MEDSA-DiEDMA micro-column, $V_m = 6.7$ μL, is equal in the HILIC and the RP mobile phase ranges (see Figures 1.2 and 1.3). The numerical values of the best-fit parameters of Equation (1.22), determined with convention 3, differ from the parameters determined

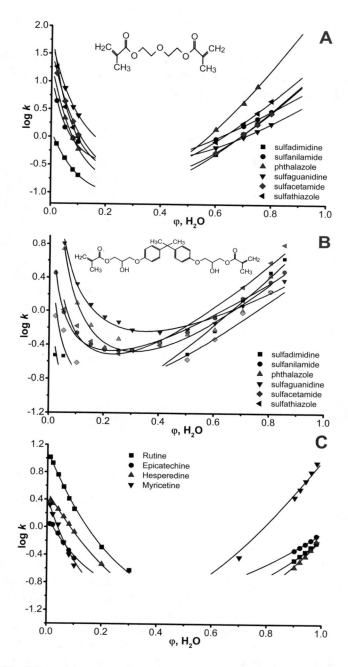

FIGURE 1.15 Effect of the volume fraction of the water, φ_{H_2O}, in buffered aqueous-organic mobile phases on the retention factors, k, of sulfonamides on monolithic MEDSA-DiEDMA (A) and MEDSA-BiGDMA (B) dual-mode micro-columns, and of flavones on a Luna HILIC (C) column. Based on the data in [7].

TABLE 1.3
Experimental Values of the Parameters *a*, *b* and *m'* for Retention Models of Equation (1.14) Sulfonamides and Derivatives of Benzoic Acids on the MEDSA-DiEDMA Monolithic Micro-Column

Comp.	*b*	*a*	*m'*	D²
		Sulfonamides		
HILIC (2–10% water/acetonitrile), $k' = \left(b + a \cdot \varphi_{H_2O}\right)^{-m'}$, $V_m = V_{r,toluene} = 8.1\ \mu L$				
SDD	0.91 ± 0.04	12.06 ± 3.63	2.16 ± 0.47	0.9998
SNA	0.01 ± 0.03	13.54 ± 0.69	1.13 ± 0.05	0.9978
SAA	0.37 ± 0.06	6.02 ± 0.46	3.68 ± 0.54	0.9999
STZ	0.00 ± 0.00	10.22 ± 0.02	1.83 ± 0.01	1.0000
HILIC (2–10% water/acetonitrile), $k' = \left(b + a \cdot \varphi_{H_2O}\right)^{-m'}$, $V_m = V_{r,polystyrene,\ Mw=1\ 800\ 000} = 6.7\ \mu L$				
SDD	0.57 ± 0.10	19.64 ± 4.80	0.98 ± 0.17	0.9997
SNA	0.00 ± 0.00	10.21 ± 0.75	1.05 ± 0.05	0.9965
SAA	0.29 ± 0.04	5.96 ± 0.25	3.11 ± 0.33	0.9999
STZ	0.00 ± 0.00	8.88 ± 0.12	1.78 ± 0.01	0.9999
RP (20–40% acetonitrile/water), $k' = \left(b + a \cdot \varphi_{ACN}\right)^{-m'}$, $V_m = V_{r,uracil} = 8.8\ \mu L$				
SDD	0.69 ± 0.15	0.99 ± 0.47	8.61 ± 4.37	0.9999
SNA	0.71 ± 0.03	0.75 ± 0.07	7.23 ± 0.69	1.0000
SAA	0.62 ± 0.10	1.19 ± 0.30	6.73 ± 1.83	0.9999
STZ	0.29 ± 0.02	1.89 ± 0.06	3.83 ± 0.17	1.0000
RP (20–40% acetonitrile/water), $k' = \left(b + a \cdot \varphi_{ACN}\right)^{-m'}$, $V_m = V_{r,polystyrene,\ Mw=1\ 800\ 000} = 6.7\ \mu L$				
SDD	0.23 ± 0.07	1.90 ± 0.16	2.94 ± 0.36	0.9999
SNA	0.41 ± 0.00	1.15 ± 0.01	3.24 ± 0.03	1.0000
SAA	0.20 ± 0.03	1.95 ± 0.07	2.64 ± 0.15	1.0000
STZ	0.05 ± 0.01	1.99 ± 0.01	2.39 ± 0.04	1.0000

Sulfonamides: SDD – Sulfadimidine, SNA – Sulfanilamide, SAA – Sulfacetamide, STZ – Sulfathiazole. D² – coefficient of determination.

according to the classical *convention 1*, with different values of V_m in the HILIC and RP mobile phase ranges. However, both conventions yield similar fitting quality and standard deviations of the parameters of the retention equations (Table 1.4). The retention factors predicted using Equation (1.22) with *convention 3* are slightly higher in comparison to the data predicted with *convention 1* (Figure 1.16). The results demonstrate that selection of the convention is not critical for the method development and optimization of the separation on the dual-mode HILIC/RP columns. Anyway, *convention 3* (inner pores) is more consistent as it employs a single V_m value for a dual-mode column over the full mobile phase range.

The LSER model (Equation (1.21)) fits the retention data measured on various dual-mode columns in the HILIC mode and in the RP mode. Different interactions

TABLE 1.4

Experimental Values of the Parameters a_1, m_{RP}, m_{HILIC} and b_1 for the HILIC/RP Dual-Mechanism Retention Model of Equation (1.22) for Sulfonamides and Aromatic Acids on the MEDSA-DiEDMA Monolithic Micro-Column

Comp.	a_1	m_{RP}	m_{HILIC}	b_1	D^2	φ_{min}
			Sulfonamides			
$V_{m,HILIC}=V_{r,toluene}=8.1\ \mu L,\ V_{m,RP}=V_{r,uracil}=8.8\ \mu L$						
SDD	0.10±0.01	7.27±0.19	6.72±0.34	6.85±0.40	0.9998	0.26
SAA	1.57±0.01	7.99±0.12	8.18±0.20	9.12±0.30	0.9999	0.33
$V_{m,HILIC}=V_{r,polystyrene\ Mw=1\ 800\ 000}=6.7\ \mu L,\ V_{m,RP}=V_{r,polystyrene\ Mw=1\ 800\ 000}=6.7\ \mu L$						
SDD	0.18±0.02	5.71±0.28	5.12±0.51	6.75±0.78	0.9995	0.24
SAA	1.73±0.05	5.86±0.36	5.05±0.46	16.46±2.41	0.9983	0.31
			Aromatic acids			
$V_{m,HILIC}=V_{r,toluene}=8.1\ \mu L,\ V_{m,RP}=V_{r,uracil}=8.8\ \mu L$						
HMB	1.91±0.08	9.08±0.62	10.40±1.22	7.79±1.28	0.9987	0.37
FMA	1.76±0.06	8.59±0.41	8.89±0.75	8.79±1.06	0.9984	0.34
PTA	0.68±0.05	8.17±0.33	6.89±0.60	9.07±1.16	0.9997	0.26
PHT	1.58±0.04	8.00±0.59	10.87±1.40	5.01±0.75	0.9986	0.39
$V_{m,HILIC}=V_{r,polystyrene\ Mw=1\ 800\ 000}=6.7\ \mu L,\ V_{m,RP}=V_{r,polystyrene\ Mw=1\ 800\ 000}=6.7\ \mu L$						
HMB	2.20±0.20	6.37±0.78	5.96±1.12	16.56±6.00	0.9933	0.37
FMA	1.96±0.06	7.34±0.37	7.02±0.60	11.12±1.52	0.9982	0.34
PTA	0.68±0.05	8.31±0.57	8.20±1.22	6.09±1.13	0.9992	0.26
PHT	1.70±0.05	6.24±0.42	7.41±0.83	7.20±1.07	0.9985	0.39

φ_{min} – the Volume Fraction of the Aqueous Part of the Mobile Phase at the Transition between the HILIC and RP Mechanisms. SDD – Sulfadimidine, SAA – Sulfacetamide. Aromatic Benzoic Acids: HMB – 4-(hydroxymethyl)benzoic Acid, FMA – 4-Formylbenzoic Acid, PTA – p-Toluic Acid, PHT – Phthalic Acid. D^2 – Coefficient of Determination.

and consequently different system response parameters apply in the two modes, and hence the dual-mode systems need separate treatment of the retention data in the RP and in the HILIC mobile phase ranges. The mobile phase composition affects the LSER characteristics of the separation phase systems (Equation (1.21)) more significantly in the HILIC mode than in the RP mode, as illustrated by the data for phenolic acids on three diol and amide columns in Table 1.5 [117]. The basicity of the stationary phase, including the adsorbed water layer (proton-acceptor interactions, parameter b) and, to a lesser extent, dipole–dipole interactions (parameter s), enhances the retention of phenolic acids in the HILIC mode range (4–10% water in acetonitrile). On the other hand, increasing molar volume (parameter v) and the

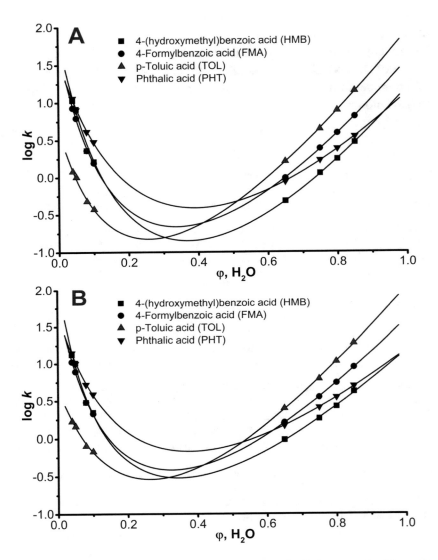

FIGURE 1.16 Effect of the phase volume convention on the U-shape graph of the volume fraction of water, φ_{H_2O}, in buffered aqueous–acetonitrile mobile phases *versus* the retention factors, k, of aromatic benzoic acids on a monolithic MEDSA-DiEDMA micro-column. (A) – *convention 1*, $V_{m,HILIC} = V_{r,toluene} = 8.1$ μL, $V_{m,RP} = V_{r,uracil} = 8.8$ μL; (B) – *convention 3*, $V_{m,HILIC} = V_{m,RP} = V_{r,polystyrene}$ Mw = 1 800 000 $= 6.7$ μL.

acidity of the stationary phase (parameter a) decreases retention in the HILIC mode, but increases retention in the RP mode range (95% water), where the dipole–dipole (s) and hydrogen-bonding interactions between the stationary phase and the solutes (b) decrease the retention. The differences in the values of the parameters for the individual columns depend on the polarity, cohesiveness and concentration of

TABLE 1.5

The System Response Parameters c, v, s, a, b and d^- of the Chirita LSER Model (Equation (1.21) for Phenolic Acids on the YMC Triart Diol-HILIC, Luna HILIC and TSK Gel Amide Columns

Column	Mode	φ_{H_2O}	c	v	s	a	b	d^-	R^2
YMC-Triart Diol-HILIC	HILIC	0.10	0.91	−1.28	0.43	−0.44	0.89	−0.56	0.9935
		0.04	1.40	−1.45	0.53	−0.47	0.90	−0.49	0.9920
	RP	0.95	−1.57	1.66	−0.39	0.19	−0.53	–	0.9240
Luna HILIC	HILIC	0.10	0.91	−1.21	0.37	−0.46	1.49	−0.39	0.9980
		0.04	1.23	−1.19	0.43	−0.48	0.67	−0.46	0.8923
	RP	0.95	−2.65	2.92	−0.59	0.65	−1.49	–	0.9344
TSK Gel Amide	HILIC	0.10	1.51	−1.63	0.45	−0.51	1.10	−0.60	0.9854
		0.04	2.45	−2.16	0.57	−0.63	1.40	−0.68	0.9906
	RP	0.95	−1.65	3.18	−0.50	0.04	−1.89	−0.50	0.9309

R^2 – Multiple Correlation Coefficient.

water adsorbed in the pores and on the differences in the structure of the stationary phases. Including parameter D^-, representing the contribution of ionic (repulsion) interactions significantly improves the fit of the LSER Equation (1.21) for phenolic acids [158].

Figure 1.17 shows several examples of the separation of phenolic antioxidants on a MEDSA-DiEDMA (A, B) and a MEDSA-BiGDMA (C, D) zwitterionic poly-methacrylate micro-column (0.32 mm i.d.) in the HILIC (A, C) and in the RP (B, D) mobile phase ranges [164]. The two columns show the almost reversed elution order in the HILIC range (80% acetonitrile) with respect to the RP range (40% acetonitrile), but there are significant differences in the retention and elution order, due to the different structures of the cross-linking agents (see Figure 1.15A and B).

The dual retention mechanism columns provide complementary selectivities in the HILIC and in the RP modes, so that they are suitable for orthogonal two-dimensional (2-D) LC-LC applications [152]. The mobile phase incompatibility challenges the possibilities of the on-line coupling of the HILIC and RP systems. However, a single dual-mode column offers the possibility for 2-D separations in a sequence of alternating sample injections and subsequent HILIC and RP gradients with decreasing and increasing concentrations of acetonitrile, allowing a sufficient equilibration time between the alternating runs [156].

Some polar columns showing a dual HILIC/RP mechanism appear useful in combined comprehensive two-dimensional (2D) RP×RP and HILIC×RP runs. For example, it is possible to run a zwitterionic polymethacrylate MEDSA-BiGDMA micro-column in the first dimension (D1), alternating in the RP mode with a highly aqueous mobile phase and in the HILIC mode at high acetonitrile concentrations [164]. In either alternating modes, an on-line coupled short (3–5 cm) core-shell or monolithic silica C18 column provides the second-dimension (D2) separation in the comprehensive 2D setup [166]. The HILIC×RP D1 period employed a gradient of

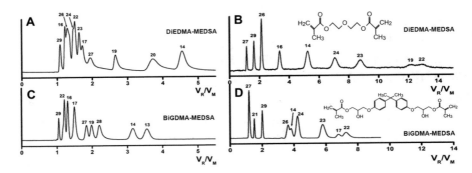

FIGURE 1.17 Dual-mode isocratic separation of flavonoid compounds on DiEDMA (179×0.32 mm) and BiGDMA (123×0.32 mm) MEDSA zwitterionic polymethacrylate micro-columns. A – HILIC separation on a MEDSA-DiEDMA (179×0.32 mm) column: flow rate 2 µL/min; mobile phase 80% acetonitrile + 20% 10 mM ammonium acetate in water; UV detection 214 nm. B – RP separation on a MEDSA-DiEDMA (168×0.32 mm) column: flow rate 5.5 µL/min; mobile phase 40% acetonitrile + 60% 10 mM ammonium acetate in water; UV detection 214 nm. C – HILIC separation on a MEDSA-BiGDMA (123×0.32 mm) column: flow rate 2.1 µL/min; mobile phase 80% acetonitrile + 20% 10 mM ammonium acetate in water; UV detection 214 nm. D – RP separation on a MEDSA-BiGDMA- (123×0.32 mm) column: flow rate 3.1 µL/min; mobile phase 40% acetonitrile + 60% 10 mM ammonium acetate in water; UV detection 214 nm. Compounds: 13 – (–)-Epicatechine, 14 – (+)-Catechine, 15 – Flavone, 16 – 7-Hydroxyflavone, 17 – Apigenine, 18 – Lutheoline, 19 – Quercetine, 20 – Rutine, 21 – Naringine, 22 – Biochanin A, 23 – Naringenine, 24 – Hesperetine, 25 – Hesperidine, 26 – Morine, 27 – Myricetine, 28 – Esculine, 29 – Esculetine, 30 – Scopoletine, 31 – 4-Hydroxycoumarine, 32 – 7-Hydroxycoumarine. Inserted structure of DiEDMA and BiGDMA cross-linkers. Based on the data in [163].

decreasing acetonitrile concentration, followed by flushing the micro-column with a highly aqueous mobile phase for five minutes. Repeated sample injection followed for the separation under reversed-phase gradient conditions with increasing concentration of acetonitrile. Figure 1.18 shows 2D chromatograms of flavones and related polyphenolic compounds, acquired with a single first-dimension MEDSA-BiGDMA micro-column employing two consecutive injections with alternating decreasing and increasing acetonitrile gradients. The dual LC×LC approach can be automated to obtain three-dimensional data (zwitterionic HILIC, zwitterionic RP and C18 RP) in a relatively short time [167].

1.5 CONCLUSIONS, FURTHER PERSPECTIVES

The last decade has witnessed the spectacular development of new, highly efficient column formats (sub-2 µm and core-shell column packing particles, monolithic and recently micro-pillar separation media). A plethora of new stationary phase chemistries have been introduced, such as mixed-mode separation materials and especially silica bonded and organic polymer materials for a relatively new HILIC technique, employing separation systems with polar columns (NP) and aqueous-organic mobile

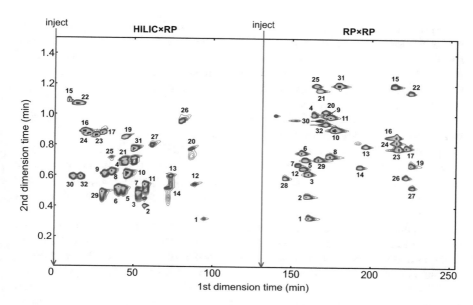

FIGURE 1.18 Dual-mode HILIC×RP + RP×RP 2D separation of phenolic acids and flavones on a MEDSA-BiGDMA micro-column (160×0.53 mm) in alternating HILIC (left from the arrow indicating the second sample injection) and RP (right) modes in the first dimension; decreasing (HILIC) and increasing (RP) gradient of acetonitrile in water. A Chromolith C18 HighResolution column (50×4.6 mm) in the second dimension. $F_{m,D1}=3$ μL/min, $F_{m,D2}=5$ mL/min. 1 – Gallic acid, 2 – Protocatechuic acid, 3 – p-Hydroxybenzoic acid, 4 – Salicylic acid, 5 – Vanillic acid, 6 – Syringic acid, 7 – 4-Hydroxyphenylacetic acid, 8 – Caffeic acid, 9 – Sinapic acid, 10 – p-Coumaric acid, 12 – Ferulic acid, 12 – Chlorogenic acid, 13 – (−)-Epicatechine, 14 – (+)-Catechine, 15 – Flavone, 16 – 7-Hydroxyflavone, 17 – Apigenine, 18 – Lutheoline, 19 – Quercetine, 20 – Rutine, 21 – Naringine, 22 – Biochanin A, 23 – Naringenine, 24 – Hesperetine, 25 – Hesperidine, 26 – Morine, 27 – Myricetine, 28 – Esculine, 29 – Esculetine, 30 – Scopoletine, 31 – 4-Hydroxycoumarine, 32 – 7-Hydroxycoumarine. Based on the data in [167].

phases (RP). The mobile phase, an equally important player in the HPLC separation systems, has attracted less attention.

Besides the trial-and error approach, the composition of the mobile phase in the isocratic LC or the gradient elution program can be adjusted using commercial tools such as DryLab optimization software, or by dedicated chemometric approaches. The approaches usually depart from a set of experimental data acquired under two to six different initial isocratic or gradient conditions with a few pure standards or with a sample itself. The optimization process employs the data interpolation or extrapolation based on a retention model describing the effects of the mobile phase composition on the retention factors, most frequently the Snyder LSS model of reversed-phase LC, or the adsorption-displacement (normal-phase) model. Besides these two-parameter models, three-parameter or more complex equations may provide a better fit for the retention data, such as the quadratic model, the mixed adsorption-partition model or the Neue–Kuss model. The three-parameter ABM model employs the parameters related to the retention in the pure weak and the pure strong solvent in binary mobile phases, regardless of the adsorption, partition or mixed (the most frequent case) retention mechanism. Several models provide equations enabling explicit calculations of the retention in gradient elution, based on the isocratic or gradient input data.

Some columns show a mixed ion-exchange-RP or ion-exchange-HILIC retention mechanism, while other polar columns provide largely prevalent RP mechanism in water-rich mobile phases with, at the same time, a HILIC mechanism at high concentrations of the organic solvent. Such a single dual-mode column often provides complementary separation selectivities for the sample repeatedly injected in alternating runs in the two mobile phase ranges, in one-dimensional or two-dimensional LC systems.

Modeling the retention, especially but not only on the dual-mode columns, meets difficulties in properly defining the ratio of the volumes of the stationary and of the mobile phases in the column. Because of the different solvation of the solid phase in polar and non-polar solvents, the hold-up volume markers provide slightly different V_m values in the HILIC and in the RP ranges. The boundary between the liquid stationary phase (adsorbed solvent) and the bulk mobile phase inside the column pores is more diffuse rather than clearly fixed, especially in HILIC systems, where the adsorbed water forms a part of the stationary phase, changes proportionally to the composition of the mobile phase and strongly differs among various polar columns. Further, in many experimental phase systems, both the partition and the adsorption effects may contribute to the retention in HPLC, which further complicates the situation. On the other hand, we need the phase volumes in the column for the thermodynamic interpretation of the chromatographic retention data. Adopting a definition of the phase volumes by convention may provide a less ambiguous solution acceptable in the method development, even though not theoretically fully rigorous. Possible conventions of the stationary phase: 1/ The inner volume of the column after subtraction of the volumes of the inner pores and of the inter-particle volume. 2/ In addition to the volumes of the intra- and inter-particle pore volumes, the volume of the inaccessible spaces (such as the volume of the core in superficially porous particles) is also subtracted from the column inner volume. 3/ The volume of the inner pores, such as in size-exclusion chromatography. The inner pore volume can be determined as the difference of the elution volumes of a small molecule (e.g., toluene) and of a large one (e.g., the polystyrene standard with $M_r > 10^6$). As the experimental determination may be inconvenient for a common analytical laboratory, the column manufacturer's information may include the information on the pore porosity, ε_i.

In the future, the development of new stationary phases, based on the silica gel or organic polymer supports, will continue, especially in the monolithic and micropillar formats. New stationary phases will provide new separation selectivities, possibly tailored according to the needs of practical problems. Our understanding of the retention models will improve, especially concerning the mixed-mode retention mechanisms. We can expect new multi-use columns with separation ranges designed for specific applications using optimized mobile phase ranges.

REFERENCES

1. Alpert, A. J. 1990. Hydrophilic-interaction chromatography for the separation of peptides, nucleic-acid and other polar compounds. *Journal of Chromatography* 499:177–196.
2. Jandera, P., and P. Janás. 2017. Recent advances in stationary phases and understanding of retention in hydrophilic interaction chromatography. A review. *Analytica Chimica Acta* 967:12–32.
3. Soukup, J., and P. Jandera. 2014. Adsorption of water from aqueous acetonitrile on silica-based stationary phases in aqueous normal-phase liquid chromatography. *Journal of Chromatography A* 1374:102–111.
4. Jandera, P., T. Hájek, and Z. Šromová. 2018. Mobile phase effects in reversed-phase and hydrophilic interaction liquid chromatography revisited. *Journal of Chromatography A* 1543:48–57.
5. Jandera, P. 2006. Can the theory of gradient liquid chromatography be useful in solving practical problems? *Journal of Chromatography A* 1126(1–2):195–218.
6. Schuster, G., and W. Lindner. 2013. Comparative characterization of hydrophilic interaction liquid chromatography columns by linear solvation energy relationships. *Journal of Chromatography A* 1273:73–94.
7. Jandera, P., and T. Hájek. 2018. Mobile phase effects on the retention on polar columns with special attention to the dual hydrophilic interaction-reversed-phase liquid chromatography mechanism, a review. *Journal of Separation Science* 41(1):145–162.
8. Melander, W. R., and C. Horváth. 1980. Reversed-phase chromatography. In: *High Performance Liquid Chromatography, Advances and Perspectives*. Edited by Csaba Horváth. 2. New York: Academic Press.
9. Engelhardt, W. H. 1977. *Hochdruck-Flussigkeits-Chromatographie*. 2nd ed. Berlin: Springer.
10. Kazakevich, Y. V., and H. M. McNair. 1995. Study of the excess adsorption of the eluent components on different reversed-phase adsorbents. *Journal of Chromatographic Science* 33(6):321–327.
11. Urban, J., and P. Jandera. 2013. Recent advances in the design of organic polymer monoliths for reversed-phase and hydrophilic interaction chromatography separations of small molecules. *Analytical and Bioanalytical Chemistry* 405(7):2123–2131.
12. McCalley, D. V., and U. D. Neue. 2008. Estimation of the extent of the water-rich layer associated with the silica surface in hydrophilic interaction chromatography. *Journal of Chromatography A* 1192(2):225–229.
13. Slaats, E. H., J. C. Kraak, W. J. T. Brugman, and H. Poppe. 1978. Study of influence of competition and solvent interaction on retention in liquid-solid chromatography by measurement of activity-coefficients in mobile phase. *Journal of Chromatography A* 149:255–270.
14. Berendsen, G. E., P. J. Schoenmakers, L. D. Galan, G. Vigh, Z. Vargapuchony, and J. Inczédy. 1980. On the determination of the hold-up time in reversed phase liquid-chromatography. *Journal of Liquid Chromatography* 3(11):1669–1686.
15. Moldoveanu, S., and V. David. 2015. Estimation of the phase ratio in reversed-phase high-performance liquid chromatography. *Journal of Chromatography A* 1381:194–201.
16. Rimmer, C. A., C. R. Simmons, and J. G. Dorsey. 2002. The measurement and meaning of void volumes in reversed-phase liquid chromatography. *Journal of Chromatography A* 965(1–2):219–232.
17. Riedo, F., and E. S. Kovats. 1982. Adsorption from liquid-mixtrures and liquid-chromatography. *Journal of Chromatography A* 239:1–28.
18. Jandera, P., T. Hájek, and M. Růžičková. 2017. Retention models on core–shell columns. *Journal of AOAC International* 100(6):1636–1646.

19. Neue, U. D. 1997. *HPLC Columns, Theory, Technology, and Practice*. New York: Wiley-VCH.
20. Poppe, H. 1997. Some reflections on speed and efficiency of modern chromatographic methods. *Journal of Chromatography A* 778(1–2):3–21.
21. Yoshida, T., and R. E. Majors. 2006. High-speed analyses using rapid resolution liquid chromatography on 1.8-mu m porous particles. *Journal of Separation Science* 29(16):2421–2432.
22. De Vos, J., K. Broeckhoven, and S. Eeltink. 2016. Advances in ultrahigh-pressure liquid chromatography technology and system design. *Analytical Chemistry* 88(1):262–278.
23. Walter, T. H., and R. W. Andrews. 2014. Recent innovations in UHPLC columns and instrumentation. *TrAC-Trends in Analytical Chemistry* 63:14–20.
24. Xiang, Y. Q., Y. S. Liu, and M. L. Lee. 2006. Ultrahigh pressure liquid chromatography using elevated temperature. *Journal of Chromatography A* 1104(1–2):198–202.
25. DeStefano, J. J., S. A. Schuster, J. M. Lawhorn, and J. J. Kirkland. 2012. Performance characteristics of new superficially porous particles. *Journal of Chromatography A* 1258:76–83.
26. Hayes, R., A. Ahmed, T. Edge, and H. F. Zhang. 2014. Core-shell particles: Preparation, fundamentals and applications in high performance liquid chromatography. *Journal of Chromatography A* 1357:36–52.
27. Aggarwal, P., H. D. Tolley, and M. L. Lee. 2012. Monolithic bed structure for capillary liquid chromatography. *Journal of Chromatography A* 1219:1–14.
28. Guiochon, G. 2007. Monolithic columns in high-performance liquid chromatography. *Journal of Chromatography A* 1168(1–2):101–168.
29. Minakuchi, H., K. Nakanishi, N. Soga, N. Ishizuka, and N. Tanaka. 1996. Octadecylsilylated porous silica rods as separation media for reversed-phase liquid chromatography. *Analytical Chemistry* 68(19):3498–3501.
30. Švec, F. 2012. Quest for organic polymer-based monolithic columns affording enhanced efficiency in high performance liquid chromatography separations of small molecules in isocratic mode. *Journal of Chromatography A* 1228:250–262.
31. Miyamoto, K., T. Hara, H. Kobayashi, H. Morisaka, D. Tokuda, K. Horie, K. Koduki, S. Makino, O. Nunez, C. Yang, T. Kawabe, T. Ikegami, H. Takubo, Y. Ishihama, and N. Tanaka. 2008. High-efficiency liquid chromatographic separation utilizing long monolithic silica capillary columns. *Analytical Chemistry* 80(22):8741–8750.
32. Nakanishi, K., and N. Tanaka. 2007. Sol-gel with phase separation. Hierarchically porous materials optimized for high-performance liquid chromatography separations. *Accounts of Chemical Research* 40(9):863–873.
33. Ali, A., F. Ali, and W. J. Cheong. 2017. Sedimentation assisted preparation of ground particles of silica monolith and their C18 modification resulting in a chromatographic phase of improved separation efficiency. *Journal of Chromatography A* 1525:79–86.
34. Cabooter, D., K. Broeckhoven, R. Sterken, A. Vanmessen, I. Vandendael, K. Nakanishi, S. Deridder, and G. Desmet. 2014. Detailed characterization of the kinetic performance of first and second generation silica monolithic columns for reversed-phase chromatography separations. *Journal of Chromatography A* 1325:72–82.
35. Arrua, R. D., M. Talebi, T. J. Causon, and E. F. Hilder. 2012. Review of recent advances in the preparation of organic polymer monoliths for liquid chromatography of large molecules. *Analytica Chimica Acta* 738:1–12.
36. Švec, F., and J. M. J. Frechet. 1995. Temperature, a simple and efficient tool for the control of pore size distribution in macroporous polymers. *Macromolecules* 28(22):7580–7582.
37. Buchmeiser, M. R. 2007. Polymeric monolithic materials: Syntheses, properties, functionalization and applications. *Polymer* 48(8):2187–2198.

38. Nischang, I., I. Teasdale, and O. Brueggemann. 2011. Porous polymer monoliths for small molecule separations: Advancements and limitations. *Analytical and Bioanalytical Chemistry* 400(8):2289–2304.
39. Urban, J. 2016. Current trends in the development of porous polymer monoliths for the separation of small molecules. *Journal of Separation Science* 39(1):51–68.
40. Nischang, I., and O. Bruggemann. 2010. On the separation of small molecules by means of nano-liquid chromatography with methacrylate-based macroporous polymer monoliths. *Journal of Chromatography A* 1217(33):5389–5397.
41. Jandera, P., M. Staňková, V. Škeříková, and J. Urban. 2013. Cross-linker effects on the separation efficiency on (poly)methacrylate capillary monolithic columns. Part I. Reversed-phase liquid chromatography. *Journal of Chromatography A* 1274:97–106.
42. Staňková, M., P. Jandera, V. Škeříková, and J. Urban. 2013. Cross-linker effects on the separation efficiency on (poly)methacrylate capillary monolithic columns. Part II. Aqueous normal-phase liquid chromatography. *Journal of Chromatography A* 1289:47–57.
43. Currivan, S., and P. Jandera. 2014. Post-polymerization modifications of polymeric monolithic columns: A review. *Chromatography* 1(1):24–53.
44. Lv, Y., Z. Lin, and F. Svec. 2012. Hypercrosslinked large surface area porous polymer monoliths for hydrophilic interaction liquid chromatography of small molecules featuring zwitterionic functionalities attached to gold nanoparticles held in layered structure. *Analytical Chemistry* 84(20):8457–8460.
45. Currivan, S., J. M. Macák, and P. Jandera. 2015. Polymethacrylate monolithic columns for hydrophilic interaction liquid chromatography prepared using a secondary surface polymerization. *Journal of Chromatography A* 1402:82–93.
46. Urban, J., F. Švec, and J. M. J. Frechet. 2010. Efficient separation of small molecules using a large surface area hypercrosslinked monolithic polymer capillary column. *Analytical Chemistry* 82(5):1621–1623.
47. Sandra, K., J. Vandenbussche, I. Vandenheede, B. Claerebout, J. Op de Beeck, P. Jacobs, W. De Malsche, G. Desmet, and P. Sandra. 2018. Peptide mapping of monoclonal antibodies and antibody-drug conjugates using micro-pillar array columns combined with mass spectrometry. *LC-GC Europe* 31(3):155–166.
48. De Malsche, W., J. O. De Beeck, S. De Bruyne, H. Gardeniers, and G. Desmet. 2012. Realization of 1 × 10(6) theoretical plates in liquid chromatography using very long pillar array columns. *Analytical Chemistry* 84(3):1214–1219.
49. McCormick, R. M., and B. L. Karger. 1980. Distribution phenomena of mobile-phase components and determination of dead volume in reversed-phase liquid-chromatography. *Analytical Chemistry* 52(14):2249–2257.
50. Poole, C. F. 2019. Influence of solvent effects on retention of small molecules in reversed-phase liquid chromatography. *Chromatographia* 82(1):49–64.
51. Teutenberg, T., K. Hollebekkers, S. Wiese, and A. Boergers. 2009. Temperature and pH-stability of commercial stationary phases. *Journal of Separation Science* 32(9):1262–1274.
52. Pesek, J. J., and M. T. Matyska. 2009. Our favorite materials: Silica hydride stationary phases. *Journal of Separation Science* 32(23–24):3999–4011.
53. Soukup, J., and P. Jandera. 2012. Hydrosilated silica-based columns: The effects of mobile phase and temperature on dual hydrophilic-reversed-phase separation mechanism of phenolic acids. *Journal of Chromatography A* 1228:125–134.
54. Zuvela, P., M. Skoczylas, J. J. Liu, T. Baczek, R. Kaliszan, M. W. Wong, and B. Buszewski. 2019. Column characterization and selection systems in reversed-phase high-performance liquid chromatography. *Chemical Reviews (Washington, DC, United States)* 119(6):3674–3729.

55. Gritti, F., Y. V. Kazakevich, and G. Guiochon. 2007. Effect of the surface coverage of endcapped C-18-silica on the excess adsorption isotherms of commonly used organic solvents from water in reversed phase liquid chromatography. *Journal of Chromatography A* 1169(1–2):111–124.

56. Vailaya, A., and C. Horvath. 1997. Solvophobic theory and normalized free energies of nonpolar substances in reversed phase chromatography. *Journal of Physical Chemistry B* 101(30):5875–5888.

57. Cole, L. A., J. G. Dorsey, and K. A. Dill. 1992. Temperature-dependence of retention in reversed-phase liquid-chromatography. 2. Mobile-phase considerations. *Analytical Chemistry* 64(13):1324–1327.

58. Poole, C. F. 2015. An interphase model for retention in liquid chromatography. *JPC-Journal of Planar Chromatography-Modern TLC* 28(2):98–105.

59. Gritti, F., A. D. Pereira, P. Sandra, and G. Guiochon. 2010. Efficiency of the same neat silica column in hydrophilic interaction chromatography and per aqueous liquid chromatography. *Journal of Chromatography A* 1217(5):683–688.

60. McCalley, D. V. 2007. Is hydrophilic interaction chromatography with silica columns a viable alternative to reversed-phase liquid chromatography for the analysis of ionisable compounds? *Journal of Chromatography A* 1171(1–2):46–55.

61. Kawachi, Y., T. Ikegami, H. Takubo, Y. Ikegami, M. Miyamoto, and N. Tanaka. 2011. Chromatographic characterization of hydrophilic interaction liquid chromatography stationary phases: Hydrophilicity, charge effects, structural selectivity, and separation efficiency. *Journal of Chromatography A* 1218(35):5903–5919.

62. Heaton, J. C., J. J. Russell, T. Underwood, R. Boughtflower, and D. V. McCalley. 2014. Comparison of peak shape in hydrophilic interaction chromatography using acidic salt buffers and simple acid solutions. *Journal of Chromatography A* 1347:39–48.

63. Strege, M. A. 1998. Hydrophilic interaction chromatography electrospray mass spectrometry analysis of polar compounds for natural product drug discovery. *Analytical Chemistry* 70(13):2439–2445.

64. McCalley, D. V. 2017. Understanding and manipulating the separation in hydrophilic interaction liquid chromatography. *Journal of Chromatography A* 1523:49–71.

65. Fountain, K. J., J. Xu, D. M. Diehl, and D. Morrison. 2010. Influence of stationary phase chemistry and mobile-phase composition on retention, selectivity, and MS response in hydrophilic interaction chromatography. *Journal of Separation Science* 33(6–7):740–751.

66. Buszewski, B., and S. Noga. 2012. Hydrophilic interaction liquid chromatography (HILIC)-a powerful separation technique. *Analytical and Bioanalytical Chemistry* 402(1):231–247.

67. Jandera, P. 2011. Stationary and mobile phases in hydrophilic interaction chromatography: A review. *Analytica Chimica Acta* 692(1–2):1–25.

68. Jiang, W., G. Fischer, Y. Girmay, and K. Irgum. 2006. Zwitterionic stationary phase with covalently bonded phosphorylcholine type polymer grafts and its applicability to separation of peptides in the hydrophilic interaction liquid chromatography mode. *Journal of Chromatography A* 1127(1–2):82–91.

69. Guo, Y., and S. Gaiki. 2005. Retention behavior of small polar compounds on polar stationary phases in hydrophilic interaction chromatography. *Journal of Chromatography A* 1074(1–2):71–80.

70. Alpert, A. J. 2008. Electrostatic repulsion hydrophilic interaction chromatography for isocratic separation of charged solutes and selective isolation of phosphopeptides. *Analytical Chemistry* 80(1):62–76.

71. Dinh, N. P., T. Jonsson, and K. Irgum. 2013. Water uptake on polar stationary phases under conditions for hydrophilic interaction chromatography and its relation to solute retention. *Journal of Chromatography A* 1320:33–47.

72. Bicker, W., J. Y. Wu, M. Lammerhofer, and W. Lindner. 2008. Hydrophilic interaction chromatography in nonaqueous elution mode for separation of hydrophilic analytes on silica-based packings with noncharged polar bondings. *Journal of Separation Science* 31(16–17):2971–2987.

73. Soukup, J., and P. Jandera. 2013. Comparison of nonaqueous hydrophilic interaction chromatography with aqueous normal-phase chromatography on hydrosilated silica-based stationary phases. *Journal of Separation Science* 36(17):2753–2759.

74. Glajch, J. L., J. J. Kirkland, K. M. Squire, and J. M. Minor. 1980. Optimization of solvent strength and selectivity for reversed-phase liquid-chromatography using an interactive mixture-design statistical technique. *Journal of Chromatography A* 199:57–79.

75. Bocian, S. 2016. Solvation processes in liquid chromatography: The importance and measurements. *Journal of Liquid Chromatography & Related Technologies* 39(16):731–738.

76. Bocian, S., J. Soukup, P. Jandera, and B. Buszewski. 2015. Thermodynamics study of solvent adsorption on octadecyl-modified silica. *Chromatographia* 78(1–2):21–30.

77. Bocian, S., M. Skoczylas, I. Gorynska, M. Matyska, J. Pesek, and B. Buszewski. 2016. Solvation processes on phenyl-bonded stationary phases–The influence of polar functional groups. *Journal of Separation Science* 39(22):4369–4376.

78. Melnikov, S. M., A. Holtzel, A. Seidel-Morgenstern, and U. Tallarek. 2011. Composition, structure, and mobility of water-acetonitrile mixtures in a silica nanopore studied by Molecular Dynamics simulations. *Analytical Chemistry* 83(7):2569–2575.

79. Wikberg, E., T. Sparrman, C. Viklund, T. Jonsson, and K. Irgum. 2011. A H-2 nuclear magnetic resonance study of the state of water in neat silica and zwitterionic stationary phases and its influence on the chromatographic retention characteristics in hydrophilic interaction high-performance liquid chromatography. *Journal of Chromatography A* 1218(38):6630–6638.

80. Greco, G., S. Grosse, and T. Letzel. 2012. Study of the retention behavior in zwitterionic hydrophilic interaction chromatography of isomeric hydroxy- and aminobenzoic acids. *Journal of Chromatography A* 1235:60–67.

81. Guo, Y., and R. Shah. 2016. Detailed insights into the retention mechanism of caffeine metabolites on the amide stationary phase in hydrophilic interaction chromatography. *Journal of Chromatography A* 1463:121–127.

82. Guo, Y., N. Bhalodia, B. Fattal, and I. Serris. 2019. Evaluating the adsorbed water layer on polar stationary phases for hydrophilic interaction chromatography (HILIC). *Separations* 6(2):19–29.

83. Vajda, P., A. Felinger, and G. Guiochon. 2013. Evaluation of surface excess isotherms in liquid chromatography. *Journal of Chromatography A* 1291:41–47.

84. Langmuir, I. 1916. The comnstituion and fundamental properties of solid and liquids. Part I. Solids. *Journal of the American Chemical Society* 38(11):2221–2295.

85. Pesek, J. J., M. T. Matyska, R. I. Boysen, Y. Z. Yang, and M. T. W. Hearn. 2013. Aqueous normal-phase chromatography using silica-hydride-based stationary phases. *TrAC-Trends in Analytical Chemistry* 42:64–73.

86. Snyder, L. R. 1968. Principles of adsorption chromatography: The separation of nonionic organic compounds. In: *Chromatographic Science*. Edited by Calvin Giddings and Roy A. Keller. New York: Marcel Dekker.

87. Jandera, P., and J. Churáček. 1974. Gradient elution in liquid chromatography: I. The influence of the composition of the mobile phase on the capacity ratio (retention volume, band width, and resolution) in isocratic elution — Theoretical considerations. *Journal of Chromatography* 91(0):207–221.

88. Snyder, L. R. 1974. Role of solvent in liquid-solid chromatography. Review. *Analytical Chemistry* 46(11):1384–1393.

89. Snyder, L. R., and J. W. Dolan. 2007. *High-Performance Gradient Elution, The Practical Application of the Linear-Solvent-Strength Model.* Hoboken, NJ: Wiley.
90. Tijssen, R., H. A. H. Billiet, and P. J. Schoenmakers. 1976. Use of solubility parameter for predicting selectivity and retention in chromatography. *Journal of Chromatography A* 122:185–203.
91. Jandera, P., and J. Kubát. 1990. Possibilities of determination and prediction of solute capacity factors in reverse-phase systems with pure water as the mobile phase. *Journal of Chromatography A* 500:281–299.
92. Jin, G., Z. Guo, F. Zhang, X. Xue, Y. Jin, and X. Liang. 2008. Study on the retention equation in hydrophilic interaction liquid chromatography. *Talanta* 76(3):522–527.
93. Neue, U. D., and H.-J. Kuss. 2010. Improved reversed-phase gradient retention modeling. *Journal of Chromatography A* 1217(24):3794–3803.
94. Tyteca, E., J. De Vos, N. Vankova, P. Cesla, G. Desmet, and S. Eeltink. 2016. Applicability of linear and nonlinear retention-time models for reversed-phase liquid chromatography separations of small molecules, peptides, and intact proteins. *Journal of Separation Science* 39(7):1249–1257.
95. Tyteca, E., A. Periat, S. Rudaz, G. Desmet, and D. Guillarme. 2014. Retention modeling and method development in hydrophilic interaction chromatography. *Journal of Chromatography A* 1337:116–127.
96. Vivo-Truyols, G., J. R. Torres-Lapasio, and M. C. Garcia-Alvarez-Coque. 2003. Error analysis and performance of different retention models in the transference of data from/to isocratic/gradient elution. *Journal of Chromatography A* 1018(2):169–181.
97. Jandera, P., M. Janderová, and J. Churáček. 1978. Gradient elution in liquid-chromatography. 8. Selection of optimal composition of mibile phase in liquid-chromatography under isocratic conditions. *Journal of Chromatography A* 148(1):79–97.
98. Jandera, P., M. Kučerová, and J. Holíková. 1997. Description and prediction of retention in normal-phase high-performance liquid chromatography with binary and ternary mobile phases. *Journal of Chromatography A* 762(1–2):15–26.
99. Beyaz, A., W. Fan, P. W. Carr, and A. P. Schellinger. 2014. Instrument parameters controlling retention precision in gradient elution reversed-phase liquid chromatography. *Journal of Chromatography A* 1371(0):90–105.
100. Jandera, P., and J. Churáček. 1981. Liquid chromatography with programmed composition of the mobile phase. *Advances in Chromatography* 19:125–260.
101. Baeza-Baeza, J. J., C. Ortiz-Bolsico, J. R. Torres-Lapasió, and M. C. García-Álvarez-Coque. 2013. Approaches to model the retention and peak profile in linear gradient reversed-phase liquid chromatography. *Journal of Chromatography A* 1284(0):28–35.
102. Jandera, P., and J. Churáček. 1974. Gradient elution in liqud-chromatography. 2. Retention characteristics (Retention volume, band width, resolution, plate number) in solvent-programmed chromatography - Theoretical considerations. *Journal of Chromatography A* 91:223–235.
103. Jandera, P., J. Churáček, and L. Svoboda. 1979. Gradient elution in liquid chromatography: X. Retention characteristics in reversed-phase gradient elution chromatography. *Journal of Chromatography A* 174(1):35–50.
104. Schoenmakers, P. J., H. A. H. Billiet, R. Tijssen, and L. Degalan. 1978. Gradient selection in reverse-phase liquid-chromatography. *Journal of Chromatography A* 149:519–537.
105. Snyder, L. R., J. W. Dolan, and J. R. Gant. 1979. Gradient elution in high-performance liquid-chromatography. 1. Theoretical basis for reversed-phase systems. *Journal of Chromatography A* 165(1):3–30.
106. Jandera, P., and M. Kucerova. 1997. Prediction of retention in gradient-elution normal-phase high-performance liquid chromatography with binary solvent gradients. *Journal of Chromatography A* 759(1–2):13–25.

107. Jandera, P., T. Hájek, and P. Česla. 2010. Comparison of various second-dimension gradient types in comprehensive two-dimensional liquid chromatography. *Journal of Separation Science* 33(10):1382–1397.

108. Vyňuchalová, K., and P. Jandera. 2013. Possibilities of retention prediction in fast gradient liquid chromatography. Part 1: Comparison of separation on packed fully porous, nonporous and monolithic columns. *Journal of Chromatography A* 1278(0):37–45.

109. Jandera, P., T. Hájek, and K. Vyňuchalová. 2014. Retention and bandwidths prediction in fast gradient liquid chromatography. Part 2. *Core–Shell Columns. Journal of Chromatography A* 1337(0):57–66.

110. Jandera, P., and T. Hájek. 2015. Possibilities of retention prediction in fast gradient liquid chromatography. Part 3: Short silica monolithic columns. *Journal of Chromatography A* 1410:76–89.

111. Schoenmakers, P. J., H. A. H. Billiet, and L. Degalan. 1981. Use of gradient elution for rapid selection of isocratic conditions in reversed-phase high-performance liquid chromatography. *Journal of Chromatography A* 205(1):13–30.

112. Kormany, R., I. Molnar, and H. J. Rieger. 2013. Exploring better column selectivity choices in ultra-high performance liquid chromatography using Quality by Design principles. *Journal of Pharmaceutical and Biomedical Analysis* 80:79–88.

113. Molnar, I. 2002. Computerized design of separation strategies by reversed-phase liquid chromatography: Development of DryLab software. *Journal of Chromatography A* 965(1–2):175–194.

114. Pirok, B. W. J., S. R. A. Molenaar, R. E. van Outersterp, and P. J. Schoenmakers. 2017. Applicability of retention modelling in hydrophilic-interaction liquid chromatography for algorithmic optimization programs with gradient-scanning techniques. *Journal of Chromatography A* 1530:104–111.

115. Tyteca, E., D. Guillarme, and G. Desmet. 2014. Use of individual retention modeling for gradient optimization in hydrophilic interaction chromatography: Separation of nucleobases and nucleosides. *Journal of Chromatography A* 1368:125–131.

116. Soukup, J., P. Janás, and P. Jandera. 2013. Gradient elution in aqueous normal-phase liquid chromatography on hydrosilated silica-based stationary phases. *Journal of Chromatography A* 1286:111–118.

117. Jandera, P., P. Janás, V. Škeříková, and J. Urban. 2017. Effect of water on the retention on diol and amide columns in hydrophilic interaction liquid chromatography. *Journal of Separation Science* 40(7):1434–1448.

118. Abraham, M. H., and M. Roses. 1994. Hydrogen-bonding. 38. Effect of solute structure and mobile-phase composition on reversed-phase high-performance liquid-chromatographic capacity factors. *Journal of Physical Organic Chemistry* 7(12):672–684.

119. Abraham, M. H., M. Roses, C. F. Poole, and S. K. Poole. 1997. Hydrogen bonding. 42. Characterization of reversed-phase high-performance liquid chromatographic C-18 stationary phases. *Journal of Physical Organic Chemistry* 10(5):358–368.

120. Chirita, R. I., C. West, S. Zubrzycki, A. L. Finaru, and C. Elfakir. 2011. Investigations on the chromatographic behaviour of zwitterionic stationary phases used in hydrophilic interaction chromatography. *Journal of Chromatography A* 1218(35):5939–5963.

121. Jandera, P., and J. Churáček. 1974. Ion-exchange chromatography of aldehydes, ketones, ethers, alcohols, polyols and saccharides. *Journal of Chromatography A* 98(1):55–104.

122. Lammerhofer, M., R. Nogueira, and W. Lindner. 2011. Multi-modal applicability of a reversed-phase/weak-anion exchange material in reversed-phase, anion-exchange, ion-exclusion, hydrophilic interaction and hydrophobic interaction chromatography modes. *Analytical and Bioanalytical Chemistry* 400(8):2517–2530.

123. Ray, S., M. Takafuji, and H. Ihara. 2012. Chromatographic evaluation of a newly designed peptide-silica stationary phase in reverse phase liquid chromatography and hydrophilic interaction liquid chromatography: Mixed mode behavior. *Journal of Chromatography A* 1266:43–52.
124. Wang, L. J., W. L. Wei, Z. N. Xia, X. Jie, and Z. Z. L. Xia. 2016. Recent advances in materials for stationary phases of mixed-mode high-performance liquid chromatography. *TrAC-Trends in Analytical Chemistry* 80:495–506.
125. Yang, Y., and X. D. Geng. 2011. Mixed-mode chromatography and its applications to biopolymers. *Journal of Chromatography A* 1218(49):8813–8825.
126. Zhang, K., and X. D. Liu. 2016. Mixed-mode chromatography in pharmaceutical and biopharmaceutical applications. *Journal of Pharmaceutical and Biomedical Analysis* 128:73–88.
127. Zhao, G. F., X. Y. Dong, and Y. Sun. 2009. Ligands for mixed-mode protein chromatography: Principles, characteristics and design. *Journal of Biotechnology* 144(1):3–11.
128. Wu, J. Y., W. G. Bicker, and W. G. Lindner. 2008. Separation properties of novel and commercial polar stationary phases in hydrophilic interaction and reversed-phase liquid chromatography mode. *Journal of Separation Science* 31(9):1492–1503.
129. Lammerhofer, M., M. Richter, J. Y. Wu, R. Nogueira, W. Bicker, and W. Lindner. 2008. Mixed-mode ion-exchangers and their comparative chromatographic characterization in reversed-phase and hydrophilic interaction chromatography elution modes. *Journal of Separation Science* 31(14):2572–2588.
130. Alpert, A. J., O. Hudecz, and K. Mechtler. 2015. Anion-exchange chromatography of phosphopeptides: Weak anion exchange versus strong anion exchange and anion-exchange chromatography versus electrostatic repulsion-hydrophilic interaction chromatography. *Analytical Chemistry* 87(9):4704–4711.
131. Aral, T., H. Aral, B. Ziyadanogullari, and R. Ziyadanogullari. 2015. Synthesis of a mixed-model stationary phase derived from glutamine for HPLC separation of structurally different biologically active compounds: HILIC and reversed-phase applications. *Talanta* 131:64–73.
132. Shen, A. J., X. L. Li, X. F. Dong, J. Wei, Z. M. Guo, and X. M. Liang. 2013. Glutathione-based zwitterionic stationary phase for hydrophilic interaction/cation-exchange mixed-mode chromatography. *Journal of Chromatography A* 1314:63–69.
133. Lowe, A. B. 2010. Thiol-ene "click" reactions and recent applications in polymer and materials synthesis. *Polymer Chemistry* 1(1):17–36.
134. Peng, X. T., T. Liu, S. X. Ji, and Y. Q. Feng. 2013. Preparation of a novel carboxyl stationary phase by "thiol-ene" click chemistry for hydrophilic interaction chromatography. *Journal of Separation Science* 36(16):2571–2577.
135. Peng, X. T., Y. Q. Feng, X. Z. Hu, and D. J. Hu. 2013. Preparation and characterization of the neomycin-bonded silica stationary phase for hydrophilic-interaction chromatography. *Chromatographia* 76(9–10):459–465.
136. Liu, X. D., and C. A. Pohl. 2010. HILIC behavior of a reversed-phase/cation-exchange/anion-exchange trimode column. *Journal of Separation Science* 33(6–7):779–786.
137. Viklund, C., A. Sjogren, K. Irgum, and I. Nes. 2001. Chromatographic interactions between proteins and sulfoalkylbetaine-based zwitterionic copolymers in fully aqueous low-salt buffers. *Analytical Chemistry* 73(3):444–452.
138. Wikberg, E., J. J. Verhage, C. Viklund, and K. Irgum. 2009. Grafting of silica with sulfobetaine polymers via aqueous reversible addition fragmentation chain transfer polymerization and its use as a stationary phase in HILIC. *Journal of Separation Science* 32(12):2008–2016.
139. Idborg, H., L. Zamani, P. O. Edlund, I. Schuppe-Koistinen, and S. P. Jacobsson. 2005. Metabolic fingerprinting of rat urine by LC/MS Part 1. Analysis by hydrophilic interaction liquid chromatography-electrospray ionization mass spectrometry. *Journal of Chromatography B-Analytical Technologies in the Biomedical and Life Sciences* 828(1–2):9–13.

140. Wade, K. L., I. J. Garrard, and J. W. Fahey. 2007. Improved hydrophilic interaction chromatography method for the identification and quantification of glucosinolates. *Journal of Chromatography A* 1154(1–2):469–472.

141. Oertel, R., V. Neumeister, and W. Kirch. 2004. Hydrophilic interaction chromatography combined with tandem-mass spectrometry to determine six aminoglycosides in serum. *Journal of Chromatography A* 1058(1–2):197–201.

142. Marrubini, G., B. E. C. Mendoza, and G. Massolini. 2010. Separation of purine and pyrimidine bases and nucleosides by hydrophilic interaction chromatography. *Journal of Separation Science* 33(6–7):803–816.

143. Takegawa, Y., K. Deguchi, T. Keira, H. Ito, H. Nakagawa, and S. Nishimura. 2006. Separation of isomeric 2-aminopyridine derivatized N-glycans and N-glycopeptides of human serum immunoglobulin G by using a zwitterionic type of hydrophilic-interaction chromatography. *Journal of Chromatography A* 1113(1–2):177–181.

144. Wohlgemuth, J., M. Karas, W. Jiang, R. Hendriks, and S. Andrecht. 2010. Enhanced glyco-profiling by specific glycopeptide enrichment and complementary monolithic nano-LC (ZIC-HILIC/RP18e)/ESI-MS analysis. *Journal of Separation Science* 33(6–7):880–890.

145. Boersema, P. J., N. Divecha, A. J. R. Heck, and S. Mohammed. 2007. Evaluation and optimization of ZIC-HILIC-RP as an alternative MudPIT strategy. *Journal of Proteome Research* 6(3):937–946.

146. Foo, H. C., J. Heaton, N. W. Smith, and S. Stanley. 2012. Monolithic poly (SPE-co-BVPE) capillary columns as a novel hydrophilic interaction liquid chromatography stationary phase for the separation of polar analytes. *Talanta* 100:344–348.

147. Takegawa, Y., H. Ito, T. Keira, K. Deguchi, H. Nakagawa, and S. I. Nishimura. 2008. Profiling of N- and O-glycopeptides of erythropoietin by capillary zwitterionic type of hydrophilic interaction chromatography/electrospray ionization mass spectrometry. *Journal of Separation Science* 31(9):1585–1593.

148. Cheng, X. D., X. T. Peng, Q. W. Yu, B. F. Yuan, and Y. Q. Feng. 2013. Preparation of a novel amino-phosphate zwitterionic stationary phase for hydrophilic interaction chromatography. *Chromatographia* 76(23–24):1569–1576.

149. Alvarez-Segura, T., J. R. Torres-Lapasio, C. Ortiz-Bolsico, and M. C. Garcia-Alvarez-Coque. 2016. Stationary phase modulation in liquid chromatography through the serial coupling of columns: A review. *Analytica Chimica Acta* 923:1–23.

150. Chalcraft, K. R., and B. E. McCarry. 2013. Tandem LC columns for the simultaneous retention of polar and nonpolar molecules in comprehensive metabolomics analysis. *Journal of Separation Science* 36(21–22):3478–3485.

151. Greco, G., S. Grosse, and T. Letzel. 2013. Serial coupling of reversed-phase and zwitterionic hydrophilic interaction LC/MS for the analysis of polar and nonpolar phenols in wine. *Journal of Separation Science* 36(8):1379–1388.

152. Jandera, P., T. Hájek, M. Staňková, K. Vyňuchalová, and P. Česla. 2012. Optimization of comprehensive two-dimensional gradient chromatography coupling in-line hydrophilic interaction and reversed phase liquid chromatography. *Journal of Chromatography A* 1268:91–101.

153. Zhang, T., D. J. Creek, M. P. Barrett, G. Blackburn, and D. G. Watson. 2012. Evaluation of coupling reversed phase, aqueous normal phase, and hydrophilic interaction liquid chromatography with Orbitrap mass spectrometry for metabolomic studies of human urine. *Analytical Chemistry* 84(4):1994–2001.

154. Falasca, S., F. Petruzziello, R. Kretz, G. Rainer, and X. Z. Zhang. 2012. Analysis of multiple quaternary ammonium compounds in the brain using tandem capillary column separation and high resolution mass spectrometric detection. *Journal of Chromatography A* 1241:46–51.

155. Cabooter, D., K. Choikhet, F. Lestremau, M. Dittmann, and G. Desmet. 2014. Towards a generic variable column length method development strategy for samples with a large variety in polarity. *Journal of Chromatography A* 1372:174–186.

156. Jandera, P., and T. Hájek. 2009. Utilization of dual retention mechanism on columns with bonded PEG and diol stationary phases for adjusting the separation selectivity of phenolic and flavone natural antioxidants. *Journal of Separation Science* 32(21):3603–3619.

157. Vlčková, H., K. Ježková, K. Štětková, H. Tomšíková, P. Solich, and L. Novaková. 2014. Study of the retention behavior of small polar molecules on different types of stationary phases used in hydrophilic interaction liquid chromatography. *Journal of Separation Science* 37(11):1297–1307.

158. Jandera, P., T. Hájek, V. Škeříková, and J. Soukup. 2010. Dual hydrophilic interaction-RP retention mechanism on polar columns: Structural correlations and implementation for 2-D separations on a single column. *Journal of Separation Science* 33(6–7):841–852.

159. Pesek, J. J., M. T. Matyska, and S. Larrabee. 2007. HPLC retention behavior on hydride-based stationary phases. *Journal of Separation Science* 30(5):637–647.

160. Pesek, J. J., M. T. Matyska, and H. Natekar. 2016. Evaluation of the dual retention properties of stationary phases based on silica hydride: Perfluorinated bonded material. *Journal of Separation Science* 39(6):1050–1055.

161. Soukup, J., and P. Jandera. 2012. The effect of temperature and mobile phase composition on separation mechanism of flavonoid compounds on hydrosilated silica-based columns. *Journal of Chromatography A* 1245:98–108.

162. Jiang, Z. J., N. W. Smith, P. D. Ferguson, and M. R. Taylor. 2007. Hydrophilic interaction chromatography using methacrylate-based monolithic capillary column for the separation of polar analytes. *Analytical Chemistry* 79(3):1243–1250.

163. Staňková, M., and P. Jandera. 2016. Dual retention mechanism in two-dimensional LC separations of barbiturates, sulfonamides, nucleic bases and nucleosides on polymethacrylate zwitterionic monolithic micro-columns. *Chromatographia* 79(11–12):657–666.

164. Jandera, P., M. Staňková, and T. Hájek. 2013. New zwitterionic polymethacrylate monolithic columns for one- and two-dimensional microliquid chromatography. *Journal of Separation Science* 36(15):2430–2440.

165. Feng, J. T., Z. M. Guo, H. Shi, J. P. Gu, Y. Jin, and X. M. Liang. 2010. Orthogonal separation on one beta-cyclodextrin column by switching reversed-phase liquid chromatography and hydrophilic interaction chromatography. *Talanta* 81(4–5):1870–1876.

166. Jandera, P., T. Hájek, and M. Staňková. 2015. Monolithic and core-shell columns in comprehensive two-dimensional HPLC: A review. *Analytical and Bioanalytical Chemistry* 407(1):139–151.

167. Hájek, T., P. Jandera, M. Staňková, and P. Česla. 2016. Automated dual two-dimensional liquid chromatography approach for fast acquisition of three-dimensional data using combinations of zwitterionic polymethacrylate and silica-based monolithic columns. *Journal of Chromatography A* 1446:91–102.

2 Temperature Effects in Reversed-Phase Liquid Chromatography

Use in Focusing, Temperature-Stable Stationary Phases, Effect on Retention, and Viscous Dissipation

Anthony R. Horner, Erin P. Shields, Michael T. Rerick, and Stephen G. Weber

CONTENTS

2.1   INTRODUCTION

This chapter focuses on recent developments in the use of temperature to improve separations by reversed-phase liquid chromatography (RPLC). Lloyd Snyder contributed enormously to chromatography in general, also being among the first to recognize and exploit the use of temperature in liquid chromatographic optimization. Snyder's early work in liquid chromatography was in normal phase separations. In typical Snyder fashion, he made quantitative comparisons of various chromatographic approaches, *viz.*, solvent programming, temperature programming, flow programming, and column coupling, to separations with a large range of retention factor (k) values.[1] He concluded that temperature programming did not seem very useful. By the late 1970s, RPLC was clearly the dominant form of LC being practiced. While both temperature and mobile phase composition had been investigated in RPLC, Snyder was the first to investigate them together.[2] He thus proposed a simple but powerful optimization scheme in which four chromatograms were generated at two temperatures and two mobile phase compositions. This matured into a successful commercial product, Drylab.[3,4] Snyder was a champion of gradient elution, and here temperature was also central to optimization. He and his group found that the effects of the gradient slope, (b) (or gradient time, t_G), and column temperature were complementary and highly effective in protein and peptide separations.[5,6] A very thorough analysis ensued[7–10] which included the effects of temperature on solute type. In particular, the latter work took advantage of Snyder's insight published nearly two decades earlier in which he put solutes into two categories: "regular" and "irregular,"[11] based on the linear relationship between $\ln k$ and ΔH°. Strategies for optimization were refined and applied to a variety of solute classes.[12–17] Later he invoked similar concepts related to solute type to explain the effects of temperature on structural isomer separations in RPLC.[18] His contribution to chromatographic optimization, particularly related to selectivity and the very practical matter of getting a good separation in a reasonable time, cannot be overstated.

Solute retention behavior as a function of temperature has been important in helping to define how reversed-phase LC works.[19–24] In a more practical vein, the Greibrokk group[25–27] and others,[12,28–31] starting about 20 years ago, showed how to control temperature in capillary columns and demonstrated high-temperature separations, and programmed temperature. Horvath's group also demonstrated temperature programming[32], and described a now oft-cited rule of thumb that a ~5° change in column temperature has about the same effect on retention in reversed-phase LC as a 1% change in the mobile phase content of the organic modifier. Importantly, the Greibrokk group also demonstrated significant benefits of on-column focusing with thermal control of the column.[26,27,33] Temperature has an effect on pK_as, so controlling the temperature can often be used to control the retention of weak acid/base solutes.[34]

Thanks to the basic understanding described in the work cited above, the use of temperature has expanded in scope. Carr[35–37] used temperature control of two columns in a series to adjust the relative roles of each column in determining retention in a process called the thermally tuned tandem column approach. Teutenberg[38] and Carr[39–41] pushed the boundaries of high flow rate and high temperatures to achieve high speed. Low temperature[42] and negative temperature gradients[43] can have advantages in particular circumstances. More recently, we[44–47] and the Desmet group[48] have advanced the instrumentation used for temperature-assisted solute focusing. Temperature continues to be an element of liquid chromatographic performance optimization for overall separation goals like speed and separation capacity[49–59] and for method development.[16,14,60]

In this chapter we will review work of ours and of others related to on-column solute focusing, stable stationary phases, and retention equations that include a temperature dependence.

2.2 ON-COLUMN SOLUTE FOCUSING

2.2.1 OVERVIEW

Solute focusing during injection results in the solute's occupying a smaller volume on column than the volume injected. Minimizing the volume on column requires the solute to have a high retention factor, k, during the injection. Further, it is also wise to avoid larger diameter columns. Capillary-scale columns are often essential for the analysis of low volume, low concentration samples because they avoid radial dilution of the sample.[61] Both temperature and solvent composition during the injection can be manipulated to temporarily increase solute retention during injection. It is this concentration enhancement effect that has made capillary-scale columns such an invaluable tool when applied to the fields of metabolomics,[62–64] proteomics,[65,66] and online monitoring of neurotransmitters.[67–69] However, the enhancement from decreasing the column diameter may not be observed if the analyte retention during injection is low or if injection volume is large in relation to the total column fluid volume.[70,71] In these cases, injecting more sample only results in a broader peak at a constant peak height. This effect is known as volume overload.[72] To maximize the concentration sensitivity of a method, inject the largest volume of sample that increases the ultimate peak width only marginally given the retention factor, k, of the solute under the conditions during the injection and the column radius.

The detrimental effects of volume overload can be countered with on-column solute focusing techniques. Commonly, this is achieved by so-called solvent focusing in which samples are injected in weak solvents, such as aqueous solutions for reversed-phase separations.[73–77] For a short time during injection the sample solvent becomes the mobile phase, temporarily generating high retention for hydrophobic solutes and a compression of the solute band relative to the volume of sample injected. Sample concentration using solvent has been quite useful particularly in coupling 2D-LC separations through active modulation. In 2017, the Stoll group developed active solvent modulation to dilute the strong ^{1}D solvent, now the injection solvent for the second dimension, with the weaker 2D separation solvent prior to the second injection for improved peak shapes.[78] Stationary-phase-assisted modulation (SPAM)[79] utilizes high-retention trapping columns between the first and second dimension to concentrate the sample before eluting with the 2D eluent allowing for improved sensitivity, smaller 2D injection volumes, and faster 2D separations.[80,81] Temperature-assisted solute focusing (TASF) can also be used to induce focusing effects alone[33,44,47,82–88] or in combination with solvent focusing[46] by manipulating column temperature. Rapid, controlled temperature changes can be realized due to the low thermal mass of capillary-scale columns[89] without the unfavorable radial temperature gradients observed in larger column diameters.[90–92]

2.2.2 Solute Band Focusing Theory

During the injection of a volume of solution, any solute with a non-zero retention factor will be found in a length of column that is smaller than the length occupied by the injection solvent. When the sample volume is injected, the front of the solute band travels at a velocity given by Equation (2.1) where u_{ave} is the average velocity of the mobile phase and k_{inj} is the retention factor of the solute during injection.

$$u_{solute} = \frac{u_{ave}}{1 + k_{inj}} \tag{2.1}$$

$$L_{iso} = \left(\frac{V_{inj}}{F} \right) \frac{u_{ave}}{1 + k_{inj}} = t_{inj} \frac{u_{ave}}{1 + k_{inj}} \tag{2.2}$$

The length of the resulting solute band following injection (L_{iso}) is given by the product of solute band velocity and injection time shown in Equation (2.2). The length of the solute band contributes to the magnitude of band broadening and thus volume overload, an effect that worsens with large injection volumes, smaller column diameters, and less analyte retention.

As retention factors depend upon both solvent and temperature, each of these parameters can induce a focusing effect through temporarily increasing solute retention during the injection process. Solute retention generally decreases with increasing temperature and organic modifier content for reversed-phase systems. In both cases, it is paramount that the leading edge of the solute band spends more time in high retention conditions than the trailing edge. By maintaining a low temperature at the head of the column during injection or injecting in a weak solvent, solute retention is increased and k_{inj} is replaced with k_{foc}. Following injection, the band is released by a rapid step increase in temperature or strong elution solvent, reducing the solute retention to a value of k_{sep}. Solute band length immediately after injection is reduced from isothermal/isocratic conditions and given by Equation (2.3).

$$L_{foc} = t_{inj} \frac{u_{ave}}{1 + k_{foc}} \tag{2.3}$$

The resulting focusing effect can be stated quantitatively as Equation (2.4).

$$\frac{L_{iso}}{L_{foc}} = \frac{1 + k_{foc}}{1 + k_{iso}} \tag{2.4}$$

Solvent focusing benefits from an additional solute band compression step following injection that is analogous to the gradient compression described by Snyder.[75,76] The effect is caused by the strong elution solvent reaching the trailing edge of the solute band before the leading edge. Several groups studied this phenomenon including Poppe,[93] Desmet,[94,95] de Jong,[96] and Weber,[84] deriving the simple expression for the observed peak following elution given by Equation (2.5).

$$\frac{L_{iso}}{L_{foc}} = \frac{k_{foc}}{k_{iso}} \tag{2.5}$$

Thus, the greatest focusing effect during injection is observed when $k_{foc} \gg k_{iso}$. The maximum effect achievable from weakening the solvent for a given solute is typically much greater than that achieved by reducing the temperature.[97] In regard to TASF, there are also limitations to the maximum separation temperature and the minimum focusing temperature that can be used for a given separation. While continuing to decrease temperature results in higher retention and stronger focusing capabilities, temperature's effect on mobile phase viscosity becomes significant.[98] High back pressures can result from cooling even a short column segment to sub-ambient temperatures, making the upper pressure limit of the high performance liquid chromatography (HPLC) pump the limiting factor for how low the focusing temperature can be. However, unlike solvent focusing, temperature focusing is not limited to one focusing event. The ability to change temperature in discrete regions along a column rapidly and easily leads to new possibilities for the use of temperature changes for focusing that would be difficult to do with solvent focusing. Thermoelectric cooling elements (TECs or Peltier devices) are an important element of the instrumentation to actively change temperature.

2.2.3   Capillary-Scale TASF

2.2.3.1   Overview

TECs or Peltier devices are capable of achieving rapid, programmed changes in temperature,[33,44–47,85,87,99–101] so they are ideally suited to controlling low thermal mass column temperature. Early work using temperature as a focusing mechanism included Holm et al.[33] and Eghbali et al.[82] who focused solute bands at the column inlet and outlet respectively. This work was followed by the Weber group utilizing one TEC to induce focusing immediately after injection at the head of the column.[44,45] This methodology successfully focused peptides[47] but was limited to one band focusing step with a dynamic temperature zone of 1 cm. Advancements in hardware and software design have led to improved TASF instruments with arrays of independently controlled TECs to expand the versatility and level of control available over column temperature.[46] Band compression can take place in multiple trap-and-release stages because the focusing effect of TASF is multiplicative. The array design allows for larger injections of more poorly retained analytes.

When being focused, the entire solute band must lie within the length of the temperature focusing zone in order to achieve the maximum focusing effect. If the solute begins to bleed off the zone, the leading edge of the solute band will experience a decrease in retention in the downstream hotter segment resulting in bandspreading. Therefore, the upper limit for V_{inj} for an analyte with a retention factor during injection of k_{foc} is given by Equation (2.6) where F is the separation flow rate, ε_T is the total column porosity, and r_{col} is the column radius:

$$\text{Upper limit } V_{inj} = L_{foc} F \left(\frac{1 + k_{foc}}{u_{ave}} \right) = L_{foc} \varepsilon_T \pi r_{col}^2 \left(1 + k_{foc} \right) \qquad (2.6)$$

Equation (2.6) states that the focusing effect is larger the longer the focusing length. However, there is a limit to how long a cooled section of a column can be. For example, imagine operating a 10-cm-long column at 80°C at 85% of maximum pump pressure. Cooling one 1-cm TEC to 20°C increases the viscosity of the 20% acetonitrile (80% water) mobile phase by more than 2.7 times, but only for one centimeter. The pressure required when cooling one centimeter becomes 99% of the pump's capacity. To lower the temperature of the full 10-cm-long column to the same 20°C without exceeding the pump's maximum pressure would require operating at only about 40% of the pump's capacity at 80°C. The 10 TEC TASF device solves this problem by allowing for focusing zones of increased and variable lengths, permitting larger injection volumes of more poorly retained analytes, while also allowing for multiple independent focusing stages. Thus, focusing can be achieved multiplicatively rather than in one single step, eliminating the need to focus at extremely low temperature or at extended lengths that significantly increase column back pressure.

2.2.3.2   Multiplicative TASF

Recently, we[85] illustrated this idea by resolving the volume overload effects of a test mixture in a targeted and multiplicative fashion. A 4 μL sample was injected onto a 121 × 0.150 mm column. The injected volume was 3.3 times greater than the column void volume. The first seven TECs were operated as three independent focusing zones for solute band compression. The first zone comprised the first four TECs, the second comprised the next two, and the third was the last of the seven TECs used. All three zones were cooled upon injection and were heated sequentially to release each focused solute band into the next focusing zone to demonstrate selectivity. The timing of these trap-and-release stages was chosen to focus ethyl paraben (PB2, $k=3.38$ at 70°C) selectively in three stages while propyl paraben (PB3, $k=9.98$) was compressed only during the first focusing stage. A visual representation of PB2 and PB3 solute band widths under isothermal conditions at each focusing stage is shown in Figure 2.1A to illustrate the band compression each focusing stage provides. A 20.5-fold increase in PB2 peak height was achieved for the three-stage focusing relative to the isothermal run. Simulations to predict peak shape based on the Neue–Kuss equation[102] were also employed to provide fast screening of temperature conditions and timing for method development as well as validation of the experimental results. The comparison of the experimental results in Figure 2.1B and predictions based on simulations in Figure 2.1C shows excellent similarity in both peak shape and retention time, demonstrating the accuracy of these simulations.

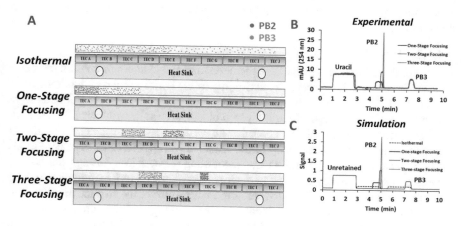

FIGURE 2.1 The experimental design and a summary of the results from the multiplicative focusing experiment using the 10 TEC TASF device are shown. In **A**, a visual representation of the PB2 (blue) and PB3 (orange) solute bands in space on the column are shown under isothermal conditions and during each stage of focusing. Focusing stages are shown immediately before the step increase in temperature releases the compressed band from the cooled trapping zone. Cooled (blue) and heated (red) zones were operated at 5°C and 70°C respectively for each of the independently controlled TECs labeled A–J. The overlaid chromatograms **B** for one- (black), two- (red), and three-stage (blue) focusing, obtained experimentally, display the peak height enhancement and selectivity for PB2 through the multiplicative focusing approach. Comparison of the experimental results to the simulation results **C** demonstrates the accuracy of the simulated predictions used to screen for experimental conditions during method development. (Reprinted (adapted) with permission from *Analytical Chemistry*, M. T. Rerick, S. R. Groskreutz, and S. G. Weber, "Multiplicative On-Column Solute Focusing Using Spatially Dependent Temperature Programming for Capillary HPLC," *Anal Chem*, 91, (2019), 2854–2860. Copyright 2019 American Chemical Society.)

2.2.4 ANALYTICAL SCALE TASF

2.2.4.1 The Effect of Column Diameter

We expanded the capabilities of TASF to an analytical scale to accommodate commercially available columns and larger sample volumes.[101] Actively changing the temperature outside a column leads to radial temperature gradients within the column. Band broadening is induced by radial temperature gradients as retention factor and viscosity are both temperature-dependent. Higher temperatures near the column wall decrease viscosity and typically decrease retention compared with the center of the column. Thus, it is imperative to avoid inducing radial temperature gradients in significant lengths of a column. The thermal entrance length, the distance required for the temperature of a flowing mobile phase to reach 99% of the wall temperature with a constant wall temperature, was used to characterize precolumn diameters that would be suitable for TASF. The small diameter of a capillary column leads to a very small thermal entrance length of 46 µm at 1.5 cm/s mobile phase velocity (150 µm inside diameter, 10 µL/min). For a 2.1 mm ID precolumn, the thermal entrance length is 9.0 mm at the same velocity. For a 1.0 mm ID precolumn, the thermal entrance length is much more acceptable at 2.0 mm. Based on this information, a 20-mm-long, 1 mm ID column should be effective as a precolumn for a 2.1 mm ID column.

2.2.4.2 Precolumn Design

The major design challenge relates to the need for rapid changes in column temperature. TECs are capable of a certain maximum power (Watts = Joules). The elements of the column each have a certain specific heat (Joule/g*K). In order to achieve a target rate of temperature change, K/s, for a given object, a certain number of TECs, each rated at a certain power, will be needed. The stainless steel portions of a 1 mm ID column dominate the heat capacity of the whole object. In our case, the calculations specified a need for eight 1-cm² TECs (accounting for heat loss through the fittings as well as heating the column). Peltier devices are flat and cannot be wrapped around the column. Therefore, we designed a column with 8 cm² of flat surface for attaching Peltier devices with minimal added mass, as shown in Figure 2.2A. Theoretically the column's temperature can be changed at a maximum rate of 27.8 K/s.

We built a 3D model in Comsol and calculated both the temperature and solute concentration within the device/column following a 50 µL injection of butylparaben onto the precolumn. Figure 2.2B shows the profiles taken 7 s after heating. The analyte band is curved because the analyte in the center of the column is cooler and thus more retained than the analyte at the walls. The second profile was taken 41 s after heating. The concentration distribution has less radial dependence but is dispersed more axially. It is important to note that the injection volume is five times the column volume.

2.2.4.3 Experimental Results

Experimentally, changes in back pressure demonstrated that the full temperature transient from 5°C (focus) to 80°C (release) takes about 10 s. Experimental van Deemter curves indicate that the reduced velocity in the precolumn at 250 µL/min

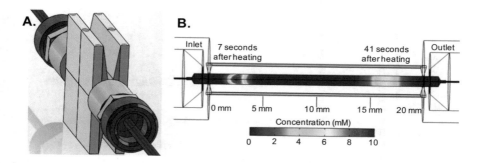

FIGURE 2.2 **A.** shows a SolidWorks rendering of the precolumn (gray). Peltier elements are shown in gold; Thermo nanoViper™ capillaries, used for fluidic connections, are shown in red and black. **B.** shows a cross section of a column simulation at different time points after injection of a 50 μL *n*-butylparaben solution. The transient heating step occurs at *t*=0 s. The two solute zones are shown 7 s and 41 s after heating. The view corresponds to a cross section where the Peltier elements are on top and bottom, and fluid flow is to the right. (Reprinted from *Journal of Chromatography A*, 1523, Groskreutz, Stephen R., Horner, Anthony R., and Weber, Stephen G., "Development of a 1.0 mm inside diameter temperature-assisted focusing precolumn for use with 2.1 mm inside diameter columns," 193–203, Copyright (2017), with permission from Elsevier.)

flow rate is about 50. Nonetheless, about 1,000 theoretical plates are generated. When operating as a precolumn to a 50×2.1 mm column, clear advantages are seen in the chromatograms from solutes across a range of modest *k* values (2.2–23.4 at the separation conditions at 65°C) as demonstrated in Figure 2.3. The effects of focusing are most significant for the ethylparaben and propylparaben peaks.

2.3 STABLE STATIONARY PHASES

2.3.1 OVERVIEW

The use of high temperature and extremes of pH require stationary phases that are stable under these conditions. This section provides a review of the materials and approaches that have been used. We have begun to use the thiol-yne reaction for modifying silica. Our results are also described.

2.3.2 SILICA SUPPORTS

Silica, SiO_2, and silicates not only make up the majority of the earth's crust, but also account for over 90% of HPLC columns' stationary phase supports.[103] Creating stable, silica-based RPLC stationary phases has been a major focus in chromatography for over 40 years, led by Unger,[104,105] Kirkland,[106–112] Carr,[113–115] and many others. Silica particles have many advantages over other types of stationary phase supports. They are rigid and able to withstand high pressures and flow rates, can be synthesized with a variety of diameters, pore sizes, and pore volumes, and can be covalently modified to give a variety of potential column chemistries.[103,116–118] However, silica does possess some drawbacks. The siloxane bond, Si–O–Si, can

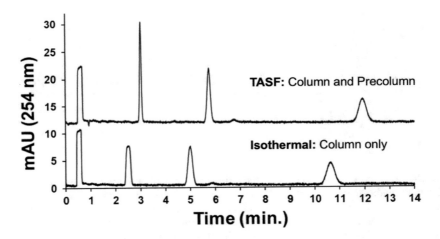

FIGURE 2.3   Example isocratic chromatograms from 50 μL injections of uracil, ethyl, propyl, and butylparaben dissolved in mobile phase. The upper trace shows a separation carried
out on a TASF precolumn interfaced with a 2.1×50 mm i.d. column. The 5°C focusing temperature was held in the precolumn for 25 s during injection before heating to 80°C. The lower
trace shows a separation where an injection is made directly into the 2.1×50 mm i.d. analytical
column maintained at 80°C (i.e., no precolumn). (Reprinted from *Journal of Chromatography
A*, 1523, Groskreutz, Stephen R., Horner, Anthony R., and Weber, Stephen G., "Development
of a 1.0 mm inside diameter temperature-assisted focusing precolumn for use with 2.1 mm
inside diameter columns," 193–203, Copyright (2017), with permission from Elsevier.)

be hydrolyzed. Increased temperature increases the hydrolysis rate,[109,119] so silica-
based stationary phases degrade faster at higher temperatures. The hydrolysis is also
increased in acidic or basic environments, limiting the pH range of early stationary
phases to pH 4–7 at temperatures less than 60°C.[103,109,116,117] The silanols are mostly
deprotonated at these pHs, so a pH below 3 is needed to help reduce poor separation
of cations. This creates the need for more pH-resistant robust silica supports to allow
separations of bases with silica-based stationary phases.[103,117]

There are three main approaches to creating stable silica stationary phases: slowing hydrolysis by sterically protecting the ligand's siloxane bond, limiting access to
the silica surface with polymers, and increasing the silica's resistance to hydrolysis.

2.3.2.1   Sterically Protected Stationary Phases

Sterically protected stationary phases are the most common stable stationary phases.
Steric hindrance slows reactions by restricting access to the reactive site. This idea
has been used to make stationary phases more resistant to hydrolysis since the late
1980s. Kirkland and others at DuPont developed a sterically protected phase that
consisted of diisopropylalkylsilane or a diisobutylalkylsilane that showed little deg-
radation over time at pH < 3.[109] These phases provide high stability from pH 2–8 and
are still sold and used today.

Bidentate ligand stationary phases were first developed concurrently with Glajch,
Farlee, and Kirkland's sterically hindered phases in the late 1980s,[109] but were not

pursued further until the late 1990s.[108] Steric hindrance and increased ligand densities, compared to the sterically hindered phases, can be achieved by using bidentate silanes, consisting of two silicon atoms usually joined by an ethyl or propyl linker chain.[108] These stationary phases exhibited reasonable stability at pH < 2 and high stability at pH 11. This allows the silica-based stationary phase to be used from pH 2–11, allowing for the analysis of many types of molecules, including neutral amines.[112] By their nature, sterically hindered phases have a reduced ligand density compared to other alkylsilane phases and endcapping is required if separations are to be carried out at high pH.

2.3.2.2 Polymer Modified Silica

For over 50 years silica particles have been modified with a polymeric layer to create chromatographic stationary phases. A polymer coating on silica particles helps prevent hydrolysis at the silica surface, increasing the stability. The layer shields the surface from acidic and basic mobile phases, extending the operable pH range of the silica particle. The polymer can adsorb onto the surface or the layer can be covalently attached to the surface. Both types of modification can be created both by growing the polymer *in situ* or by adsorbing/reacting a polymeric ligand.[120,121]

Horvath was first to protect the surface of silica particles by polymerizing divinylbenzene and styrene on the surface of the particles. This resulted in a thick polymer layer that filled pores and decreased chromatographic performance.[122] Besides solution-based polymerization, more recent techniques involve adsorbing the monomers and radical initiator to the surface and polymerizing the dried mixture.[123] This reduces some of the negative aspects, but still results in decreased performance. The poor performance of these techniques has shifted the research focus to other methods of polymer modification.[120]

The adsorption of preformed polymers to the silica surface can help reduce the problems of polymer thickness and clogged pores that plague the surface polymerization techniques. The preformed polymer stationary phases are usually made by mixing the polymers with the silica particles and causing adsorption by evaporating the solvent. Often these polymers are then crosslinked to create a more stable surface.[121] The main polymers used for this technique include polybutadiene,[105,124] polystyrene-divinylbenzene,[125] and polymethacrylates.[126,127] The adsorption of polymers generally creates a patchwork covering of the silica particle. This can result in poor performance by the solutes that interact with the silica surface,[124] and this limits their effectiveness in creating a highly stable silica stationary phase. The technique is still in use and is especially important for coating other metal oxide particles, which will be discussed later.

Covalently attached polymer coatings can also increase silica stability. In the mid-1970s, Wheals used vinylsilane-modified silica to covalently build various acrylate-based polymeric layers from the surface-bound vinyl groups.[128] He noted reduced silanol activity and increased stability with multiple mobile phases, and that the polymer adhered to the particle longer than an adsorbed polymer. Others have built polymeric coatings from the surface,[126] but without controlled reactions the pores can clog and thick polymer layers can form reducing the performance of the stationary phases.[120]

Most procedures for a chemically bonded, polymer-coated stationary phase work by grafting the preformed polymer to the surface of the silica. This technique gives reproducible polymer layers that provide an abundance of column chemistries and increases the stability of the silica support.[116,120,121] In 1983, Kurganov first reported use of a chiral polystyrene copolymerized with vinyltriethoxysilane covalently attached to a silica particle.[129] The vinylsilane anchor is common for the many alkene-based polymers used, like styrene- or methacrylate-based polymers.[130,131] Chlorophenylsilanes were used in Carr's hypercrosslinked stationary phases to anchor the polymer to the surface.[113,132–134]

The foregoing examples illustrate that post-polymerization modification or grafting polymers to the surface of silica particles can lead to a large variety of column chemistries. Polymers made with peptides, polysaccharides, ethers, sulfonate, or amine groups, along with styrene, butadiene, and divinylbenzene, have all been used to create unique and stable stationary phases.[135]

Crosslinking the polymers is not always needed. Ihara has covalently attached poly(4-vinylpyridine)[136] and poly(2-N-carbazolylethyl acrylate)[137] to silica. They adhere to the silica surface with hydrogen bonding and greatly reduce the mobile phase's and solutes' interaction with the silica. This helps increase the silica's stability.

Polymers based on poly(alkylsiloxanes) are also used to coat silica particles and enhance their stability. An early commercial phase used poly(dimethylsiloxane)-coated particles to reduce the silanol activity and the rate of hydrolysis.[131] The use of longer chain alkyl groups, like methyloctyl[138] and methyloctadecyl[139] siloxane, helped to create stationary phases with similar chromatographic behavior to traditional RPLC columns. Polysiloxanes can be created *in situ* by using either trichloro- or trialkoxysilanes. In the presence of a small amount of water the silanes will polymerize, forming a high surface coverage phase, over 4 μmol/m^2, while increasing the stability of the RPLC stationary phase.[116,140]

2.3.2.3 Robust Silica Particles

Over 50 years ago, Stöber introduced a process to synthesize monodisperse nonporous silica particles with diameters from 50 nm to 2.0 μm.[141] Shortly thereafter, methods were found to create monodisperse porous particles 2.0–10 μm in diameter by coalescing the small 10–300 nm sols made by the Stöber process into larger porous particles.[142,143] The methods and chemicals used then improved,[144] giving the highly pure silica supports used today.

The surface area and porosity of the silica particle impacts the stability of the stationary phase. Silica particles that have a high surface area (over 300 m^2/g) and high porosity are less hydrolytically stable than those with lower surface areas and porosities. Particles made to have a very high surface area are usually made by coalescing soluble or fumed silicates. This results in thin pore walls are prone to dissolution. The lower surface area and lower porosity silicas are made using larger, more insoluble sols and have thicker more uniform pore walls, creating a support that is more resistant to dissolution.[110,111]

Reactions with silica are slowed by the addition of carbon substituents to a silicon atom. The silicon atom has more electron density when bonded to carbon versus a silicon atom bonded to oxygen. The increased electron density reduces electrophilicity of the silicon atom, slowing hydrolysis and other reactions. The carbon substituents also sterically hinder a nucleophile's attack on the silicon atom. This is evident in the evolution of chemically bonded reversed-phase stationary phases, with alkylsilanes, Si–C, instead of the silicic esters, Si–O–R, originally used for stationary phases.[145,146] In the mid-1970s, Unger helped develop organosilica hybrid particles that were synthesized using the standard tetraethoxysilane and an organotriethoxysilane, with benzyl- or 1,2-diol-3-propoxypropyl organic groups.[104] It took nearly 25 years for an organosilica-based stationary phase to make it to market when Waters released their methyltrialkoxysilane hybrid phase, which showed high stability up to pH 11.5.[147] A few years later, Waters released an ethylene-bridged particle that showed even higher base stability and reduced silanol activity.[148] Continued research into organosilica particles has developed a stable high flow-through phase,[149] and particles with alkyl ligands, C18, and others mixed into the particle.[150]

2.3.3 OTHER STATIONARY PHASE SUPPORTS

Organic polymer particles are often used as an alternative to silica. They can be hydrophilic or hydrophobic and can be functionalized covalently with stable carbon–carbon bonds. This gives polymeric phases their main advantage: a very broad pH range, generally from pH 1–13, enabling their use for many applications. Hydrophobic polymers, like poly(divinylbenzene), poly(styrene-divinylbenzene), and poly(alkylmethacrylate), are used as reversed phases or are functionalized to have a hydrophilic nature, e.g., for ion exchange. Hydrophilic polymers, such as poly(vinylalcohol) or poly(hydroxyalkylmethacrylate),[116] or hydrophobic polymers functionalized with hydrophilic ligands,[151,152] may be used for ion exchange or size exclusion chromatography. The hydrophilic surface eliminates any secondary retention characteristics that may interfere with the techniques.

While polymer-based stationary phases provide excellent pH stability, they have many disadvantages that keep them from widespread use. Polymer particles have low mechanical strength and deform at high pressures, limiting columns using porous polymers to about 400 bar. This limits their use. Polymers have the tendency to swell or shrink in various solvents. Hydrophobic polymers swell in nonpolar solvents and shrink in more polar solvents, and vice versa with hydrophilic polymers. This can cause problems because of the associated change in permeability and in retention characteristics with varying mobile phase composition. The particles' natural pore structure decreases their efficiency for some molecules as well. The particles have larger pores with smaller pores inside. It has been noted that flat, rigid molecules that can fit into the small pores have slow mass transfer in the small pores, decreasing the efficiency of the column for those molecules.[116,153]

Metal oxides other than silica have been used in chromatography for a long time. Alumina, Al_2O_3, zirconia, ZrO_2, and titania, TiO_2, are the most common metal oxides that are in use. The primary benefits of these metal oxides are their high mechanical strength with excellent temperature and pH stability. They can be used regularly at 200°C and from pH 1–13, conditions much more extreme than those in which silica can be used.[153,154] Alumina has been used in normal phase TLC and chromatography for decades, but has not been adequately developed for RPLC.[153,154] Titania phases are the least used support mainly due to the lack of availability of particles that permit their use in HPLC columns.[154] Zirconia supports have been studied and modified more than the others, and can be found with a variety of particle sizes and pore diameters.[114,154] The surface chemistries for the metal oxide particles are complex with hydroxyls, Lewis acid, and Lewis basic sites.[154] The approaches using and blocking these active sites are similar for all the metal oxides, so zirconia will be the focus below.

Unlike silica, zirconia and the other metal oxides cannot undergo simple covalent modification. There have been a few positive results silanizing the hydroxyl groups,[155,156] but little has come from them because of the weak Zi–O–Si bond. One way to counter this is to polymerize tetraalkoxyzirconium onto the surface of a silica particle creating a zirconized silica particle. Then poly(alkylsiloxane) is polymerized on the surface to give a reversed-phase, highly pH-resistant stationary phase.[157–159] The polymer coatings increase the binding strength between the zirconium and silicon esters, but these phases can still have strong Lewis acid sites and silanols exposed, reducing their efficiency and stability.

The most common way to modify metal oxide surfaces is to adsorb polymers on the surface,[153] much like the adsorption of polymers on silica. Polybutadiene-modified zirconia is the most common type recorded.[153,160] The polymer is adsorbed to the surface from solution by evaporation of the solvent and then crosslinked to add to the stability.[160] This gives the particles a hydrophobic reversed-phase characteristic, but the patchy coverage leaves Lewis acid sites that are detrimental to the chromatography of the bases and ionizable analytes. For better separations of ions, several ion-exchange phases have been synthesized using a zirconia support. Anion exchange phases can be made using a polyethyleneimine coating, allowing the controlled separation of anions.[161] Likewise, cation exchange phases have been made by adsorbing a poly(acrylic acid) anhydride or reacting the polyethyleneimine phase with succinic anhydride.[162] The versatility of the polymeric coverings and improved coverages allow zirconia to be a highly stable useful alternative to silica. However, the Lewis acid sites, lack of stable covalent bonds, and lack of availability of a wide range of particle size and porosity options keep zirconia and other metal oxide phases from dominating the market.

Carbon-based supports offer high pH resistance and unique retention mechanisms. The most common form is a porous graphitic carbon support developed and evaluated by Knox in the early 1980s.[163] Over the next 30 years, several commercially available stationary phases with a porous graphitic support arose from this work.[164] The synthesis involved pyrolyzing an organic phenol polymer in a highly porous silica particle. The silica was then dissolved, and the porous carbon was graphitized in a furnace. This gives a very hydrophobic surface that has strong π–π interactions and stereoselectivity.[163,164] With stability in a broad pH range, these particles are often a useful stationary phase. However, they have a pressure limit of about 550 bar, can easily be fouled by large molecules, and require high organic content mobile phases,[116] limiting their usefulness.

Stationary phases have been developed recently that utilize diamonds and nanodiamonds to help improve the stability of the supports and offer unique selectivities.

Polystyrene-divinylbenzene particles were synthesized with oxidized nanodiamonds creating a hybrid particle structure. Hybrid particles that withstood higher pressures than the pure polymer were used as a reversed phase and were also modified to act as an anion exchange phase.[165] Core-shell particles were produced by coating a diamond core with alternating layers of polyallylamine and nanodiamonds. The final amine-rich layer was functionalized with epoxyoctadecane to create a C18-like stationary phase that is stable at high pressure and across a wide pH range.[166] Another application of this process used a carbon core to produce a very stable and efficient stationary phase.[167] Nanodiamonds have also been used to coat silica particles to increase their stability and various column chemistries.[168] Both oxidized and hydrogenated nanodiamonds can be bonded to silica using a radical reaction. The oxidized nanodiamonds produce a hydrophilic stationary phase, while the hydrogenated ones act like a reversed-phase stationary phase.[168]

Of particular interest for high-temperature separations on a customizable stationary phase is the Carr group's hypercrosslinked stationary phases, which can be synthesized with a variety of column chemistries and which show excellent stability at high temperature and low pH.[113,133,169,170] These are crosslinked polystyrene block copolymer phases that provide a foundation for the creation of numerous column chemistries by adding ligands onto the polymer surface with the Friedel–Crafts reaction. However, this is a complex, multi-step reaction that can create active sites from adsorbed metal catalysts.

2.3.4 STATIONARY PHASES BASED ON THE THIOL-YNE REACTION

We liked Carr's hypercrosslinked phase for its stability at extreme conditions, but the synthesis is very complex. Thus, we developed a very simple way to create a crosslinked phase that has the potential to be an outstanding basis for a wide variety of chemistries. Click chemistry is useful for stationary phase production,[171] but the thiol-yne[172] has only been used once, to our knowledge, for modifying particles, by Zou's group.[173] It has also been used for monoliths.[174–176]

We used a simple multi-step process using 1,4-diethynylbenzene (DEB) and 1,6-hexanedithiol as the monomers to create a crosslinked stationary phase along the surface of thiol-functionalized silica. Proposed structures are shown in Figure 2.4. The Tanaka characterization[177] reveals that the crosslinked phase showed a low phase ratio, methylene selectivity typical of a reversed phase, and extremely high shape selectivity compared to other commercial phases. The crosslinked phase showed good stability in tests at low and high pH. At pH 0.5 and 70°C the phase suffers only about 10% change in k over 144 h, compared to about 25% change with an acid-stable SB-C18 phase. At pH 12.0 and 70°C the crosslinked phase showed a very small pressure increase and about a 15% increase in the k of butylparaben over three hours of exposure, but the SB-C18's performance dropped significantly and clogged two times over the three hours.[178]

In the process of developing this pH-stable, highly crosslinked stationary phase we discovered a new charge transfer stationary phase.[179] The first step in the preparation of the crosslinked phase is to attach 1,4-diethynylbenzene (DEB) to thiol-functionalized silica particles. Upon preparation of that phase, we noticed that the color

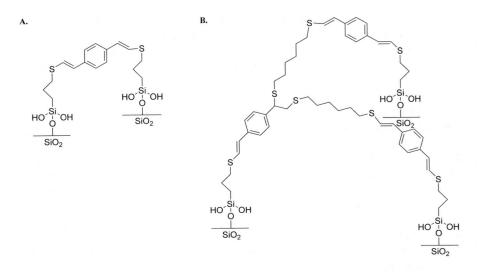

FIGURE 2.4 Possible structure of crosslinked stationary phase. A. The bidentate DEB ligand that is likely found on the silica. B. Example of the dithiol crosslinker with the majority of the DEB only substituted one time. The surface is likely a mixture of A and B. (Reprinted from *Journal of Chromatography A,* 1598, Erin P Shields and Stephen G Weber, "A pH-stable, crosslinked stationary phase based on the thiol-yne reaction," 132–140, Copyright (2019), with permission from Elsevier.)

of the particles was different when the modified particles were wet with aromatic solvents compared to nonaromatic solvents. This color change was still apparent upon crosslinking the DEB with 1,6-hexanedithiol to create the crosslinked thiol-yne (CTY) stationary phase. The CTY phase has high shape selectivity as shown by the α-value for the flat triphenylene over the bulkier *o*-terphenyl, $\alpha_{T/OT}=4.91\pm0.08$, almost twice that of the most shape-selective reversed-phase column.[180] The difference of the two compounds' entropy contributions to retention free energy, $\Delta(T\Delta S^{\circ})$ at 298 K, is only -0.1 ± 0.9 kJ/mol (statistically indistinguishable from zero), leading us to believe that the shape selectivity is not consistent with the slot model. To test the hypothesis that the DEB–thiol adduct was responsible for the observed behavior, we prepared a low coverage DEB phase (<2 μmol/m^2) which, unlike higher density, crosslinked, or polymeric phases, should not display shape selectivity based on "slots." With the low-coverage DEB phase, the shape selectivity remained with $\alpha_{T/OT}=3.23\pm0.01$. We discovered further that the DEB ligand, shown with an electron-rich resonance structure in Figure 2.5, has electron-donating characteristics. The selectivity for nitrobenzene compared to benzene on a commercial C18 stationary phase is 0.64 ± 0.01, while on the DEB phase it is remarkably higher, 1.83 ± 0.10. This shows that the thiol-yne-based DEB phase is an *electron-donating charge transfer* stationary phase.

This is not unprecedented. It turns out that Sander et al. has described an alternative shape selectivity mechanism based on charge transfer.[181] Lucy has attributed at least some of the high selectivity seen with a dinitroaniline-based phase in a normal

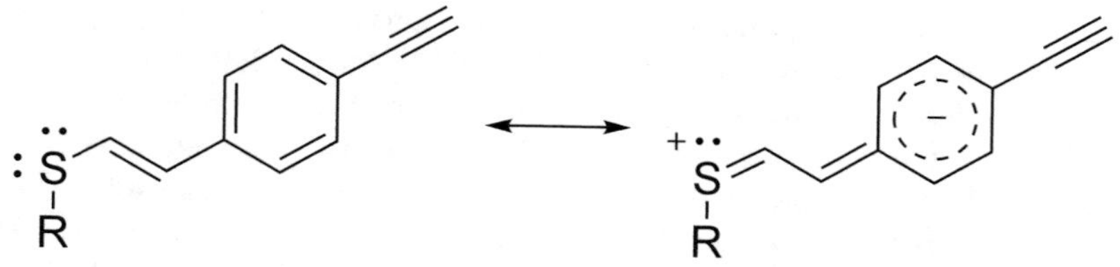

FIGURE 2.5 The DEB ligand structure with electron-donating resonance structure. (Reprinted from *Journal of Chromatography A*, 1591, Erin P Shields and Stephen G Weber, "A liquid chromatographic charge transfer stationary phase based on the thiol-yne reaction," 1–6, Copyright (2019), with permission from Elsevier.)

phase determination of $\alpha_{T/OT}$ to charge transfer: the stationary phase is an electron acceptor (EA).[182] Lindner's group thoroughly characterized naphthalimide phases for shape selectivity.[183] Most charge transfer stationary phases have an electron poor aromatic system to create an EA stationary phase.[181,182,184–186] We are aware of only three ED phases – all based on pyrene – that have been examined.[181,184,187] The new phase we created is unique.

2.4 THE INFLUENCE OF TEMPERATURE ON RETENTION

2.4.1 NEUTRAL ANALYTES

2.4.1.1 Background on Retention Models for Neutral Analytes

Generally, there are several forms that retention models have taken, namely, thermodynamic, empirical, and correlations with experimental or calculated properties of an analyte such as linear solvation energy relationships (LSERs) or quantitative structure-retention relationships (QSRRs). Thermodynamic models serve to demonstrate how and why chromatography works,[21,22,188–191] but are not typically used to predict retention, partially due to the difficulty in measuring the phase ratio. This is due to the heterogeneous nature of the stationary phase which comprises alkyl chains, silanols, organic modifier, water, and buffer components.[20,192,193] The first empirical model, linear solvent strength theory, has been used to model retention, typically in gradient separations. However, it does not explicitly include temperature dependence. In addition, its simplicity arises from assuming a linear dependence of retention-free energy on mobile phase composition.[194] We have investigated other models based at least partly on theoretically defined dependencies that have been developed. Neue improved on linear solvent strength to include a nonlinear dependence of $\ln(k)$ on mobile phase composition[195] and later expanded with Kuss to create the Neue–Kuss expression (NK) which includes a temperature dependence.[102] The Pappa-Louisi group has developed two models, one based on partitioning of solute (PL-P) and one based on adsorption of solute (PL-A). Both include a temperature and mobile phase composition dependence.[196] The last category comprises correlations with solvent or solute properties. Correlating properties of analytes and stationary phases with retention has provided significant insight into selectivity. Carr initiated a set of investigations using linear solvent strength models.[192] The relationship between retention and certain physicochemical properties has been heavily investigated with QSRRs.[197] The latter relate chromatographic retention to chemical structure by using experimental or theoretically predicted molecular descriptors to predict retention.[198] Historically, these relationships have been observed between retention factor and nonpolar surface area of solute,[190] logP and water solubility,[199,200] molecular connectivity,[201] and the presence/abundance of certain chemical moieties: methylene group,[202–206] polyethyleneglycol,[207,208] and aromatic rings.[205] The descriptors are then chosen and QSRR can be used to predict retention times for new analytes and to gain insight into the mechanism of retention. Experimentation or computation is required to determine these molecular descriptors. Haddad successfully extended the QSRRs for use with the hydrophobic subtraction model (HSM). This allowed for this approach to be used with different stationary phases[198] for which HSM parameters were available. Additionally, UNIFAC (*UNI*QUAC *f*unctional-group *a*ctivity *c*oefficients), a group-contribution method typically used to predict activity coefficients in liquid mixtures without electrolytes,[209,210] has also been invoked in a chromatography context. These calculated activity coefficients have been correlated with retention data in RPLC.[211] They have also been used to calculate activity coefficients in stationary and mobile phases, which were subsequently used to calculate partition coefficients.[212] Park et al. demonstrated that UNIFAC is useful for predicting elution order, relative solvent strengths,[213] and the effects of methylene contribution,[214] but showed the inaccuracies in its ability to quantitatively predict retention.[213,214]

2.4.1.2   Comparing Retention Models for Neutral Analytes

Knowledge of retention behavior over a wide range of conditions allows for both the simulation and optimization of advanced chromatographic techniques including gradient separations, TASF, multidimensional liquid chromatography, and solvent focusing. Recently, we assessed three retention models for their ability to recapitulate experimentally determined retention as a function of temperature and solvent composition. We further assessed their ability to predict the retention enthalpy of these compounds.[215] The first equation (Equation (2.7)) was a four-parameter expression developed by Neue and Kuss (NK) as a function of temperature (T) and mobile phase composition (v/v%, ϕ).[102]

$$\ln(k) = \ln\left(k_{0,T}\right) + \frac{D}{T} + 2\ln\left(1 + a\phi\right) - \left(1 + \frac{D}{T}\right)\frac{B_T\phi}{1 + a\phi} \tag{2.7}$$

This model allows for curvature in a plot of $\ln(k)$ vs. mobile phase composition, but $\ln(k)$ has a linear dependence on the inverse of absolute temperature. The other two were developed by Pappa-Louisi. One is based on a partitioning model (PL-P) and has six parameters (Equation (2.8)). Note that $\ln(k)$ has a quadratic dependence on mobile phase composition and a linear dependence on the inverse of temperature.

$$\ln(k) = \frac{1}{T}\left(A\phi^2 + B\phi + C\right) + D\phi^2 + E\phi + F \tag{2.8}$$

The other expression from Pappa-Louisi et al. is based on an adsorption model (PL-A) and also has six parameters (Equation (2.9)).

$$\ln(k) = A + \frac{B}{T} - \frac{\phi\left(C + \dfrac{D}{T}\right)e^{E+\frac{F}{T}}}{1 + \phi\left(e^{E+\frac{F}{T}} - 1\right)} \tag{2.9}$$

The dependence of ln(k) on mobile phase composition and temperature in this expression is complex.

The derivative of each of these models with respect to the inverse of the absolute temperature provides an apparent retention enthalpy as shown in Equation (2.10).

$$\Delta H^{\circ} = -R \frac{d \ln(k)}{d \dfrac{1}{T}} \tag{2.10}$$

Apparent enthalpy based on the NK expression (Equation (2.11)) consists of three parameters and is only dependent on mobile phase composition.

$$\Delta H^{\circ}_{NK}(\phi) = -RD\left(1 - \frac{B_T \phi}{1 + a\phi}\right) \tag{2.11}$$

Apparent enthalpy from the PL-P expression (Equation (2.12)) also consists of three parameters and is quadratically dependent on mobile phase composition.

$$\Delta H^{\circ}_{PLP}(\phi) = -R\left(A\phi^2 + B\phi + C\right) \tag{2.12}$$

Apparent enthalpy from the PL-A expression (Equation (2.13a–c)) is more complicated, containing five parameters and being dependent on both mobile phase composition and temperature.

$$\Delta H^{\circ}_{PLA}(\phi, T) = R\left(\frac{DX\phi + FXY\phi}{X\phi - \phi + 1} - \frac{FX^2 Y\phi^2}{(X\phi - \phi + 1)^2} - B\right) \tag{2.13a}$$

$$X = e^{\left(E + \frac{F}{T}\right)} \tag{2.13b}$$

$$Y = C + \frac{D}{T} \tag{2.13c}$$

The dataset used to compare these models consists of retention factors taken at different mobile phase compositions and temperatures for 101 compounds containing alkylbenzenes, alkylbenzoates, hydroxyalkylbenzoates, alkylphenones, alkylhydroxyphenones, mono- and disubstituted benzenes, NSAIDs and related compounds, polycyclic aromatic hydrocarbons, and polycyclic heterocycles (see http://stephen-weber-chemistry.squarespace.com/useful-links/). All measurements were done in acid conditions and all solutes were neutral. All experimentally determined k values were repeated typically four times. The within-day experimental uncertainty in $\ln(k)$ was 0.2%, and 2.1% for between-day measurements, demonstrating that this dataset has a small experimental uncertainty, allowing it to be used to test these retention models. ("Between day" incorporates a range from next day, same column to many months later, different column.)

The residual uncertainty from the fits of $\ln(k)$ over the experimental range of T and ϕ for the 101 compounds were NK: 3.9%, PL-P: 4.3%, and PL-A: 2.7%. As these models have different numbers of parameters, we wanted to be sure that the benefit did not come from overfitting. The Akaike information criterion uses the log likelihood function to award a model for a good fit while penalizing for the number of parameters. The NK model was the best for one solute, while the PL-P model was best for 23 solutes and the PL-A for 77 solutes.

Apparent enthalpy was measured using the slope of the van't Hoff plots taken within a day and compared to that predicted by Equations (2.11–2.13). The NK showed an error of 2.01 kJ/mol, which equates to a 21.3% error, while the PL-P showed a 1.42 kJ/mol error (12.8%) and the PL-A showed a 1.08 kJ/mol error (10.7%).

The PL-A model appears to be the most accurate with a prediction uncertainty at the same level as day-to-day reproducibility and the lowest uncertainty in enthalpy of retention as well.

2.4.2 IONIZABLE ANALYTES

2.4.2.1 Overview
The temperature dependence of the retention of ionizable analytes is complicated. In the following description we consider an ionizable solute. It will be referred to as having neutral and ionized forms with a single pK_a for simplicity. In a buffered solution containing such an ionizable solute several changes occur as temperature changes, namely: 1) buffer pK_a shifts; 2) solute pK_a shifts; and 3) the retention factors of the ionized and neutral species change. The buffer pK_a shifts to a degree related to its enthalpy of ionization when temperature changes. Thus, a change in temperature induces a shift in mobile phase pH. Buffer pH determines the relative proportion of the acid and base forms of a solute in the chromatographic system, and thus buffer pH has a significant effect on ionizable solute retention. This was first observed by Horvath et al.,[216] who made the important observation that the ionization enthalpy of the buffer contributes to the observed enthalpy of a separation. Solute pK_a also shifts, affecting the distribution of ionized and neutral species at a certain pH. Tran et al.[49] explored the effect of temperature at different pH values, temperatures, mobile phase compositions, and stationary phases. While the pH shift with temperature was not accounted for and different buffers were used for the pH values tested, they noted different selectivity trends with temperature for analytes containing amino groups which tend to have large positive ionization enthalpies. McCalley noted unique changes in selectivity for amine containing compounds as well.[217] The foregoing

effects on retention are most significant if the pH is near the pK_a of the solute and the pK_a of the buffer is similar to that of the solute. Finally, retention factors of the neutral and ionized forms of the solute shift depending on the apparent enthalpy of adsorption/partitioning of each of the neutral and ionized compounds.

2.4.2.2 Effect of Buffer Ionization Enthalpy

The most interesting aspect of these effects of temperature is that the chromatographer has some control over the effect of a temperature change by choosing buffers with particular properties. We demonstrate the effect of temperature for a representative solute in three different buffer systems that are effective in the pH range where carboxylic acids ionize. We consider a solute with a pK_a and ionization enthalpy that are representative of a carboxylic acid solute. The pK_a is 4.76 (same as acetic acid), and the ionization enthalpy is -0.50 kJ/mol. The three buffers represent different ionization enthalpies. The enthalpy of ionization of acetic acid in water is small (-0.41 kJ/mol), so the pH does not significantly shift with temperature. A buffer based on oxalate ($HOx^- \rightarrow Ox^{2-}$) is representative of a buffer in the desired pK_a range with an acid having a significant negative enthalpy of dissociation (-7.00 kJ/mol). The acid is favored as temperature increases (dissociation is greater at low temperature and pH increases with increasing temperature). Piperazine ($PipH_2^{2+} \rightarrow PipH^+$) has a large positive ionization enthalpy (31.11 kJ/mol) so pH decreases with increasing temperature.[218] The shifts in buffer pH as temperature is increased from 25°C to 65°C are shown in Figure 2.6A. The changes in k at different mobile phase pH (25°C) as temperature is changed to 65°C are shown in Figure 2.6B. Retention factors of the neutral and ionized forms are shown by the arrows at the k axis. The retention behavior of the solute at 25°C in all three buffer systems is given by the curve with square symbols. It is identical in all cases. (Ion pairing is assumed not to affect retention.) When the temperature is raised, retention at the high and low ends of the pH range is the same in each buffer system because the solute exists in predominantly a single form at the extremes of pH. However, the pH dependences of the retention of the solute in the three buffers are significantly different. Notably, when the initial pH at 25°C is greater than 5, retention *increases* with increasing temperature in the piperazine buffer with its positive enthalpy of ionization.

2.4.2.3 What Is the pH of a Mobile Phase?

There are challenges in making pH measurements in mobile phases. When the pH probe is calibrated in aqueous solutions and measurements are made in aqueous solutions, the pH is stated as being the $_w^w pH$. The superscript is for the calibration solution (w for water) and the subscript is for the sample. It is common in HPLC work to prepare an aqueous buffer at a specific pH, and then combine it with the organic portion of the mobile phase. This provides a reproducible mobile phase, but the shifts in pK_as and pH must be estimated. Equations have been developed to predict changes in pK_a[219–222] and pH[223–226] with the addition of the organic portion of the mobile phase; however, these expressions are specific to classes of compounds. Measurements can be made directly in the mobile phase with calibration either in a mixed solvent or in water-based buffers. To make measurements in a mixed solvent, the calibration standards must be made in the exact mobile phase composition

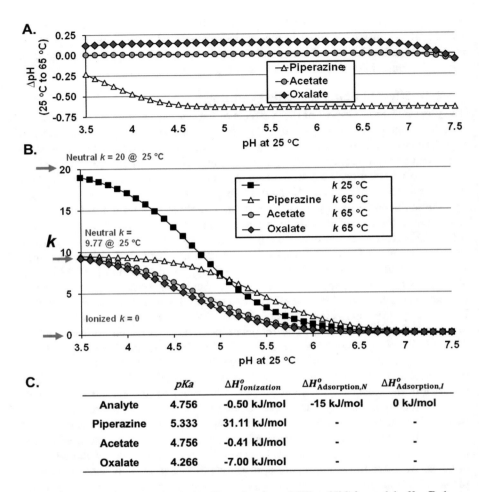

FIGURE 2.6 **A** shows the change in pH moving from 25°C to 65°C for each buffer. **B** shows the retention factor as a function of the initial pH at 25°C for neutral forms, ionized forms, and the average retention factors in buffered mobile phase at 25°C and 65°C. **C** shows the thermodynamic properties for analyte and buffers.

in which the measurements will be made. The result is called $_s^s$pH (s for solvent). This is a tedious process and somewhat difficult to work with, but it is the most accurate method for measuring the activity of the proton in the solution if liquid junction potentials are controlled. A balance of accuracy and ease is found in the $_w^s$pH scale, in which pH electrodes are calibrated in aqueous buffers and measurements are made in mobile phase. This procedure is easy to carry out in the lab and will work on any compound or class of compounds. Following this measurement, $_s^s$pH can be estimated for different buffers at different mobile phase compositions as shown by Espinosa et al.,[227] who reported "correction factors." Canals et al. developed an expression for this correction factor in methanol mobile phase mixtures at room temperature[228] and Gagliardi et al. developed the expression for acetonitrile mobile phase mixtures ranging from 15°C to 60°C.[229] Analyte pK_a in mixed solvents can be measured using the techniques described above and these methods have been automated.[230] Mobile phase pH gradients have also been used to estimate analyte pK_a chromatographically.[231]

2.4.2.4 Retention Prediction

As discussed above, analyte pK_a, buffer pK_a, and mobile phase pH depend on organic cosolvent fraction and temperature. Retention factors have been successfully predicted using aqueous buffer pH and aqueous solute pK_a prior to the addition of the organic modifier.[232] Mobile phase organic gradients have used pH and pK_a shifts as another tool in the separation of ionizable solutes.[231,233–236] Under conditions of constant ϕ (30% ACN, v/v%), the pH and temperature dependence of retention was modeled by Gagliardi et al.[237] They quantitatively described the response of ionizable analytes to temperature changes depending on buffer type. This work was extended to include methanol mobile phases, and a simpler expression focusing on only the equilibria which significantly affect retention was developed.[238] The prediction of analyte retention as a function of mobile phase pH and temperature has been extended to include mobile phase composition as well.[239–241] Agrafiotou studied 22 analytes in 12 buffers at a variety of pHs, temperature, and mobile phase compositions. They focused on developing simple models for the prediction of retention and ultimately arrived at an eight-parameter expression. Pous-Torres et al. modeled these effects using a dataset of 11 analytes and one buffer system, at a range of acetonitrile compositions, pHs, and temperatures. They used an eight-parameter expression to predict retention and subsequently focused on minimizing the number of experiments required to predict retention as a function of pH, mobile phase composition, and temperature.[240,241] A unique extension of this work explores the use of 1) pH gradients and 2) mobile phase organic and pH gradients in combination. The theory is complicated, but could be useful in the fast yet selective separation of medical or environmental samples.[231]

2.4.2.5 Conclusion

Understanding the effect of temperature changes on the retention of ionizable analytes where mobile phase pH is near the pK_a of the analyte is quite complicated because the buffer pK_a, analyte pK_a, and retention factors of the ionized and neutral forms of the solutes are all temperature-dependent. While this leads to complexity, as a practical matter, buffer selection allows for an extra parameter of control over solute selectivity.

2.4.3 TEMPERATURE-DEPENDENT CHANGES IN STATIONARY PHASES

There have been several publications demonstrating the intriguing properties of temperature-dependent stationary phases where the stationary phase itself changes at different temperatures. Almost all of these phases are based on poly(isopropyl acrylamide)[242–250] or related polymers.[250–257] The intriguing nature of this system is that retention (reversed phase) increases when temperature increases. For example,

Cao et al.[258] used a traveling cold spot to elute proteins in a preparative separation with an *N*-isopropyl acrylamide-based cation exchange gel (fast protein liquid chromatography, FPLC). However, the very poor plate count typically observed makes this phase less useful than it might otherwise be.

Pesek et al. demonstrated temperature-dependent retention due to the phase transition of a liquid crystalline phase.[259] Sander, Wise, and coworkers[260–268] have convincingly demonstrated the existence of the temperature dependence of shape selectivity using polycyclic aromatic hydrocarbon mixtures with reversed phases of a particular sort. Phases whose structure allows for self-association of adjacent alkyl ligands at low temperatures tend to show shape selectivity.

2.5 THE EFFECTS OF VISCOUS DISSIPATION

2.5.1 Overview

Chromatographic efficiency decreases above specific inlet pressures while average eluent temperature increases.[269] This heating is caused by viscous dissipation, which is a process that causes fluid temperature to increase because of shear forces in the fluid creating friction. The effects of viscous dissipation are greater for high fluid velocities and small fluid through-ways, e.g., narrow ID open tubes or small particle-packed columns. Heat is generated throughout the column and is lost through the column outlet and the walls. Heat loss through the walls creates a radial temperature gradient with the center of the column hotter than the walls. This is a problem in chromatography as higher temperatures in the center of the column decrease mobile phase viscosity and typically decrease the solute retention factor, so the velocity of the solute is higher in the center of the column than at the walls. This causes band broadening. Approaches to solving this problem are described briefly below. The most successful approach is to have column walls that do not conduct heat axially or radially. Such an adiabatic column would have no radial temperature gradient.[270] Under adiabatic conditions, the temperature increase is proportional to the pressure drop divided by mobile phase heat capacity.[271]

2.5.2 Temperature Profiles within a Column

The radial temperature gradient induced by viscous dissipation was first studied in chromatographic packed beds by Lin and Horvath.[272] They studied thermostatted columns and demonstrated that the radial temperature gradients can be softened by having the inlet mobile phase temperature lower than the column wall temperature. This increases the thermal entrance length but flattens the radial temperature profile. They also studied adiabatic columns and demonstrated that a radial temperature gradient exists because the steel walls conduct heat. Poppe et al.[270,273] developed a model to predict the temperature profile within a column under viscous dissipation effects, which was later extended by Dapremont et al.[274] to account for different column wall and inlet mobile phase temperatures. Kaczmarski et al. developed models to predict the heat generated by viscous dissipation and the resulting chromatographic effects[275–277] Using a 2.1 mm × 5 cm column packed with 1.7 μm particles C18 particles with pure acetonitrile mobile phase at about 1,000 bar, it was shown that the difference between the column wall and column center is about 6 K and that the mobile phase temperature increases by about 20 K as it travels down the column.[276]

2.5.3 COUNTERACTING BAND BROADENING CAUSED BY VISCOUS DISSIPATION

Poppe et al.[270] proposed four potential solutions: 1) Adiabatic columns would minimize radial temperature gradients; however, heat will be conducted along the column wall; he suggested using a chemically inert and mechanically strong material with low thermal conductivity for the column wall; 2) the sample could be injected in the central region of a column only so that it is not affected by a radial temperature gradient; 3) high column pressures could be avoided; however, column performance will suffer; 4) thermostatted, smaller-diameter columns could be used.[270] It has been shown that chromatographic peak shape and band broadening can be improved by thermostatting columns and decreasing the inlet temperature,[278,279] identical to Horvath's approach to softening the radial temperature gradient.[272] Desmet further demonstrated that viscous dissipation causes chromatographic plate height to increase with column radius to the sixth power.[280] A fifth solution was presented by Broeckhoven et al.,[281] who intermittently cooled the chromatographic system by using shorter connected columns of equivalent length to a larger column. This proved to be promising as it decreased the maximum temperature reached on the column and increased peak resolution.

2.5.4 ADIABATIC COLUMNS

The use of adiabatic columns has proven the most successful of all proposed solutions, but the deleterious effect on band spreading still exists at very high pressures. Gritti et al.[282] developed an adiabatic column using a vacuum jacket and minimized thermal radiation by wrapping the column and inner walls of the jacket with aluminum tape. This minimized heat leakage to 1%. Use of this column jacket was shown to decrease the A term in the van Deemter equation by 0.8 ± 0.1 reduced plate height units. In these columns, the temperature at the center of the column is only 0.01 K higher than the column walls at about 900 bar, with 1.8 μm particles demonstrating that the thermal conductivity along the column wall does not cause a significant radial temperature gradient.[283] Broeckhoven et al. modeled the effects of viscous dissipation, discussed the limitations, and showed that the conducting heat along the column wall and the end fittings becomes an issue at pressures over 1,250 bar with sub-2 μm particles. He stated that a change in column hardware is needed before operating pressures can be increased further for analytical-scale columns.[284] An additional concern of adiabatic columns is that a steeper axial temperature gradient is present when compared with traditional thermostatted columns. Band broadening could be caused by higher temperatures downstream causing the front of the peak to be moving faster than the tail. This effect was explored by K. Horvath et al., who showed that theoretical plates and chromatographic resolution are not significantly affected by an axial temperature gradient.[285]

ACKNOWLEDGMENTS

The authors would like to thank the National Science Foundation (Grant CHE 1608757) and the National Institutes of Health (R01 GM044842 and R01 MH104386) for financial support.

REFERENCES

1. Snyder, L. R., Comparisons of normal elution, coupled-columns, and solvent, flow, or temperature programming in liquid chromatography. *J. Chromatogr. Sci.* 1970, *8*(12), 692–706.
2. Gant, J. R.; Dolan, J. W.; Snyder, L. R., Systematic approach to optimizing resolution in reversed-phase liquid chromatography, with emphasis on the role of temperature. *J. Chromatogr.* 1979, *185*(1), 153–177.
3. Dolan, J. W.; Lommen, D. C.; Snyder, L. R., High-performance liquid chromatographic computer simulation based on a restricted multi-parameter approach. I. Theory and verification. *J. Chromatogr.* 1990, *535*(1–2), 55–74.
4. Dolan, J. W.; Lewis, J. A.; Raddatz, W. D.; Snyder, L. R., Multiparameter computer simulation for HPLC method development. *Am. Lab. (Shelton, Connecticut)* 1992, *24*(5), 40D, 40F-40J, 40L.
5. Chloupek, R. C.; Hancock, W. S.; Marchylo, B. A.; Kirkland, J. J.; Boyes, B. E.; Snyder, L. R., Temperature as a variable in reversed-phase high-performance liquid chromatographic separations of peptide and protein mixtures. II. Selectivity effects observed in the separation of several peptide and protein mixtures. *J. Chromatogr. A* 1994, *686*(1), 45–59.
6. Hancock, W. S.; Chloupek, R. C.; Kirkland, J. J.; Snyder, L. R., Temperature as a variable in reversed-phase high-performance liquid chromatographic separations of peptide and protein samples. I. Optimizing the separation of a growth hormone tryptic digest. *J. Chromatogr. A* 1994, *686*(1), 31–43.
7. Zhu, P. L.; Dolan, J. W.; Snyder, L. R.; Djordjevic, N. M.; Hill, D. W.; Lin, J. T.; Sander, L. C.; Van Heukelem, L., Combined use of temperature and solvent strength in reversed-phase gradient elution. IV. Selectivity for neutral (non-ionized) samples as a function of sample type and other separation conditions. *J. Chromatogr. A* 1996, *756*(1–2), 63–72.
8. Zhu, P. L.; Dolan, J. W.; Snyder, L. R.; Hill, D. W.; Van Heukelem, L.; Waeghe, T. J., Combined use of temperature and solvent strength in reversed-phase gradient elution. III. Selectivity for ionizable samples as a function of sample type and pH. *J. Chromatogr. A* 1996, *756*(1–2), 51–62.
9. Zhu, P. L.; Dolan, J. W.; Snyder, L. R., Combined use of temperature and solvent strength in reversed-phase gradient elution. II. Comparing selectivity for different samples and systems. *J. Chromatogr. A* 1996, *756*(1–2), 41–50.
10. Zhu, P. L.; Snyder, L. R.; Dolan, J. W.; Djordjevic, N. M.; Hill, D. W.; Sander, L. C.; Waeghe, T. J., Combined use of temperature and solvent strength in reversed-phase gradient elution. I. Predicting separation as a function of temperature and gradient conditions. *J. Chromatogr. A* 1996, *756*(1–2), 21–39.
11. Snyder, L. R., Temperature-induced selectivity in separations by reversed-phase liquid chromatography. *J. Chromatogr.* 1979, *179*(1), 167–172.
12. Dolan, J. W.; Snyder, L. R.; Wolcott, R. G.; Haber, P.; Baczek, T.; Kaliszan, R.; Sander, L. C., Reversed-phase liquid chromatographic separation of complex samples by optimizing temperature and gradient time. III. Improving the accuracy of computer simulation. *J. Chromatogr. A* 1999, *857*(1–2), 41–68.
13. Dolan, J. W.; Snyder, L. R.; Djordjevic, N. M.; Hill, D. W.; Waeghe, T. J., Reversed-phase liquid chromatographic separation of complex samples by optimizing temperature and gradient time. II. Two-run assay procedures. *J. Chromatogr. A* 1999, *857*(1–2), 21–39.

14. Dolan, J. W.; Snyder, L. R.; Djordjevic, N. M.; Hill, D. W.; Waeghe, T. J., Reversed-phase liquid chromatographic separation of complex samples by optimizing temperature and gradient time. I. Peak capacity limitations. *J. Chromatogr. A* 1999, *857*(1–2), 1–20.

15. Snyder, L. R.; Dolan, J. W., Systematic approaches to HPLC method development for reversed-phase separation. *Chem. Anal. (Warsaw)* 1998, *43*(4), 495–512.

16. Dolan, J. W.; Snyder, L. R.; Saunders, D. L.; Van Heukelem, L., Simultaneous variation of temperature and gradient steepness for reversed-phase high-performance liquid chromatography method development. II. The use of further changes in conditions. *J. Chromatogr. A* 1998, *803*(1–2), 33–50.

17. Dolan, J. W.; Snyder, L. R.; Djordjevic, N. M.; Hill, D. W.; Saunders, D. L.; Van Heukelem, L.; Waeghe, T. J., Simultaneous variation of temperature and gradient steepness for reversed-phase high-performance liquid chromatography method development I. Application to 14 different samples using computer simulation. *J. Chromatogr. A* 1998, *803*(1–2), 1–31.

18. Snyder, L. R.; Dolan, J. W., Reversed-phase separation of achiral isomers by varying temperature and either gradient time or solvent strength. *J. Chromatogr. A* 2000, *892*(1–2), 107–121.

19. Melander, W.; Campbell, D. E.; Horvath, C., Enthalpy-entropy compensation in reversed-phase chromatography. *J. Chromatogr.* 1978, *158*, 215–225.

20. Carr, P. W.; Doherty, R. M.; Kamlet, M. J.; Taft, R. W.; Melander, W.; Horvath, C., Study of temperature and mobile-phase effects in reversed-phase high-performance liquid chromatography by the use of the solvatochromic comparison method. *Anal. Chem.* 1986, *58*(13), 2674–2680.

21. Dill, K. A., The mechanism of solute retention in reversed-phase liquid chromatography. *J. Phys. Chem.* 1987, *91*(7), 1980–1988.

22. Dorsey, J. G.; Dill, K. A., The molecular mechanism of retention in reversed-phase liquid chromatography. *Chem. Rev.* 1989, *89*(2), 331–346.

23. Cole, L. A.; Dorsey, J. G., Temperature dependence of retention in reversed-phase liquid chromatography. 1. Stationary-phase considerations. *Anal. Chem.* 1992, *64*(13), 1317–1323.

24. Cole, L. A.; Dorsey, J. G.; Dill, K. A., Temperature dependence of retention in reversed-phase liquid chromatography. 2. Mobile-phase considerations. *Anal. Chem.* 1992, *64*(13), 1324–1327.

25. Trones, R.; Iveland, A.; Greibrokk, T., High-temperature liquid chromatography on packed capillary columns with nonaqueous mobile phases. *J. Microcolumn Sep.* 1995, *7*(5), 505–512.

26. Molander, P.; Gundersen, T. E.; Haas, C.; Greibrokk, T.; Blomhoff, R.; Lundanes, E., Determination of retinoids by packed-capillary liquid chromatography with large-volume on-column focusing and temperature optimization. *J. Chromatogr. A* 1999, *847*(1–2), 59–68.

27. Molander, P.; Thommesen, S. J.; Bruheim, I. A.; Trones, R.; Greibrokk, T.; Lundanes, E.; Gundersen, T. E., Temperature-programmed packed capillary liquid chromatography separation with large volume on-column focussing of retinyl esters. *J. High Resolut. Chromatogr.* 1999, *22*(9), 490–494.

28. Houdiere, F.; Fowler, P. W. J.; Djordjevic, N. M., The combination of column temperature gradient and mobile phase flow gradient in microcolumn and capillary column high-performance liquid chromatography. *Anal. Chem.* 1997, *69*(13), 2589–2593.

29. Djordjevic, N. M.; Houdiere, F.; Fowler, P., High temperature and temperature programming in capillary HPLC. *Biomed. Chromatogr.* 1998, *12*(3), 153–154.

30. Djordjevic, N. M.; Fowler, P. W. J.; Houdiere, F., High temperature and temperature programming in high-performance liquid chromatography: Instrumental considerations. *J. Microcolumn Sep.* 1999, *11*(6), 403–413.

31. Kennedy, R. T.; German, I.; Thompson, J. E.; Witowski, S. R., Fast analytical-scale separations by capillary electrophoresis and liquid chromatography. *Chem. Rev. (Washington, DC)* 1999, *99*(10), 3081–3131.

32. Chen, M. H.; Horvath, C., Temperature programming and gradient elution in reversed-phase chromatography with packed capillary columns. *J. Chromatogr. A* 1997, *788*(1–2), 51–61.

33. Holm, A.; Molander, P.; Lundanes, E.; Greibrokk, T., Novel column oven concept for cold spot large volume sample enrichment in high throughput temperature gradient capillary liquid chromatography. *J. Sep. Sci.* 2003, *26*(12–13), 1147–1153.

34. Gagliardi, L. G.; Tascon, M.; Castells, C. B., Effect of temperature on acid-base equilibria in separation techniques. A review. *Anal. Chim. Acta* 2015, *889*, 35–57.

35. Mao, Y.; Carr, P. W., Application of the thermally tuned tandem column concept to the separation of several families of environmental toxicants. *Anal. Chem.* 2000, *72*(13), 2788–2796.

36. Mao, Y.; Carr, P. W., Separation of selected basic pharmaceuticals by reversed-phase and ion-exchange chromatography using thermally tuned tandem columns. *Anal. Chem.* 2001, *73*(18), 4478–4485.

37. Mao, Y.; Carr, P. W., Separation of barbiturates and phenylthiohydantoin amino acids using the thermally tuned tandem column concept. *Anal. Chem.* 2001, *73*(8), 1821–1830.

38. Teutenberg, T.; Goetze, H. J.; Tuerk, J.; Ploeger, J.; Kiffmeyer, T. K.; Schmidt, K. G.; Kohorst, W.; Rohe, T.; Jansen, H. D.; Weber, H., Development and application of a specially designed heating system for temperature-programmed high-performance liquid chromatography using subcritical water as the mobile phase. *J. Chromatogr. A* 2006, *1114*(1), 89–96.

39. Yan, B.; Zhao, J.; Brown, J. S.; Blackwell, J.; Carr, P. W., High-temperature ultrafast liquid chromatography. *Anal. Chem.* 2000, *72*(6), 1253–1262.

40. Thompson, J. D.; Carr, P. W., High-speed liquid chromatography by simultaneous optimization of temperature and eluent composition. *Anal. Chem.* 2002, *74*(16), 4150–4159.

41. Xiang, Y.; Yan, B.; Yue, B.; McNeff, C. V.; Carr, P. W.; Lee, M. L., Elevated-temperature ultrahigh-pressure liquid chromatography using very small polybutadiene-coated nonporous zirconia particles. *J. Chromatogr. A* 2003, *983*(1–2), 83–89.

42. Jinno, K.; Hirata, Y., Investigation of the low-temperature effect in normal-phase micro high-performance liquid chromatography. *HRC CC J. High Resolut. Chromatogr. Chromatogr. Commun.* 1982, *5*(2), 85–90.

43. Skuland, I. L.; Andersen, T.; Trones, R.; Eriksen, R. B.; Greibrokk, T., Determination of polyethylene glycol in low-density polyethylene by large volume injection temperature gradient packed capillary liquid chromatography. *J. Chromatogr. A* 2003, *1011*(1–2), 31–36.

44. Groskreutz, S. R.; Weber, S. G., Temperature-assisted on-column solute focusing: A general method to reduce pre-column dispersion in capillary high performance liquid chromatography. *J. Chromatogr. A* 2014, *1354*, 65–74.

45. Groskreutz, S. R.; Horner, A. R.; Weber, S. G., Temperature-based on-column solute focusing in capillary liquid chromatography reduces peak broadening from pre-column dispersion and volume overload when used alone or with solvent-based focusing. *J. Chromatogr. A* 2015, *1405*, 133–139.

46. Groskreutz, S. R.; Weber, S. G., Temperature-assisted solute focusing with sequential trap/release zones in isocratic and gradient capillary liquid chromatography: Simulation and experiment. *J. Chromatogr. A* 2016, *1474*, 95–108.

47. Wilson, R. E.; Groskreutz, S. R.; Weber, S. G., Improving the sensitivity, resolution, and peak capacity of gradient elution in capillary liquid chromatography with large-volume injections by using temperature-assisted on-column solute focusing. *Anal. Chem. (Washington, DC, US)* 2016, *88*(10), 5112–5121.

48. De Vos, J.; Desmet, G.; Eeltink, S., Enhancing detection sensitivity in gradient liquid chromatography via post-column refocusing and strong-solvent remobilization. *J. Chromatogr. A* 2016, *1455*, 86–92.
49. Tran, J. V.; Molander, P.; Greibrokk, T.; Lundanes, E., Temperature effects on retention in reversed phase liquid chromatography. *J. Sep. Sci.* 2001, *24*(12), 930–940.
50. Molander, P.; Olsen, R.; Lundanes, E.; Greibrokk, T., The impact of column inner diameter on chromatographic performance in temperature gradient liquid chromatography. *Analyst (Cambridge, UK)* 2003, *128*(11), 1341–1345.
51. Guillarme, D.; Heinisch, S.; Rocca, J. L., Effect of temperature in reversed phase liquid chromatography. *J. Chromatogr. A* 2004, *1052*(1–2), 39–51.
52. Cabooter, D.; Heinisch, S.; Rocca, J. L.; Clicq, D.; Desmet, G., Use of the kinetic plot method to analyze commercial high-temperature liquid chromatography systems. I: Intrinsic performance comparison. *J. Chromatogr. A* 2007, *1143*(1–2), 121–133.
53. Clicq, D.; Heinisch, S.; Rocca, J. L.; Cabooter, D.; Gzil, P.; Desmet, G., Use of the kinetic plot method to analyze commercial high-temperature liquid chromatography systems. II. Practically constrained performance comparison. *J. Chromatogr. A* 2007, *1146*(2), 193–201.
54. McNeff, C. V.; Yan, B.; Stoll, D. R.; Henry, R. A., Practice and theory of high temperature liquid chromatography. *J. Sep. Sci.* 2007, *30*(11), 1672–1685.
55. Heinisch, S.; Desmet, G.; Clicq, D.; Rocca, J.-L., Kinetic plot equations for evaluating the real performance of the combined use of high temperature and ultra-high pressure in liquid chromatography. Application to commercial instruments and 2.1 and 1 mm I.D. columns. *J. Chromatogr. A* 2008, *1203*(2), 124–136.
56. Teutenberg, T., Potential of high temperature liquid chromatography for the improvement of separation efficiency—A review. *Anal. Chim. Acta* 2009, *643*(1–2), 1–12.
57. Carr, P. W.; Wang, X.; Stoll, D. R., Effect of pressure, particle size, and time on optimizing performance in liquid chromatography. *Anal. Chem.* 2009, *81*(13), 5342–5353.
58. Carr, P. W.; Stoll, D. R.; Wang, X., Perspectives on recent advances in the speed of high-performance liquid chromatography. *Anal. Chem.* 2011, *83*(6), 1890–1900.
59. Blumberg, L. M.; Desmet, G., Kinetic performance factor - A measurable metric of separation-time-pressure tradeoff in liquid and gas chromatography. *J. Chromatogr. A* 2018, *1567*, 26–36.
60. Teutenberg, T., *High-Temperature Liquid Chromatography: A User's Guide for Method Development.* RSC Pub: Cambridge, 2010, p. 210.
61. Novotny, M. V.; Ishii, D., *Microcolumn Separations: Columns, Instrumentation, and Ancillary Techniques.* Elsevier; Distributors for the U.S. and Canada, Elsevier Science Pub. Co.: Amsterdam; New York, 1985, p. xi, 336 p.
62. Filla, L. A.; Yuan, W.; Feldman, E. L.; Li, S.; Edwards, J. L., Global metabolomic and isobaric tagging capillary liquid chromatography-tandem mass spectrometry approaches for uncovering pathway dysfunction in diabetic mouse aorta. *J. Proteome Res.* 2014, *13*(12), 6121–6134.
63. Gao, X.; Zhang, Q.; Meng, D.; Isaac, G.; Zhao, R.; Fillmore, T. L.; Chu, R. K.; Zhou, J.; Tang, K.; Hu, Z.; Moore, R. J.; Smith, R. D.; Katze, M. G.; Metz, T. O., A reversed-phase capillary ultra-performance liquid chromatography-mass spectrometry (UPLC-MS) method for comprehensive top-down/bottom-up lipid profiling. *Anal. Bioanal. Chem.* 2012, *402*(9), 2923–2933.
64. Theodoridis, G.; Gika, H. G.; Wilson, I. D., LC-MS-based methodology for global metabolite profiling in metabonomics/metabolomics. *TrAC Trends Anal. Chem.* 2008, *27*(3), 251–260.
65. Ishihama, Y., Proteomic LC–MS systems using nanoscale liquid chromatography with tandem mass spectrometry. *J. Chromatogr. A* 2005, *1067*(1), 73–83.

66. Haskins, W. E.; Wang, Z.; Watson, C. J.; Rostand, R. R.; Witowski, S. R.; Powell, D. H.; Kennedy, R. T., Capillary LC-MS2 at the attomole level for monitoring and discovering endogenous peptides in microdialysis samples collected in vivo. *Anal. Chem.* 2001, *73*(21), 5005–5014.

67. Cepeda, D. E.; Hains, L.; Li, D.; Bull, J.; Lentz, S. I.; Kennedy, R. T., Experimental evaluation and computational modeling of tissue damage from low-flow push-pull perfusion sampling in vivo. *J. Neurosci. Methods* 2015, *242*, 97–105.

68. Wong, J.-M. T.; Malec, P. A.; Mabrouk, O. S.; Ro, J.; Dus, M.; Kennedy, R. T., Benzoyl chloride derivatization with liquid chromatography-mass spectrometry for targeted metabolomics of neurochemicals in biological samples. *J. Chromatogr. A* 2016, *1446*, 78–90.

69. Maes, K.; Van Liefferinge, J.; Viaene, J.; Van Schoors, J.; Van Wanseele, Y.; Bechade, G.; Chambers, E. E.; Morren, H.; Michotte, Y.; Vander Heyden, Y.; Claereboudt, J.; Smolders, I.; Van Eeckhaut, A., Improved sensitivity of the nano ultra-high performance liquid chromatography-tandem mass spectrometric analysis of low-concentrated neuropeptides by reducing aspecific adsorption and optimizing the injection solvent. *J. Chromatogr. A* 2014, *1360*, 217–228.

70. Kucera, P., *Microcolumn High-Performance Liquid Chromatography.* Elsevier; Distributors for the U.S. and Canada, Elsevier Science Pub. Co.: Amsterdam; New York, 1984, p. xvi, 302 p.

71. Gilar, M.; McDonald, T. S.; Johnson, J. S.; Murphy, J. P.; Jorgenson, J. W., Wide injection zone compression in gradient reversed-phase liquid chromatography. *J. Chromatogr. A* 2015, *1390*, 86–94.

72. van Deemter, J. J.; Zuiderweg, F. J.; Klinkenberg, A., Longitudinal diffusion and resistance to mass transfer as causes of nonideality in chromatography. *Chem. Eng. Sci.* 1995, *50*(24), 3869–3882.

73. Stoll, D. R.; Talus, E. S.; Harmes, D. C.; Zhang, K., Evaluation of detection sensitivity in comprehensive two-dimensional liquid chromatography separations of an active pharmaceutical ingredient and its degradants. *Anal. Bioanal. Chem.* 2015, *407*(1), 265–277.

74. Snyder, L. R., Linear elution adsorption chromatography: VII. gradient elution theory. *J. Chromatogr. A* 1964, *13*, 415–434.

75. Snyder, L. R., Principles of gradient elution. *Chromatogr. Rev.* 1965, *7*, 1–51.

76. Snyder, L. R.; Saunders, D. L., Optimized solvent programming for separations of complex samples by liquid-solid adsorption chromatography in columns. *J. Chromatogr. Sci.* 1969, *7*(4), 195–208.

77. Lankelma, J.; Poppe, H., Determination of methotrexate in plasma by on-column concentration and ion-exchange chromatography. *J. Chromatogr. A* 1978, *149*, 587–598.

78. Stoll, D. R.; Shoykhet, K.; Petersson, P.; Buckenmaier, S., Active solvent modulation: A valve-based approach to improve separation compatibility in two-dimensional liquid chromatography. *Anal. Chem.* 2017, *89*(17), 9260–9267.

79. Vonk, R. J.; Gargano, A. F. G.; Davydova, E.; Dekker, H. L.; Eeltink, S.; de Koning, L. J.; Schoenmakers, P. J., Comprehensive two-dimensional liquid chromatography with stationary-phase-assisted modulation coupled to high-resolution mass spectrometry applied to proteome analysis of Saccharomyces cerevisiae. *Anal. Chem.* 2015, *87*(10), 5387–5394.

80. Baglai, A.; Blokland, M. H.; Mol, H. G. J.; Gargano, A. F. G.; van der Wal, S.; Schoenmakers, P. J., Enhancing detectability of anabolic-steroid residues in bovine urine by actively modulated online comprehensive two-dimensional liquid chromatography – High-resolution mass spectrometry. *Anal. Chim. Acta* 2018, *1013*, 87–97.

81. Jakobsen, S. S.; Christensen, J. H.; Verdier, S.; Mallet, C. R.; Nielsen, N. J., Increasing flexibility in two-dimensional liquid chromatography by pulsed elution of the first dimension: A proof of concept. *Anal. Chem.* 2017, *89*(17), 8723–8730.
82. Eghbali, H.; Sandra, K.; Tienpont, B.; Eeltink, S.; Sandra, P.; Desmet, G., Exploring the possibilities of cryogenic cooling in liquid chromatography for biological applications: A proof of principle. *Anal. Chem.* 2012, *84*(4), 2031–2037.
83. Young, C.; Podtelejnikov, A. V.; Nielsen, M. L., Improved reversed phase chromatography of hydrophilic peptides from spatial and temporal changes in column temperature. *J. Proteome Res.* 2017, *16*(6), 2307–2317.
84. Groskreutz, S. R.; Weber, S. G., Quantitative evaluation of models for solvent-based, on-column focusing in liquid chromatography. *J. Chromatogr. A* 2015, *1409*, 116–124.
85. Rerick, M. T.; Groskreutz, S. R.; Weber, S. G., Multiplicative on-column solute focusing using spatially dependent temperature programming for capillary HPLC. *Anal. Chem.* 2019, *91*(4), 2854–2860.
86. Greibrokk, T.; Andersen, T., Temperature programming in liquid chromatography. *J. Sep. Sci.* 2001, *24*(12), 899–909.
87. Collins, D.; Nesterenko, E.; Connolly, D.; Vasquez, M.; Macka, M.; Brabazon, D.; Paull, B., Versatile capillary column temperature control using a thermoelectric array based platform. *Anal. Chem.* 2011, *83*(11), 4307–4313.
88. Gritti, F., Combined solvent- and non-uniform temperature-programmed gradient liquid chromatography. I – A theoretical investigation. *J. Chromatogr. A* 2016, *1473*, 38–47.
89. Verstraeten, M.; Pursch, M.; Eckerle, P.; Luong, J.; Desmet, G., Modelling the thermal behaviour of the low-thermal mass liquid chromatography system. *J. Chromatogr. A* 2011, *1218*(16), 2252–2263.
90. Gritti, F.; Sanchez, C. A.; Farkas, T.; Guiochon, G., Achieving the full performance of highly efficient columns by optimizing conventional benchmark high-performance liquid chromatography instruments. *J. Chromatogr. A* 2010, *1217*(18), 3000–3012.
91. Gritti, F.; Guiochon, G., Performance of new prototype packed columns for very high pressure liquid chromatography. *J. Chromatogr. A* 2010, *1217*(9), 1485–1495.
92. Gritti, F.; Guiochon, G., Measurement of the axial and radial temperature profiles of a chromatographic column: Influence of thermal insulation on column efficiency. *J. Chromatogr. A* 2007, *1138*(1), 141–157.
93. Poppe, H.; Paanakker, J.; Bronckhorst, M., Peak width in solvent-programmed chromatography: I. General description of peak broadening in solvent-programmed elution. *J. Chromatogr. A* 1981, *204*, 77–84.
94. De Vos, J.; Desmet, G.; Eeltink, S., A generic approach to post-column refocusing in liquid chromatography. *J. Chromatogr. A* 2014, *1360*, 164–171.
95. De Vos, J.; Eeltink, S.; Desmet, G., Peak refocusing using subsequent retentive trapping and strong eluent remobilization in liquid chromatography: A theoretical optimization study. *J. Chromatogr. A* 2015, *1381*, 74–86.
96. Hilhorst, M. J.; Somsen, G. W.; de Jong, G. J., Sensitivity enhancement in capillary electrochromatography by on-column preconcentration. *Chromatographia* 2000, *53*(3), 190–196.
97. Michelle Hong, C.; Horváth, C., Temperature programming and gradient elution in reversed-phase chromatography with packed capillary columns. *J. Chromatogr. A* 1997, *788*(1–2), 51–61.
98. Billen, J.; Broeckhoven, K.; Liekens, A.; Choikhet, K.; Rozing, G.; Desmet, G., Influence of pressure and temperature on the physico-chemical properties of mobile phase mixtures commonly used in high-performance liquid chromatography. *J. Chromatogr. A* 2008, *1210*(1), 30–44.

99. Im, K.; Park, H.-W.; Kim, Y.; Chung, B.; Ree, M.; Chang, T., Comprehensive two-dimensional liquid chromatography analysis of a block copolymer. *Anal. Chem.* 2007, *79*(3), 1067–1072.

100. Wu, X.; Langan, T. J.; Durney, B. C.; Holland, L. A., Thermally responsive phospholipid preparations for fluid steering and separation in microfluidics. *Electrophoresis* 2012, *33*(17), 2674–2681.

101. Groskreutz, S. R.; Horner, A. R.; Weber, S. G., Development of a 1.0 mm inside diameter temperature-assisted focusing precolumn for use with 2.1 mm inside diameter columns. *J. Chromatogr. A* 2017, *1523*, 193–203.

102. Neue, U. D.; Kuss, H.-J., Improved reversed-phase gradient retention modeling. *J. Chromatogr. A* 2010, *1217*(24), 3794–3803.

103. Berthod, A., Silica: Backbone material of liquid chromatographic column packings. *J. Chromatogr. A* 1991, *549*, 1–28.

104. Unger, K. K.; Becker, N.; Roumeliotis, P., Recent developments in the evaluation of chemically bonded silica packings for liquid chromatography. *J. Chromatogr. A* 1976, *125*(1), 115–127.

105. Hanson, M.; Unger, K. K.; Schomburg, G., Non-porous polybutadiene-coated silicas as stationary phases in reversed-phase chromatography. *J. Chromatogr. A* 1990, *517*, 269–284.

106. Kirkland, J. J., Stability of silica-based, monofunctional C18 bonded-phase column packing for HPLC at high pH. *J. Chromatogr. Sci.* 1996, *34*(7), 309–313.

107. Kirkland, J. J., Development of some stationary phases for reversed-phase HPLC. *J. Chromatogr. A* 2004, *1060*(1–2), 9–21.

108. Kirkland, J. J.; Adams, J. B.; van Straten, M. A.; Claessens, H. A., Bidentate silane stationary phases for reversed-phase high-performance liquid chromatography. *Anal. Chem.* 1998, *70*(20), 4344–4352.

109. Kirkland, J. J.; Glajch, J. L.; Farlee, R. D., Synthesis and characterization of highly stable bonded phases for high-performance liquid chromatography column packings. *Anal. Chem.* 1989, *61*(1), 2–11.

110. Kirkland, J. J.; Henderson, J. W.; DeStefano, J. J.; van Straten, M. A.; Claessens, H. A., Stability of silica-based, endcapped columns with pH 7 and 11 mobile phases for reversed-phase high-performance liquid chromatography. *J. Chromatogr. A* 1997, *762*(1), 97–112.

111. Kirkland, J. J.; van Straten, M. A.; Claessens, H. A., High pH mobile phase effects on silica-based reversed-phase high-performance liquid chromatographic columns. *J. Chromatogr. A* 1995, *691*(1–2), 3–19.

112. Kirkland, J. J.; van Straten, M. A.; Claessens, H. A., Reversed-phase high-performance liquid chromatography of basic compounds at pH 11 with silica-based column packings. *J. Chromatogr. A* 1998, *797*(1), 111–120.

113. Trammell, B. C.; Ma, L.; Luo, H.; Hillmyer, M. A.; Carr, P. W., An ultra acid stable reversed stationary phase. *J. Am. Chem. Soc.* 2003, *125*(35), 10504–10505.

114. Carr, P. W.; Rigney, M. O.; Funkenbusch, E. F.; Coleman, P. L.; Hanggi, D. A., High-stability porous zirconium oxide spherules for use as chromatographic column support. EP331283A1, 1989.

115. Zhao, J.; Carr, P. W., Synthesis and evaluation of an aromatic polymer-coated zirconia for reversed-phase liquid chromatography. *Anal. Chem.* 1999, *71*(22), 5217–5224.

116. Anderson, D. J., High-performance liquid chromatography (advances in packing materials). *Anal. Chem.* 1995, *67*(12), 475–486.

117. Nawrocki, J., The silanol group and its role in liquid chromatography. *J. Chromatogr. A* 1997, *779*(1), 29–71.

118. Kirkland, J. J.; Truszkowski, F. A.; Ricker, R. D., Atypical silica-based column packings for high-performance liquid chromatography. *J. Chromatogr. A* 2002, *965*(1), 25–34.

119. Lenher, V.; Merrill, H. B., The solubility of silica. *J. Am. Chem. Soc.* 1917, *39*(12), 2630–2638.

120. Hanson, M.; Kurganov, A.; Unger, K. K.; Davankov, V. A., Polymer-coated reversed-phase packings in high-performance liquid chromatography. *J. Chromatogr. A* 1993, *656*(1), 369–380.

121. Schomburg, G., Polymer coating of surfaces in column liquid chromatography and capillary electrophoresis. *TrAC Trends Anal. Chem.* 1991, *10*(5), 163–169.

122. Horvath, C. G.; Lipsky, S. R., Use of liquid ion exchange chromatography for the separation of organic compounds. *Nature* 1966, *211*(5050), 748–749.

123. Airapetyan, S. S.; Khachatryan, A. G., Chromatographic properties of silica gel packing materials as influenced by polymeric coating. *Russ. J. Appl. Chem.* 2003, *76*(11), 1864–1866.

124. Hanson, M.; Eray, B.; Unger, K.; Neimark, A. V.; Schmid, J.; Albert, K.; Bayer, E., A model for polybutadiene coatings on porous silica. *Chromatographia* 1993, *35*(7), 403–409.

125. Fischer, G.; Skogsberg, U.; Bachmann, S.; Yüksel, H.; Steinbrecher, S.; Plies, E.; Albert, K., Synthesis, characterization, and evaluation of divinylbenzene-coated spherical nonporous silica. *Chem. Mater.* 2003, *15*(23), 4394–4400.

126. Hanson, M.; Unger, K. K.; Mant, C. T.; Hodges, R. S., Polymer-coated reversed-phase packings with controlled hydrophobic properties: I. Effect on the selectivity of protein separations. *J. Chromatogr. A* 1992, *599*(1), 65–75.

127. Hayrapetyan, S. S.; Khachatryan, H. G.; Neue, U. D., A detailed discussion of the influence of the amount of deposited polymer on the retention properties of polymer-coated silicas. *J. Sep. Sci.* 2006, *29*(6), 801–809.

128. Wheals, B. B., Chemically bonded phases for liquid chromatography modification of silica with vinyl monomers. *J. Chromatogr. A* 1975, *107*(2), 402–406.

129. Kurganov, A. A.; Tevlin, A. B.; Davankov, V. A., High-performance ligand-exchange chromatography of enantiomers: Studies on polystrene-type chiral phases bonded to microparticulate silicas. *J. Chromatogr. A* 1983, *261*, 223–233.

130. Buchmeiser, M. R., New synthetic ways for the preparation of high-performance liquid chromatography supports. *J. Chromatogr. A* 2001, *918*(2), 233–266.

131. Takeuchi, T.; Hu, W.; Haraguchi, H.; Ishii, D., Evaluation of the stability of polymer-coated silica-based packing materials for high-performance liquid chromatography. *J. Chromatogr. A* 1990, *517*, 257–262.

132. Trammell, B. C.; Boissel, C. A.; Carignan, C.; O'Shea, D. J.; Hudalla, C. J.; Neue, U. D.; Iraneta, P. C., Development of an accelerated low-pH reversed-phase liquid chromatography column stability test. *J. Chromatogr. A* 2004, *1060*(1–2), 153–163.

133. Trammell, B. C.; Ma, L.; Luo, H.; Jin, D.; Hillmyer, M. A.; Carr, P. W., Highly cross-linked self-assembled monolayer stationary phases: An approach to greatly enhancing the low pH stability of silica-based stationary phases. *Anal. Chem.* 2002, *74*(18), 4634–4639.

134. Zhang, Y.; Huang, Y.; Carr, P. W., Optimization of the synthesis of a hyper-crosslinked stationary phases: A new generation of highly efficient, acid-stable hyper-crosslinked materials for HPLC. *J. Sep. Sci.* 2011, *34*(12), 1407–1422.

135. Petro, M.; Berek, D., Polymers immobilized on silica gels as stationary phases for liquid chromatography. *Chromatographia* 1993, *37*(9–10), 549–561.

136. Ihara, H.; Fukui, M.; Mimaki, T.; Shundo, A.; Dong, W.; Derakhshan, M.; Sakurai, T.; Takafuji, M.; Nagaoka, S., Poly(4-vinylpyridine) as a reagent with silanol-masking effect for silica and its specific selectivity for PAHs and dinitropyrenes in a reversed phase. *Anal. Chim. Acta* 2005, *548*(1), 51–57.

137. Gautam, U. G.; Sawada, T.; Gautam, M. P.; Takafuji, M.; Ihara, H., Poly(2-N-carbazolylethyl acrylate)-modified silica as a new polymeric stationary phase for reversed-phase high-performance liquid chromatography. *J. Chromatogr. A* 2009, *1216*(44), 7422–7426.

138. Ohmacht, R.; Kele, M.; Matus, Z., Polymer coated stationary phases for liquid chromatography. *Chromatographia* 1989, *28*(1–2), 19–23.

139. Hetem, M. J. J.; De Haan, J. W.; Claessens, H. A.; Cramers, C. A.; Deege, A.; Schomburg, G., Characterization and stability of silanized and polymer-coated octadecyl reversed phases. *J. Chromatogr.* 1991, *540*(1–2), 53–76.

140. Rimmer, C. A.; Sander, L. C.; Wise, S. A.; Dorsey, J. G., Synthesis and characterization of C13 to C18 stationary phases by monomeric, solution polymerized, and surface polymerized approaches. *J. Chromatogr. A* 2003, *1007*(1–2), 11–20.

141. Stöber, W.; Fink, A.; Bohn, E., Controlled growth of monodisperse silica spheres in the micron size range. *J. Colloid Interface Sci.* 1968, *26*(1), 62–69.

142. Unger, K.; Schick-Kalb, J.; Krebs, K.-F., Preparation of porous silica spheres for column liquid. *J. Chromatogr. A* 1973, *83*, 5–9.

143. Kohlschütter, H. W.; Mihm, U., Das kugelförmige und poröse Korn von Silicagel als Reaktionsprodukt. *Kolloid-Zeitschrift und Zeitschrift für Polymere* 1971, *243*(2), 148–152.

144. Unger, K. K.; Kinkel, J. N.; Anspach, B.; Giesche, H., Evaluation of advanced silica packings for the separation of biopolymers by high-performance liquid chromatography: I. Design and properties of parent silicas. *J. Chromatogr. A* 1984, *296*, 3–14.

145. Kirkland, J. J.; DeStefano, J. J., Controlled surface porosity supports with chemically-bonded organic stationary phases for gas and liquid chromatography. *J. Chromatogr. Sci.* 1970, *8*(6), 309–314.

146. Kirkland, J. J., Columns for modern analytical liquid chromatography. *Anal. Chem.* 1971, *43*(12), 36A–48a.

147. Cheng, Y.-F.; Walter, T. H.; Lu, Z.; Iraneta, P.; Alden, B. A.; Gendreau, C.; Neue, U. D.; Grassi, J. M.; Carmody, J. L.; O'Gara, J. E.; Fisk, R. P., Hybrid organic-inorganic particle technology: Breaking through traditional barriers of HPLC separations. *Lc Gc* 2000, *18*(11), 1162, 1164, 1166, 1168, 1170, 1172.

148. Hudalla, C.; Alden, B.; Walter, T.; Walsh, D.; Bouvier, E.; Iraneta, P.; Lawrence, N.; Wyndham, K., Synthesis and applications of BEH particles in liquid chromatography. *Lc Gc* 2012, *30*(4), 20–29.

149. Li, J.; Xu, L.; Shi, Z.-g., Waxberry-like hierarchically porous ethyl-bridged hybrid silica microsphere: A substrate for enzyme catalysis and high-performance liquid chromatography. *J. Chromatogr. A* 2019, *1587*, 79–87.

150. Yue, X.-Y.; Jiang, D.-D.; Shu, L.; Chen, S.; Nie, Z.-R.; Wei, Q.; Cui, S.-P.; Li, Q.-Y., Mesoporous C18-bonded ethyl-bridged organic-inorganic hybrid silica: A facile one-pot synthesis and liquid chromatographic performance. *Micropor. Mesopor. Mater.* 2016, *236*, 277–283.

151. Leonard, M.; Fournier, C.; Dellacherie, E., Polyvinyl alcohol-coated macroporous polystyrene particles as stationary phases for the chromatography of proteins. *J. Chromatogr. B Biomed. Sci. Appl.* 1995, *664*(1), 39–46.

152. Huber, C. G.; Kleindienst, G.; Bonn, G. K., Application of micropellicular poly-styrene/divinylbenzene stationary phases for high-performance reversed-phase liquid chromatography electrospray-mass spectrometry of proteins and peptides. *Chromatographia* 1997, *44*(7), 438–448.

153. Nawrocki, J.; Dunlap, C.; Li, J.; Zhao, J.; McNeff, C. V.; McCormick, A.; Carr, P. W., Part II. Chromatography using ultra-stable metal oxide-based stationary phases for HPLC. *J. Chromatogr. A* 2004, *1028*(1), 31–62.

154. Nawrocki, J.; Dunlap, C.; McCormick, A.; Carr, P. W., Part I. Chromatography using ultra-stable metal oxide-based stationary phases for HPLC. *J. Chromatogr. A* 2004, *1028*(1), 1–30.

155. Wirth, H. J.; Eriksson, K. O.; Holt, P.; Aguilar, M.; Hearn, M. T. W., High-performance liquid chromatography of amino acids, peptides and proteins CXXIX. Ceramic-based particles as chemically stable chromatographic supports. *J. Chromatogr. A* 1993, *646*(1), 129–141.

156. Trüdinger, U.; Müller, G.; Unger, K. K., Porous zirconia and titania as packing materials for high-performance liquid chromatography. *J. Chromatogr. A* 1990, *535*, 111–125.

157. Goraieb, K.; Collins, C. H., Evaluation of a doubly zirconized silica-based stationary phase for HPLC. *Chromatographia* 2013, *76*(15), 899–908.

158. Silva, C. R.; Airoldi, C.; Collins, K. E.; Collins, C. H., A new generation of more pH stable reversed phases prepared by silanization of zirconized silica. *J. Chromatogr. A* 2008, *1191*(1), 90–98.

159. Faria, A. M.; Silva, C. R.; Collins, C. H.; Jardim, I. C. S. F., Development of a polymer-coated stationary phase with improved chemical stability in alkaline mobile phases. *J. Sep. Sci.* 2008, *31*(6–7), 953–960.

160. Li, J.; Carr, P. W., Retention characteristics of polybutadiene-coated zirconia and comparison to conventional bonded phases. *Anal. Chem.* 1996, *68*(17), 2857–2868.

161. McNeff, C.; Zhao, Q.; Carr, P. W., High-performance anion exchange of small anions with polyethyleneimine-coated porous zirconia. *J. Chromatogr. A* 1994, *684*(2), 201–211.

162. Hu, Y.; Carr, P. W., Synthesis and characterization of new zirconia-based polymeric cation-exchange stationary phases for high-performance liquid chromatography of proteins. *Anal. Chem.* 1998, *70*(9), 1934–1942.

163. Knox, J. H.; Kaur, B.; Millward, G. R., Structure and performance of porous graphitic carbon in liquid chromatography. *J. Chromatogr. A* 1986, *352*, 3–25.

164. West, C.; Elfakir, C.; Lafosse, M., Porous graphitic carbon: A versatile stationary phase for liquid chromatography. *J. Chromatogr. A* 2010, *1217*(19), 3201–3216.

165. Huang, Z.; Yao, P.; Zhu, Q.; Wang, L.; Zhu, Y., The polystyrene-divinylbenzene stationary phase hybridized with oxidized nanodiamonds for liquid chromatography. *Talanta* 2018, *185*, 221–228.

166. Saini, G.; Jensen, D. S.; Wiest, L. A.; Vail, M. A.; Dadson, A.; Lee, M. L.; Shutthanandan, V.; Linford, M. R., Core-shell diamond as a support for solid-phase extraction and high-performance liquid chromatography. *Anal. Chem.* 2010, *82*(11), 4448–4456.

167. Wiest, L. A.; Jensen, D. S.; Hung, C.-H.; Olsen, R. E.; Davis, R. C.; Vail, M. A.; Dadson, A. E.; Nesterenko, P. N.; Linford, M. R., Pellicular particles with spherical carbon cores and porous nanodiamond/polymer shells for reversed-phase HPLC. *Anal. Chem.* 2011, *83*(14), 5488–5501.

168. Xue, Z.; Vinci, J. C.; Colón, L. A., Nanodiamond-decorated silica spheres as a chromatographic material. *ACS Appl. Mater. Interfaces* 2016, *8*(6), 4149–4157.

169. Trammell, B. C.; Ma, L.; Luo, H.; Hillmyer, M. A.; Carr, P. W., Synthesis and characterization of hypercrosslinked, surface-confined, ultra-stable silica-based stationary phases. *J. Chromatogr. A* 2004, *1060*(1–2), 61–76.

170. Zhang, Y.; Carr, P. W., Novel ultra stable silica-based stationary phases for reversed phase liquid chromatography--Study of a hydrophobically assisted weak acid cation exchange phase. *J. Chromatogr. A* 2011, *1218*(6), 763–777.

171. Arnold, R. M.; Patton, D. L.; Popik, V. V.; Locklin, J., A dynamic duo: Pairing click chemistry and postpolymerization modification to design complex surfaces. *Acc. Chem. Res.* 2014, *47*(10), 2999–3008.

172. Hoyle, C. E.; Bowman, C. N., Thiol-ene click chemistry. *Angew. Chem. Int. Ed.* 2010, *49*(9), 1540–1573.

173. Huang, G.; Ou, J.; Wang, H.; Ji, Y.; Wan, H.; Zhang, Z.; Zou, H.; Huang, G.; Wang, H.; Zhang, Z.; Huang, G.; Wang, H.; Peng, X., Synthesis of a stationary phase based on silica modified with branched octadecyl groups by Michael addition and photoinduced thiol-yne click chemistry for the separation of basic compounds. *J. Sep. Sci.* 2016, *39*(8), 1461–1470.

174. Ma, S.; Zhang, H.; Li, Y.; Li, Y.; Zhang, N.; Ou, J.; Ye, M.; Wei, Y., Fast preparation of hybrid monolithic columns via photo-initiated thiol-yne polymerization for capillary liquid chromatography. *J. Chromatogr. A* 2018, *1538*, 8–16.

175. Dao, T. T. H.; Guerrouache, M.; Carbonnier, B., Thiol-yne click adamantane monolithic stationary phase for capillary electrochromatography. *Chinese J. Chem.* 2012, *30*(10), 2281–2284.

176. Liu, Z.; Ou, J.; Lin, H.; Wang, H.; Liu, Z.; Dong, J.; Zou, H., Preparation of monolithic polymer columns with homogeneous structure via photoinitiated thiol-yne click polymerization and their application in separation of small molecules. *Anal. Chem.* 2014, *86*(24), 12334–12340.

177. Kimata, K.; Iwaguchi, K.; Onishi, S.; Jinno, K.; Eksteen, R.; Hosoya, K.; Araki, M.; Tanaka, N., Chromatographic characterization of silica C18 packing materials. Correlation between a preparation method and retention behavior of stationary phase. *J. Chromatogr. Sci.* 1989, *27*(12), 721–728.

178. Shields, E. P.; Weber, S. G., A pH-stable, crosslinked stationary phase based on the thiol-yne reaction. *J. Chromatogr. A* 2019, *1598*, 132–140.

179. Shields, E. P.; Weber, S. G., A liquid chromatographic charge transfer stationary phase based on the thiol-yne reaction. *J. Chromatogr. A* 2019, *1591*, 1–6.

180. Euerby, M. R.; Petersson, P., Chromatographic classification and comparison of commercially available reversed-phase liquid chromatographic columns using principal component analysis. *J. Chromatogr. A* 2003, *994*(1–2), 13–36.

181. Sander, L. C.; Parris, R. M.; Wise, S. A.; Garrigues, P., Shape discrimination in liquid chromatography using charge-transfer phases. *Anal. Chem.* 1991, *63*(22), 2589–2597.

182. Jiang, P.; Lucy, C. A., Retentivity, selectivity and thermodynamic behavior of polycyclic aromatic hydrocarbons on charge-transfer and hypercrosslinked stationary phases under conditions of normal phase high performance liquid chromatography. *J. Chromatogr. A* 2016, *1437*, 176–182.

183. Horak, J.; Maier, N. M.; Lindner, W., Investigations on the chromatographic behavior of hybrid reversed-phase materials containing electron donor–acceptor systems: II. Contribution of π–π aromatic interactions. *J. Chromatogr. A* 2004, *1045*(1), 43–58.

184. Kimata, K.; Hosoya, K.; Araki, T.; Tanaka, N.; Barnhart, E. R.; Alexander, L. R.; Sirimanne, S.; McClure, P. C.; Grainger, J.; Patterson, D. G. Jr., Electron-acceptor and electron-donor chromatographic stationary phases for the reversed-phase liquid chromatographic separation and isomer identification of polychlorinated dibenzo-p-dioxins. *Anal. Chem.* 1993, *65*(18), 2502–2509.

185. Matyska, M. T.; Pesek, J. J.; Grandhi, V., Charge-transfer-like stationary phase for HPLC prepared via hydrosilation on silica hydride. *J. Sep. Sci.* 2002, *25*(12), 741–748.

186. Yu, Q.-W.; Shi, Z.-G.; Lin, B.; Wu, Y.; Feng, Y.-Q., HPLC separation of fullerenes on two charge-transfer stationary phases. *J. Sep. Sci.* 2006, *29*(6), 837–843.

187. Kayillo, S.; Dennis, G. R.; Shalliker, R. A., Retention of polycyclic aromatic hydrocarbons on propyl-phenyl stationary phases in reversed-phase high performance liquid chromatography. *J. Chromatogr. A* 2007, *1148*(2), 168–176.

188. Jaroniec, M.; Martire, D. E., A general model of liquid—solid chroamtography with mixed mobile phases involving concurrent adsorption and partition effects. *J. Chromatogr. A* 1986, *351*, 1–16.

189. Dorsey, J. G.; Cooper, W. T.; Wheeler, J. F.; Barth, H. G.; Foley, J. P., Liquid chromatography: Theory and methodology. *Anal. Chem.* 1994, *66*(12), 500–546.

190. Horváth, C.; Melander, W.; Molnár, I., Solvophobic interactions in liquid chromatography with nonpolar stationary phases. *J. Chromatogr. A* 1976, *125*(1), 129–156.

191. Antle, P. E.; Goldberg, A. P.; Snyder, L. R., Characterization of silica-based reversed-phase columns with respect to retention selectivity: Solvophobic effects. *J. Chromatogr. A* 1985, *321*, 1–32.

192. Sadek, P. C.; Carr, P. W.; Doherty, R. M.; Kamlet, M. J.; Taft, R. W.; Abraham, M. H., Study of retention processes in reversed-phase high-performance liquid chromatography by the use of the solvatochromic comparison method. *Anal. Chem.* 1985, *57*(14), 2971–2978.

193. Gritti, F.; Guiochon, G., Adsorption mechanisms and effect of temperature in reversed-phase liquid chromatography. Meaning of the classical Van't Hoff plot in chromatography. *Anal. Chem.* 2006, *78*(13), 4642–4653.

194. Snyder, L. R.; Dolan, J. W., The linear-solvent-strength model of gradient elution. *Adv. Chromatogr.* 1998, *38*, 115–187.

195. Neue, U. D., Nonlinear retention relationships in reversed-phase chromatography. *Chromatographia* 2006, *63*(Suppl.), S45–S53.

196. Pappa-Louisi, A.; Nikitas, P.; Papachristos, K.; Zisi, C., Modeling the combined effect of temperature and organic modifier content on reversed-phase chromatographic retention: Effectiveness of derived models in isocratic and isothermal mode retention prediction. *J. Chromatogr. A* 2008, *1201*(1), 27–34.

197. Kaliszan, R., QSRR: Quantitative structure-(chromatographic) retention relationships. *Chem. Rev. (Washington, DC, US)* 2007, *107*(7), 3212–3246.

198. Wen, Y.; Talebi, M.; Amos, R. I. J.; Szucs, R.; Dolan, J. W.; Pohl, C. A.; Haddad, P. R., Retention prediction in reversed phase high performance liquid chromatography using quantitative structure-retention relationships applied to the Hydrophobic Subtraction Model. *J. Chromatogr. A* 2018, *1541*, 1–11.

199. Hafkenscheid, T. L.; Tomlinson, E., Estimation of aqueous solubilities of organic non-electrolytes using liquid chromatographic retention data. *J. Chromatogr. A* 1981, *218*, 409–425.

200. Hafkenscheid, T. L.; Tomlinson, E., Relationships between hydrophobic (lipophilic) properties of bases and their retention in reversed-phase liquid chromatography using aqueous methanol mobile phases. *J. Chromatogr. A* 1984, *292*(2), 305–317.

201. Karger, B. L.; Gant, J. R.; Martkopf, A.; Weiner, P. H., Hydrophobic effects in reversed-phase liquid chromatography. *J. Chromatogr. A* 1976, *128*(1), 65–78.

202. Tchapla, A.; Heron, S.; Colin, H.; Guiochon, G., Role of temperature in the behavior of a homologous series in reversed phase liquid chromatography. *Anal. Chem.* 1988, *60*(14), 1443–1448.

203. Carr, P. W.; Tan, L. C.; Park, J. H., Revisionist look at solvophobic driving forces in reversed-phase liquid chromatography. III. Comparison of the behavior of nonpolar and polar solutes. *J. Chromatogr. A* 1996, *724*(1–2), 1–12.

204. Carr, P. W.; Li, J.; Dallas, A. J.; Eikens, D. I.; Tan, L. C., Revisionist look at solvophobic driving forces in reversed-phase liquid chromatography. *J. Chromatogr. A* 1993, *656*(1–2), 113–133.

205. McGuffin, V. L.; Chen, S.-H., Molar enthalpy and molar volume of methylene and benzene homologues in reversed-phase liquid chromatography. *J. Chromatogr. A* 1997, *762*(1–2), 35–46.

206. Andersen, S. I.; Birdi, K. S., Retention and thermodynamics of homologous series in reversed-phase liquid chromatography. *Prog. Colloid Polym. Sci.* 1990, *82*, 52–61 (Surfactants Macromol.: Self-Assem. Interfaces Bulk).

207. Nguyen Viet, C.; Trathnigg, B., Determination of thermodynamic parameters in reversed phase chromatography for polyethylene glycols and their methyl ethers in different mobile phases. *J. Sep. Sci.* 2010, *33*(4–5), 464–474.

208. Lochmuller, C. H.; Moebus, M. A.; Liu, Q.; Jiang, C.; Elomaa, M., Temperature effect on retention and separation of poly(ethylene glycol)s in reversed-phase liquid chromatography. *J. Chromatogr. Sci.* 1996, *34*(2), 69–76.

209. Fredenslund, A.; Jones, R. L.; Prausnitz, J. M., Group-contribution estimation of activity coefficients in nonideal liquid mixtures. *AIChE J.* 1975, *21*(6), 1086–1099.

210. Kang, J. W.; Abildskov, J.; Gani, R.; Cobas, J., Estimation of mixture properties from first- and second-order group contributions with the UNIFAC model. *Ind. Eng. Chem. Res.* 2002, *41*(13), 3260–3273.

211. Petrovic, S. M.; Lomic, S.; Sefer, I., Utilization of the functional group contribution concept in liquid chromatography on chemically bonded reversed phases. *J. Chromatogr.* 1985, *348*(1), 49–65.

212. Dasko, L., Application of the UNIFAC method for assessment of retention in reversed-phase liquid chromatography. *J. Chromatogr.* 1991, *543*(2), 267–275.

213. Park, J. H.; Lee, J. E.; Jang, M. D.; Li, J. J.; Carr, P. W., UNIFAC model as a heuristic guide for estimating retention in chromatography. *J. Chromatogr.* 1991, *586*(1), 1–9.

214. Park, J. H.; Jang, M. D.; Chae, J. J.; Kim, H. C.; Suh, J. K., UNIFAC model as a heuristic guide for estimating retention in reversed-phase liquid chromatography. *J. Chromatogr. A* 1993, *656*(1–2), 69–79.

215. Horner, A. R.; Wilson, R. E.; Groskreutz, S. R.; Murray, B. E.; Weber, S. G., Evaluation of three temperature- and mobile phase-dependent retention models for reversed-phase liquid chromatographic retention and apparent retention enthalpy. *J. Chromatogr. A* 2018.

216. Melander, W. R.; Stoveken, J.; Horvath, C., Mobile phase effects in reversed-phase chromatography. II. Acidic amine phosphate buffers as eluents. *J. Chromatogr.* 1979, *185*(1), 111–127.

217. McCalley, D. V., Effect of temperature and flow-rate on analysis of basic compounds in high-performance liquid chromatography using a reversed-phase column. *J. Chromatogr. A* 2000, *902*(2), 311–321.

218. Goldberg, R. N.; Kishore, N.; Lennen, R. M., Thermodynamic quantities for the ionization reactions of buffers. *J. Phys. Chem. Ref. Data* 2002, *31*(2), 140.

219. Espinosa, S.; Bosch, E.; Roses, M., Retention of ionizable compounds in high-performance liquid chromatography 14. Acid-base pK values in acetonitrile-water mobile phases. *J. Chromatogr. A* 2002, *964*(1–2), 55–66.

220. Bosch, E.; Espinosa, S.; Roses, M., Retention of ionizable compounds on high-performance liquid chromatography. III. Variation of pK values of acids and pH values of buffers in acetonitrile-water mobile phases. *J. Chromatogr. A* 1998, *824*(2), 137–146.

221. Rived, F.; Canals, I.; Bosch, E.; Roses, M., Acidity in methanol-water. *Anal. Chim. Acta* 2001, *439*(2), 315–333.

222. Rived, F.; Roses, M.; Bosch, E., Dissociation constants of neutral and charged acids in methyl alcohol. The acid strength resolution. *Anal. Chim. Acta* 1998, *374*(2–3), 309–324.

223. Subirats, X.; Bosch, E.; Roses, M., Retention of ionisable compounds on high-performance liquid chromatography XVIII: pH variation in mobile phases containing formic acid, piperazine, tris, boric acid or carbonate as buffering systems and acetonitrile as organic modifier. *J. Chromatogr. A* 2009, *1216*(12), 2491–2498.

224. Subirats, X.; Bosch, E.; Roses, M., Retention of ionisable compounds on high-performance liquid chromatography XVII. Estimation of the pH variation of aqueous buffers with the change of the methanol fraction of the mobile phase. *J. Chromatogr. A* 2007, *1138*(1–2), 203–215.

225. Subirats, X.; Bosch, E.; Roses, M., Retention of ionizable compounds on high-performance liquid chromatography. XIX. pH variation in mobile phase containing formic acid, piperazine and tris as buffer systems and methanol as organic modifier. *J. Chromatogr. A* 2009, *1216*(8), 5445–5448.

226. Subirats, X.; Bosch, E.; Roses, M., Retention of ionizable compounds on high-performance liquid chromatography. XV. Estimation of the pH variation of aqueous buffers with the change of the acetonitrile fraction of the mobile phase. *J. Chromatogr. A* 2004, *1059*(1–2), 33–42.

227. Espinosa, S.; Bosch, E.; Roses, M., Retention of ionizable compounds on HPLC. 5. pH scales and the retention of acids and bases with acetonitrile-water mobile phases. *Anal. Chem.* 2000, *72*(21), 5193–5200.

228. Canals, I.; Oumada, F. Z.; Rosés, M.; Bosch, E., Retention of ionizable compounds on HPLC. 6. pH measurements with the glass electrode in methanol–water mixtures. *J. Chromatogr. A* 2001, *911*(2), 191–202.

229. Gagliardi, L. G.; Castells, C. B.; Rafols, C.; Roses, M.; Bosch, E., Delta conversion parameter between pH scales (SWpH and SSpH) in acetonitrile/water mixtures at various compositions and temperatures. 2007, (0003-2700 (Print)).

230. Padró, J. M.; Acquaviva, A.; Tascon, M.; Gagliardi, L. G.; Castells, C. B., Effect of temperature and solvent composition on acid dissociation equilibria, I: Sequenced determination of compounds commonly used as buffers in high performance liquid chromatography coupled to mass spectroscopy detection. *Anal. Chim. Acta* 2012, *725*, 87–94.

231. Kaliszan, R.; Wiczling, P., Gradient reversed-phase high-performance chromatography of ionogenic analytes. *TrAC, Trends Anal. Chem.* 2011, *30*(9), 1372–1381.

232. Subirats, X.; Bosch, E.; Roses, M., Retention of ionisable compounds on high-performance liquid chromatography XVI. Estimation of retention with acetonitrile/water mobile phases from aqueous buffer pH and analyte pKa. *J. Chromatogr. A* 2006, *1121*(2), 170–177.

233. Wiczling, P.; Kaliszan, R., Influence of pH on retention in linear organic modifier gradient RP HPLC. *Anal. Chem.* 2008, *80*(20), 7855–7861.

234. Andres, A.; Tellez, A.; Roses, M.; Bosch, E., Chromatographic models to predict the elution of ionizable analytes by organic modifier gradient in reversed phase liquid chromatography. *J. Chromatogr. A* 2012, *1247*, 71–80.

235. Andres, A.; Roses, M.; Bosch, E., Prediction of the chromatographic retention of acid-base compounds in pH buffered methanol-water mobile phases in gradient mode by a simplified model. *J. Chromatogr. A* 2015, *1385*, 42–48.

236. Andres, A.; Roses, M.; Bosch, E., Gradient retention prediction of acid-base analytes in reversed phase liquid chromatography: A simplified approach for acetonitrile-water mobile phases. *J. Chromatogr. A* 2014, *1370*, 129–134.

237. Gagliardi, L. G.; Castells, C. B.; Rafols, C.; Roses, M.; Bosch, E., Effect of temperature on the chromatographic retention of ionizable compounds. II. Acetonitrile-water mobile phases. *J. Chromatogr. A* 2005, *1077*(2), 159–169.

238. Gagliardi, L. G.; Castells, C. B.; Rafols, C.; Roses, M.; Bosch, E., Modeling retention and selectivity as a function of pH and column temperature in liquid chromatography. *Anal. Chem.* 2006, *78*(16), 5858–5867.

239. Agrafiotou, P.; Rafols, C.; Castells, C.; Bosch, E.; Roses, M., Simultaneous effect of pH, temperature and mobile phase composition in the chromatographic retention of ionizable compounds. *J. Chromatogr. A* 2011, *1218*(30), 4995–5009.

240. Pous-Torres, S.; Torres-Lapasió, J. R.; Baeza-Baeza, J. J.; García-Álvarez-Coque, M. C., Combined effect of solvent content, temperature and pH on the chromatographic behaviour of ionisable compounds: II. Benefits of the simultaneous optimisation. *J. Chromatogr. A* 2008, *1193*(1), 117–128.

241. Pous-Torres, S.; Torres-Lapasio, J. R.; Baeza-Baeza, J. J.; Garcia-Alvarez-Coque, M. C., Combined effect of solvent content, temperature and pH on the chromatographic behavior of ionizable compounds. *J. Chromatogr. A* 2007, *1163*(1–2), 49–62.

242. Hosoya, K.; Kimata, K.; Araki, T.; Tanaka, N.; Frechet, J. M. J., Temperature-controlled high-performance liquid chromatography using a uniformly sized temperature-responsive polymer-based packing material. *Anal. Chem.* 1995, *67*(11), 1907–1911.

243. Kanazawa, H.; Yamamoto, K.; Matsushima, Y.; Takai, N.; Kikuchi, A.; Sakurai, Y.; Okano, T., Temperature-responsive chromatography using poly(N-isopropylacrylamide)-modified silica. *Anal. Chem.* 1996, *68*(1), 100–105.

244. Kanazawa, H.; Kashiwase, Y.; Yamamoto, K.; Matsushima, Y.; Kikuchi, A.; Sakurai, Y.; Okano, T., Temperature-responsive liquid chromatography. 2. Effects of hydrophobic groups in N-Isopropylacrylamide copolymer-modified silica. *Anal. Chem.* 1997, *69*(5), 823–830.

245. Go, H.; Sudo, Y.; Hosoya, K.; Ikegami, T.; Tanaka, N., Effects of mobile-phase composition and temperature on the selectivity of poly(N-isopropylacrylamide)-bonded silica gel in reversed-phase liquid chromatography. *Anal. Chem.* 1998, *70*(19), 4086–4093.

246. Song, Y. X.; Wang, J. Q.; Su, Z. X.; Chen, D. Y., High-performance liquid chromatography on silica modified with temperature-responsive polymers. *Chromatographia* 2001, *54*(3/4), 208–212.

247. Ayano, E.; Okada, Y.; Sakamoto, C.; Kanazawa, H.; Okano, T.; Ando, M.; Nishimura, T., Analysis of herbicides in water using temperature-responsive chromatography and an aqueous mobile phase. *J. Chromatogr. A* 2005, *1069*(2), 281–285.

248. Roohi, F.; Fatoglu, Y.; Titirici, M.-M., Thermo-responsive columns for HPLC: The effect of chromatographic support and polymer molecular weight on the performance of the columns. *Anal. Methods* 2009, *1*(1), 52–58.

249. Ayano, E.; Suzuki, Y.; Nishio, T.; Nagata, Y.; Kanazawa, H.; Nagase, K.; Okano, T., Liquid chromatography-mass spectrometric analysis of dehydroepiandrosterone and related steroids utilizing a temperature-responsive stationary phase. *Chromatography* 2014, *35*(3), 131–138.

250. Poplewska, I.; Muca, R.; Strachota, A.; Piatkowski, W.; Antos, D., Adsorption behavior of proteins on temperature-responsive resins. *J. Chromatogr. A* 2014, *1324*, 181–189.

251. Kurata, K.; Shimoyama, T.; Dobashi, A., Enantiomeric separation using temperature-responsive chiral polymers composed of L-valine diamide derivatives in aqueous liquid chromatography. *J. Chromatogr. A* 2003, *1012*(1), 47–56.

252. Kanazawa, H.; Ayano, E.; Sakamoto, C.; Yoda, R.; Kikuchi, A.; Okano, T., Temperature-responsive stationary phase utilizing a polymer of proline derivative for hydrophobic interaction chromatography using an aqueous mobile phase. *J. Chromatogr. A* 2006, *1106*(1–2), 152–158.

253. Yagi, H.; Yamamoto, K.; Aoyagi, T., New liquid chromatography method combining thermo-responsive material and inductive heating via alternating magnetic field. *J. Chromatogr. B Anal. Technol. Biomed. Life Sci.* 2008, *876*(1), 97–102.

254. Miserez, B.; Lynen, F.; Wright, A.; Euerby, M.; Sandra, P., Thermoresponsive poly(N-vinylcaprolactam) as stationary phase for aqueous and green liquid chromatography. *Chromatographia* 2010, *71*(1–2), 1–6.

255. Techawanitchai, P.; Aoyagi, T.; Yamamoto, K.; Ebara, M., Surface design with self-heating smart polymers for on-off switchable traps. *Sci. Technol. Adv. Mater.* 2011, *12*(4), 044609.

256. Nishio, T.; Kanazashi, R.; Nojima, A.; Kanazawa, H.; Okano, T., Effect of polymer containing a naphthyl-alanine derivative on the separation selectivity for aromatic compounds in temperature-responsive chromatography. *J. Chromatogr. A* 2012, *1228*, 148–154.

257. Satti, A. J.; Espeel, P.; Martens, S.; Van Hoeylandt, T.; Du Prez, F. E.; Lynen, F., Tunable temperature responsive liquid chromatography through thiolactone-based immobilization of poly(N-isopropylacrylamide). *J. Chromatogr. A* 2015, *1426*, 126–132.

258. Cao, P.; Muller, T. K. H.; Ketterer, B.; Ewert, S.; Theodosiou, E.; Thomas, O. R. T.; Franzreb, M., Integrated system for temperature-controlled fast protein liquid chromatography. II. Optimized adsorbents and 'single column continuous operation'. *J. Chromatogr. A* 2015, *1403*, 118–131.

259. Pesek, J. J.; Vidensek, M. A.; Miller, M., Synthesis of chemically bonded liquid crystals for high-performance liquid chromatography. New phases via the organochlorosilane pathway. *J. Chromatogr.* 1991, *556*(1–2), 373–381.

260. Wise, S. A.; Sander, L. C., Factors affecting the reversed-phase liquid chromatographic separation of polycyclic aromatic hydrocarbon isomers. *J. High Resolut. Chromatogr.* 1985, *8*(5), 248–255.

261. Sander, L. C.; Wise, S. A., Subambient temperature modification of selectivity in reversed-phase liquid chromatography. *Anal. Chem.* 1989, *61*(15), 1749–1754.

262. Bell, C. M.; Sander, L. C.; Wise, S. A., Temperature dependence of carotenoids on C18, C30 and C34 bonded stationary phases. *J. Chromatogr. A* 1997, *757*(1–2), 29–39.

263. Sander, L. C.; Pursch, M.; Wise, S. A., Shape selectivity for constrained solutes in reversed-phase liquid chromatography. *Anal. Chem.* 1999, *71*(21), 4821–4830.

264. Sander, L. C.; Wise, S. A., The influence of column temperature on selectivity in reversed-phase liquid chromatography for shape-constrained solutes. *J. Sep. Sci.* 2001, *24*(12), 910–920.

265. Lippa, K. A.; Sander, L. C.; Wise, S. A., Chemometric studies of polycyclic aromatic hydrocarbon shape selectivity in reversed-phase liquid chromatography. *Anal. Bioanal. Chem.* 2004, *378*(2), 365–377.

266. Lippa, K. A.; Sander, L. C.; Mountain, R. D., Molecular dynamics simulations of alkyl-silane stationary-phase order and disorder. 2. Effects of temperature and chain length. *Anal. Chem.* 2005, *77*(24), 7862–7871.

267. Meyer, C.; Pascui, O.; Reichert, D.; Sander, L. C.; Wise, S. A.; Albert, K., Conformational temperature dependence of a poly(ethylene-co-acrylic acid) stationary phase investigated by nuclear magnetic resonance spectroscopy and liquid chromatography. *J. Sep. Sci.* 2006, *29*(6), 820–828.

268. Srinivasan, G.; Sander, L. C.; Muller, K., Effect of surface coverage on the conformation and mobility of C18-modified silica gels. *Anal. Bioanal. Chem.* 2006, *384*(2), 514–524.

269. Endele, R.; Halasz, I. n.; Unger, K., Influence of the particle size (5–35 μm) of spherical silica on column efficiencies in high-pressure liquid chromatography. *J. Chromatogr. A* 1974, *99*, 377–393.

270. Poppe, H.; Kraak, J. C.; Huber, J. F. K.; van den Berg, J. H. M., Temperature gradients in HPLC columns due to viscous heat dissipation. *Chromatographia* 1981, *14*(9), 515–523.

271. Martin, M.; Eon, C.; Guiochon, G., Study of the pertinency of pressure in liquid chromatography: I. Theoretical analysis. *J. Chromatogr. A* 1974, *99*, 357–376.

272. Lin, H.-J.; Horváth, S., Viscous dissipation in packed beds. *Chem. Eng. Sci.* 1981, *36*(1), 47–55.

273. Poppe, H.; Kraak, J. C., Influence of thermal conditions on the efficiency of high-performance liquid chromatographic columns. *J. Chromatogr. A* 1983, *282*, 399–412.

274. Dapremont, O.; Cox, G. B.; Martin, M.; Hilaireau, P.; Colin, H., Effect of radial gradient of temperature on the performance of large-diameter high-performance liquid chromatography columns. I. Analytical conditions. *J. Chromatogr. A* 1998, *796*(1), 81–99.

275. Kaczmarski, K.; Gritti, F.; Guiochon, G., Prediction of the influence of the heat generated by viscous friction on the efficiency of chromatography columns. *J. Chromatogr. A* 2008, *1177*(1), 92–104.

276. Kaczmarski, K.; Gritti, F.; Kostka, J.; Guiochon, G., Modeling of thermal processes in high pressure liquid chromatography. II. Thermal heterogeneity at very high pressures. *J. Chromatogr. A* 2009, *1216*(38), 6575–6586.

277. Kaczmarski, K.; Kostka, J.; Zapala, W.; Guiochon, G., Modeling of thermal processes in high pressure liquid chromatography. I. Low pressure onset of thermal heterogeneity. *J. Chromatogr. A* 2009, *1216*(38), 6560–6574.

278. Mayr, G.; Welsch, T., Influence of viscous heat dissipation on efficiency in high-speed high-performance liquid chromatography. *J. Chromatogr. A* 1999, *845*(1–2), 155–163.

279. Welsch, T.; Schmid, M.; Kutter, J.; Kalman, A., Temperature of the eluent: A neglected tool in high-performance liquid chromatography? *J. Chromatogr. A* 1996, *728*(1–2), 299–306.
280. Desmet, G., Theoretical calculation of the retention enthalpy effect on the viscous heat dissipation band broadening in high performance liquid chromatography columns with a fixed wall temperature. *J. Chromatogr. A* 2006, *1116*(1–2), 89–96.
281. Broeckhoven, K.; Billen, J.; Verstraeten, M.; Choikhet, K.; Dittmann, M.; Rozing, G.; Desmet, G., Towards a solution for viscous heating in ultra-high pressure liquid chromatography using intermediate cooling. *J. Chromatogr. A* 2010, *1217*(13), 2022–2031.
282. Gritti, F.; Gilar, M.; Jarrell, J. A., Quasi-adiabatic vacuum-based column housing for very high-pressure liquid chromatography. *J. Chromatogr. A* 2016, *1456*, 226–234.
283. Gritti, F.; Gilar, M.; Jarrell, J. A., Achieving quasi-adiabatic thermal environment to maximize resolution power in very high-pressure liquid chromatography: Theory, models, and experiments. *J. Chromatogr. A* 2016, *1444*, 86–98.
284. Broeckhoven, K.; Desmet, G., Considerations for the use of ultra-high pressures in liquid chromatography for 2.1 mm inner diameter columns. *J. Chromatogr. A* 2017, *1523*, 183–192.
285. Horvath, K.; Horvath, S.; Lukacs, D., Effect of axial temperature gradient on chromatographic efficiency under adiabatic conditions. *J. Chromatogr. A* 2017, *1483*, 80–85.

3 Kosmotropic Chromatography of Proteins
Theory and Practice

Carlos Calleja-Amador, J. F. Ogilvie,
and Rigoberto Blanco

CONTENTS

3.1   INTRODUCTION

During the late 1980s and early 1990s, there was a special interest in studying what was thought to be the ultimate effort to complete a theoretical and practical understanding of the chromatographic separation of proteins. As a result of these studies, fundamental achievements in the theory and practice of hydrophobic interaction, ion-pairing and reversed-phase chromatography were made. This work was conducted by various scientists such as Lloyd Snyder, Csaba Horváth, Georges Guiochon, Dan Martire, Fred Regnier and Barry Karger, among others. One of the most important subjects was an understanding of the practical consequences of the salting out effect.

Salting out is caused by different salts that decrease protein solubility in water. Salts that increase protein solubility are said to promote salting in. Salting out is a consequence of changes in the structure of proteins in an aqueous solution. It is useful in the chromatographic analysis of proteins using aqueous mobile phases and hydrophobic stationary phases [1].

Water has a unique, low entropy structure, governed by hydrogen bonding. It is known that ions are hydrated in aqueous solutions. Ion hydration causes changes in the structure of the aqueous environment. For example, the presence of ions is related to protein unfolding and protein separations through salting in and salting out processes. Hydrated ions have strong interactions with water molecules, and can change the structure of water. When hydrated ions increase the structuring of water, they promote protein salting out and are called "order makers" or "kosmotropes." Other ions decrease the structure of water and promote protein salting in. These are called "disorder makers" or "chaotropes." Both terms, "kosmotropes" and "chaotropes," originated from the Hofmeister series, which orders ions in terms of their ability to stabilize or destabilize proteins [2].

Kosmotropes are usually small ions with a high charge density. Chaotropes are large ions with a lower charge density. A general rule that helps to distinguish one type from the other is that kosmotropic ions show radii below 106 pm for monovalent cations, and below 178 pm for singly charged anions [2].

This review outlines the main theoretical aspects related to the effect of kosmotropic salts on chromatographic separations. It also describes the relation of the salting out effect to the Hofmeister series, and some practical high performance liquid chromatography (HPLC) applications. This review is intended not to be exhaustive but to provide a description of various chromatographic behaviors observed in the hydrophobic interaction chromatography (HIC) of proteins; and the most common analytical strategies for method development are discussed. Exhaustive reviews on specific subjects such as electrostatic interactions, Hofmeister effects on biological systems and the behavior of non-aqueous systems are available elsewhere [1, 2, 3]. Many publications on chromatography are related to kosmotropic effects that are applied mainly in the pharmaceutical industry [4], but they are omitted here with the explicit intention of addressing kosmotropicity as an important phenomenon in protein separations.

The objectives of this work are to review some theoretical models proposed to explain the kosmotropic effect, and to describe strategies for method development in order to improve kosmotropic chromatography. The amount of literature published on this subject is increasing. To date, most models proposed are far from being unifying rationales that explain the results and observations.

3.2 KOSMOTROPICITY IN CHROMATOGRAPHY

"Kosmotropicity" is a term to describe the effect of an aqueous solute capable of acting as an agent that increases the structure of water. The opposite effect is termed "chaotropicity." The most common kosmotropic agents are ammonium sulfate, potassium phosphate and sodium sulfate. The first attempts to explain kosmotropicity treated the electrostatic interactions between an ionic solute and its hydration sphere, mainly the anion. These interactions were supposed to favor an increase in water structure. More structured water consequently expelled a non-electrolytic analyte from the increasingly organized eluent, thereby decreasing its solubility [5]. However, recent evidence indicates that this explanation might be incorrect.

Kosmotropic chromatography, originally referred to as "salting out chromatography," is based on a decreased solubility of an analyte in the mobile phase caused by the presence of salts of specific types. The altered analyte solubility also causes an increased interaction with the stationary phase by a mechanism that is not fully understood, as is discussed in further sections. The salting out effect has hence been applied in order to modify the retention factor in separations of small and large molecules. Some separations use a single concentration of the salt, whereas others implement a salt gradient [5]. Its practical application resides in an empirical relation between the retention factor and the salt concentration. Salts commonly used as kosmotropic agents have been observed to follow a trend known as the Hofmeister series [6].

The Hofmeister series is related to kosmotropicity and chaotropicity. Štrop [6] and Loeser [7] observed that different salts showed the same tendency as that observed in the Hofmeister series in HIC and in reversed-phase liquid chromatography (RPLC), respectively. In general, chaotropic salts act by increasing the solubility of proteins (salting in), and kosmotropic salts act by decreasing it (salting out) [6, 7].

Several authors have tried to explain kosmotropicity in terms of electrostatic interactions [8, 9, 10], thermodynamic considerations [11], quantum-mechanical explanations [12, 13] and the simulation through molecular dynamics [14, 15, 16]. As evidence subsequently showed, structure-making in aqueous solutions by kosmotropes is not a consequence of the interactions with water molecules. Several results indicate that the main interaction occurs between a salt and an analyte, not because of a direct relationship with the hydration sphere of the salt. Anions have been observed to interact with proteins more strongly than cations because they are more polarizing [15]. This result opens a new way of describing the mechanism of interaction of kosmotropes with proteins.

3.3 KOSMOTROPIC SALTS IN HYDROPHOBIC LIQUID CHROMATOGRAPHY

This section addresses three types of chromatographic modes used in protein separations with kosmotropic additives: conventional reversed-phase liquid chromatography (RPLC), hydrophobic interaction chromatography (HIC) and ion-pairing chromatography (IPC). The use of kosmotropic salts is briefly presented in agreement with the nature of each technique.

A liquid-chromatographic system has a complication arising from a combination of several parameters. In RPLC these parameters include: the sample containing the analytes that must be separated, an aqueous solution to which non-ionic organic compounds (that can be polar or non-polar) might be added, the various salts that are used for the buffer eluent, the hydrophobic or hydrophilic nature of the adsorbate, and a variety of their concentrations. This complication has made it extremely difficult to propose a theoretical framework able to correlate all the experimental observations, and to make predictions. One such observation yet to be explained is the use of salts to modify the retention of the analyte upon a change in the analyte solubility.

The solubility of a non-polar analyte in an aqueous solution of an electrolyte might change from the salting effect. If solubility decreases with increasing concentration of the electrolyte in the mobile phase, salting out occurs. If solubility increases, then the non-polar analyte is said to be salted in [5, 17]. For the purpose of the definition, electrolytes and non-polar analytes are compounds that have large and small solubilities in water, respectively. The influence of an electrolyte in an aqueous solution of a non-polar analyte is expressed mathematically according to the Setschenow equation [5, 17]:

$$\log \frac{s_0}{s} = \log f_c = k * c_s \tag{3.1}$$

in which s_0 and s are the solubilities of the non-polar analyte in water and the electrolyte in the solution, respectively; c_s is the concentration of the electrolyte in moles per liter; f_c is the activity coefficient of the non-polar analyte; and k is called the "salting parameter." A positive value of k indicates salting out whereas a negative value indicates salting in. Because s_0 is constant in an aqueous solution, it follows from Equation (3.1) that the amount of electrolyte added is directly proportional to the amount of non-polar analyte precipitated and vice versa.

The effect predicted by the Setschenow equation is particularly evident when kosmotropic salts are used with the specific purpose of salting out proteins in HIC. This HPLC mode uses hydrophobicity as a property to purify and to separate macro biomolecules. It comprises chromatographic techniques that have in common the addition of salts to the mobile phase to help modulate or to modify hydrophobicity, and hence the retention of analytes on the hydrophobic sites of the stationary phase. The targets might be small molecules, non-ionic or ionic small molecules or biomolecules (proteins and DNA) that differ in their hydrophobicity. These techniques originated in the 1950s. The mechanism of the loss of solubility has since been related to the effects of ions, and was the first to be described as "salting out chromatography." Biomolecules are the target analytes to be salted out from the mobile phase in order to modulate their retention on the stationary phase as a function of salt concentration [18]. Figure 3.1 shows a classical separation of four biomolecules using HIC. This separation uses ammonium sulfate as a salting out additive and ammonium acetate as a buffer. The use of ammonium sulfate along with the acetate buffer can increase the retention factors, improving resolution and the time for the chromatographic test [18]. Figure 3.1 shows the separation of a mixture of biomolecules using 3 M ammonium sulfate [18].

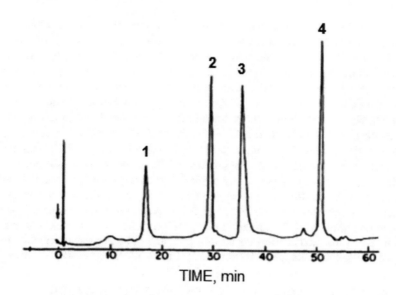

FIGURE 3.1 Hydrophobic interaction chromatogram of a protein mixture in linear gradient from 3 M ammonium sulfate + 0.5 M ammonium acetate, pH 6.0 to 0.5 M ammonium acetate, pH 6.0. Peak **1** corresponds to cytochrome-c, **2** ribunuclease, **3** lysozyme and **4** (α)-chymotrysinogen A. Reproduced from [18] with permission of the publisher.

Figure 3.1 is an example of the role that salting ions play in HIC separations. Among possible patterns, ions exhibit attractive forces with water molecules and with the organic modifiers added to the mobile phase. They also show adsorption on the stationary phase, but their effect is more strongly related to concentration. This effect arises from interactions of two types – non-specific [9, 19] and specific [4]. At small concentrations, non-specific interactions between the solute and the salting out ions are related to the charge of the ions. At any concentration, the ions produce double layer shielding [20, 21]. Other effects arise from electrostatic interactions between ions and solute molecules, or between the solute molecules and the stationary phases. These effects are independent of the type of salt, but dependent on concentration.

Specific effects of non-ionized species, such as hydrophobic interactions, appear at concentrations greater than 100 mM [21] and have been related to ions in the Hofmeister series through their increasing capacity to salt out proteins from aqueous solutions [15]. Other effects are the polarization induced by anions over adjacent water molecules, for instance, interference with the hydrophobic hydration of macromolecules and direct binding to the macromolecule. At concentrations below 100 mM in the presence of salts, hydrophobicity is of concern, because at that concentration IPC takes place.

IPC is a reversed-phase alternative that uses the addition of an ion-pairing reagent (IPR) to the mobile phase. The IPR contributes with counterions to form neutral ion pairs with the analyte ions. Ion pairs having a neutral electric charge are attracted to the stationary phase, retained and separated [22]. The formation of ion pairs was proposed by Manning in 1969 as counterion interaction based on an infinitely long charged row. For polyelectrolyte solutions, ion pairs form at a limit of zero concentration [23]. Manning's conclusions support empirical observations in IPC about the retention behavior in a reversed-phase mode. In an IPC system at a given ionic strength, the addition of an IPR favors counterion interaction. As a consequence, the inverse of the product of the charges of the polyion and its counterion remain approximately constant. This causes deviations in the order of the salting out of various salts observed in RPLC and HIC, as confirmed by Florez and Kazakevich [24].

These authors studied the separation of ionic analytes using RPLC with kosmotropic and chaotropic additives. They observed that, under the conditions of their experiments, the salting out effect of a positively charged analyte had no relation to the effect of the additives (chaotropic or kosmotropic) on the separation of ionic analytes. This is the opposite behavior of that expected in protein separations with the same technique [24]. The relation of the salting effect to the Hofmeister series is discussed in the next section for RPLC, HIC and IPC. For the lattermost, recent evidence shows that salts fail to follow the Hofmeister trend in an IPC regime.

3.4 KOSMOTROPIC SALTS AND THE HOFMEISTER SERIES IN THREE CHROMATOGRAPHIC REGIMES

The effect of kosmotropic salts on the salting out of proteins from the mobile phase in RPLC follows the Hofmeister series. This behavior is more evident under HIC conditions and not evident in IPC. An important generalization about the behavior of kosmotropic salts is that large concentrations are necessary for their retention effects to be manifested in the stationary phase. These effects are generally in compliance with the salting out or salting in potential of ions. The trend was first described by Hofmeister in 1888 and the series contains both chaotropic and kosmotropic ions [5]. Here the concept of large concentrations of kosmotropic ions is negligible. It has been pointed out that salting out effects arising from Hofmeister behavior are observed in biological systems at concentrations of about 100 mM under physiological conditions. This opposes chromatographic systems using hydrophobic interaction, where concentrations can, however, be as great as 3 M and even 6 M [18].

A classification between chaotropic (anions) and kosmotropic (cations) ions arose initially according to an erroneous belief that anions have no hydration layer. Such an absence of a hydration layer allowed anions to affect the structure of the water more directly, for example by interrupting the network of hydrogen bonds. Anions could also cause protein unfolding and promote salting in effects. In contrast, cations that have a well-defined hydration layer would help reinforce the structure of the water and strengthen and stabilize the salting out effects of proteins and other macromolecules [15]. Although these effects occur commonly in multiple processes, the mechanisms that explain this behavior are poorly understood [1].

It is accepted that ions exert two opposing effects on proteins and charged macromolecules. One effect is the electrostatic interaction between the charges of proteins (or macromolecules) and the positive or negative ion. These interactions are non-specific in nature and favor a distribution of counterions around or near the surface of proteins (or macromolecules). The second effect is the solvation of ions located outside the double electric layer, which is a specific interaction. This solvation favors ion hydration in the bulk solvent. Under these solvation conditions, movement of ions from the bulk solvent to the surface of the stationary phase is unfavorable. Even though this charge distribution decreases the electrostatic interaction of the protein with the stationary phase, retention can be modulated by controlling salt type and concentration [1, 25].

During chromatography a protein can exist either in the mobile phase or be adsorbed in the stationary phase (folded in the mobile phase or unfolded in the stationary phase). Differences arise from the surface which the proteins are interacting with, and from the presence of salts that can alter the equilibrium in any direction (non-specific effects in the mobile phase and specific effects in the stationary phase). The way in which the salts displace the solubility equilibrium to favor protein salting out is not clear. The interaction involved in the solubility (and in the salting out) of proteins was studied by Kirkwood [1]. This author proposed that a salt ion has a repulsive interaction with a charge inside a low relative permittivity cavity in the protein. At low salt concentrations, solvation interactions dominate (known as Debye–Hückel interactions). The salts displace the equilibria toward the dissolution phase, causing salting in due to an attractive interaction with the charge inside the cavity. At large concentrations, the hydration effect of the ions drives the equilibria toward repulsive interaction, causing the protein to be displaced from water to the solid stationary phase. That is, decreasing the solubility and causing salting out. Kirkwood's findings were studied more rigorously in terms of the work required for the attractive and repulsive interactions [25]. The chemical potential of the protein depends on two kinds of work. One is the amount of work necessary to charge the salt ions around the protein's low dielectric cavity. The other is the work necessary to charge the protein in the presence of the electrically charged salt ions [1]. The latter is the salting in work. The former is the salting out work, also called the "Kirkwood term" [25]. The results allowed the relation of the Kirkwood term (W^{κ}) in Equation (3.2) to the work of charging the ions around the cavity of small relative permittivity of the protein [25]. This term, plus the term corresponding to the Debye–Hückel interactions (ΔG_{DH}), yielded the following expression for the solubility of a protein [25]:

$$-k_B T \ln\left(\frac{S}{S_0}\right) = \Delta G_{DH} + W^k \tag{3.2}$$

In Equation (3.2) k_B is the Boltzmann constant, T is the temperature in K, S is the solubility of the protein in g/L in the presence of the salt, S_0 is the solubility of the protein in g/L in the absence of the salt, ΔG_{DH} is Gibbs energy of the interactions of Debye–Hückel type and W^k is the Kirkwood term. Equation (3.2) describes a diphasic behavior with increasing salt concentration. It shows that, initially, increasing the concentration of a salt, measured according to the ionic strength, I, increases the solubility of a protein (salting in) favored by the term ΔG_{DH}. Further increasing the salt concentration results in the solubility reaching a maximum and beginning to decrease, as the term W^k becomes dominant. Most salts at small concentration ($I < \sim 1$ M) have a salting in effect on proteins, consistent with the term ΔG_{DH}, and show a salting out behavior as concentration increases [1, 25]. An example of this behavior is shown in the next section in Figure 3.3.

However, Kirkwood's treatment does not explain observations about the behavior of electrolytes with themselves. In order to account for the behavior of electrolytes, electrostatic interactions must be considered: for instance, the formation of cation–anion pairs, and the interaction of the ion pair with the solvent. Although two 1:1 electrolytes are expected to behave similarly, it is known that the pH or electrolyte activity depends on the specific type of cation–anion pair and its interactions [26]. An example of such interactions are those of the cation–anion pair with solvent molecules and also with the adsorbent. Local interactions between an ion and a solute (or part of it) are also possible. Another possible interaction is with an interface, which means that an ion might be specifically adsorbed. All these possible interactions change the surface tension of the solvent. In consequence, surface tension changes with the surface concentration of the ion pairs. This has been confirmed by X-ray diffraction [26]. At the interface between the liquid and vapor of a saturated solution, anion and cation concentrations are different from those in the bulk solvent [27]. It was observed that there is an excess of anions with respect to cations. X-ray experiments with potassium halides showed that the ratio of iodide to potassium was greater than the ratio of bromide to potassium, revealing specific effects. This trend notably follows the observed effects of the Hofmeister series [27].

Electrostatic theories fail to explain several observations. For example, how the salting effects alter when the electrolyte is altered, or how a Born radius varies with temperature (a small Born radius is related to weak electrostatic interactions in bulk water) [26]. The paradigm used in electrostatic theories fails to take into account the existence of dipole–dipole interactions, or the short-range interactions of van der Waals forces. In order to take these interactions into account, the following approach has been proposed [26].

The solvent is considered to be a continuum that is characterized with a relative permittivity and lacks its own structure in the calculations. Short range forces in the solvent can be attractive or repulsive and can affect the solution. Calculations in a solution phase require a model of many bodies to describe all possible interactions. Among the most relevant interactions are those of Debye (induction), Keesom (orientation) and London (dispersion). Debye and Keesom interactions arise from interactions between permanent dipoles, ions and induced dipoles. London forces are quantum mechanical in nature, and relate to polarizability and ionization energy of the ions, reflecting their specific nature. In summary, the problem is manifold because of the presence of the structure of the solvent [26].

When the solvent is water, such effects follow the Hofmeister series [26]. Other kinds of kosmotropic salts that follow the Hofmeister series include ionic liquids [28–35]. In the following sections, we try to present several theories which attempt to explain mechanistically the interaction of ions in the Hofmeister series with different analytes. This will shed some light on the HIC of macromolecules.

3.4.1 Thermodynamic Aspects of the Hofmeister Series

Kosmotropic ions are supposed to induce local water structuring through hydration. This structuring depends on the size of the ions and on electrostatic and quantum-mechanical interactions. The short-range interactions involving the ionic species depend on van der Waals forces between the hydrated ions and the surfaces (due to the superposition of the hydration layers). The long-range electrostatic interactions depend on the nature of the solvent and its interaction with the molecules that might be present, such as proteins, other biomolecules and impurities (dissolved gases) [26, 27].

The efficiency of the most common ions as promoters of salting out (kosmotropicity) follows these decreasing orders of the Hofmeister series [26, 27]:

For anions,

$$OH^- > SO_4^{-2}, CO_3^{-2} > ClO_4^- > BrO_3^- > Cl^- > H_3CCOO^- > IO_3^-, IO_4^- > Br^-, I^- > SCN^- > NO_3^-$$

For cations,

$$Na^+ > K^+ > Li^+ > Ba^2 > Rb^+ > Ca^2 > Ni^2 > Co^2 > Mg^2 > Fe^2 > Zn^2 >$$

$$Cs^+ > Mn^2 > Al^3 > Fe^3, Cr^3 > NH_4^+ > H^+$$

This order is not rigorous but is subject to variation due to the nature of the system and the type of effect studied. A pictorial view of the Hofmeister series for anions and cations, adapted from reference [36], appears in Figure 3.2.

Figure 3.2 shows the anions and cations in decreasing order of kosmotropic effect from top to bottom. The effect has been studied by several authors [17] and has special interest in the effect on protein stability under specific conditions [26]. As an example, Figure 3.3 shows the effect of salt concentration on thermal denaturation of poly(N-isopropylacrylamide), PNIPAM. The effect is shown on cooling PNIPAM in the presence of different salts [37].

The maximum chaotropic effect is shown by SCN^-, followed by I^-, Br^- and NO_3^-. The kosmotropic ions F^-, $H_2PO_4^-$, $S_2O_3^{2-}$, SO_4^{2-} and CO_3^{2-} show an approximately linear dependence of protein stability versus salt concentration. In the case of chaotropes, the stabilization of protein structure shows an increase in temperature at small salt concentrations. The open circles correspond to biphasic behavior points on a temperature gradient [37]. The trends observed are in agreement with Equation (3.2).

The terms "kosmotropes" and "chaotrope" conform to a supposition that chaotropic ions break the structure of water formed through hydrogen bonds. Chaotropes also destabilize the folded proteins promoting salting in. On the other hand, kosmotropic ions are strongly hydrated. These ions stabilize the folded proteins and macromolecules and produce salting out. This presumption is not substantiated, as experiments have shown no relation between ions and water structure whatsoever, as is discussed below.

As has been shown, ions do not affect the properties of bulk water. Instead, their salting effect on the solubilization of proteins seems to be due to specific ion–protein interactions, even at small salt concentrations [38]. For example, the preferential accumulation of anions around a positively charged protein follows the order $SO_4^{2-} > SCN^- > I^- > Cl^-$. Except for SO_4^{2-}, the observed order is almost

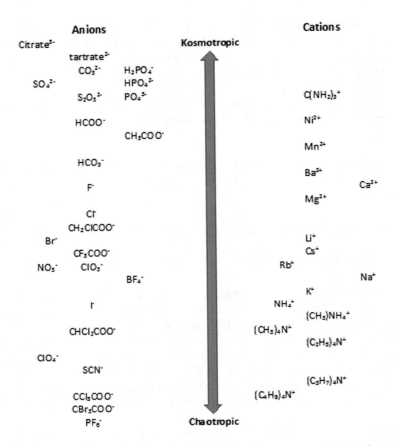

FIGURE 3.2 Hofmeister series for anions and cations in aqueous solution. Adapted from the scheme by Mazzini et al. [36].

the inverse of the Hofmeister series [38]. These selective effects are observed at concentrations of 50 mN and are manifested as dependent on the effective charge of the anion–protein interaction, but not on the cation–protein interaction [38]. This result demonstrates a progressively greater binding of the monovalent anion to the protein. The SO_4^{2-} ion, despite being strongly hydrated, interacts directly with the surface of a protein [38]. Polyvalent cations can bind strongly to acidic residues of the protein and reverse the net charge of the protein [39]. Figure 3.4 was adapted from reference [14]. It shows a possible mechanism according to which interactions of this kind arise between ions and a protein, with no direct interaction affecting the structure of water [14].

Figure 3.4 [14] indicates the possible ways in which cations and anions interact with water and protein dipoles. For example, kosmotropic anions polarize water molecules that are hydrogen-bonded to proteins on the positive end of the dipole. Zhang and Cremer obtained a similar result for the interaction of PNIPAM in Figure 3.3 in the presence of kosmotropic anions [15]. Anions polarize the

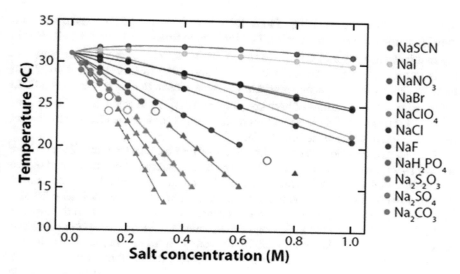

FIGURE 3.3 Effect of different sodium salts on the temperature of thermal denaturation on cooling of the macromolecule poly(N-isopropylacrylamide). At concentration 0.1 M, the specific effects of anions are manifested. For the chaotropic anions I⁻ and SCN⁻, the characteristic stabilizing effect is manifested. The five kosmotropic ions show a linear dependence on the salt concentration. Taken from [37] with permission of the publisher.

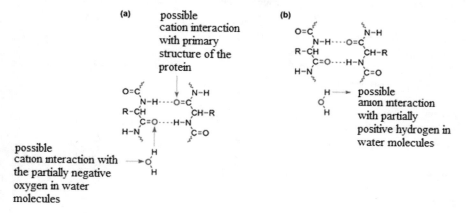

FIGURE 3.4 Possible mechanism of ion effects over the primary structure of a protein adapted from Xie and Gao [14]. (a) The cation can "solvate" carbonyl directly or indirectly by increasing the availability of water hydrogen, and (b) the anion competes with the carbonyl for hydration.

water molecules hydrogen-bonded to the amide moieties. Conversely, Xie and Gao included also the cation interaction with the oxygen atoms in water [14]. Experimental results support the possibility that ions do not affect the structural properties of the bulk water [14].

The time it takes for water dipoles to change their orientation in ionic solutions has been measured by spectroscopic techniques [3]. Measurements showed that, outside the first solvation shells of ions, there is no influence of the ions on the rotational dynamics of water molecules. The correlation times for the water molecules in the first hydration layer of Cl^-, Br^-, I^- and ClO_4^- were smaller than for molecules in the bulk (in increasing order of ionic radii). These experiments clearly showed that these anions have no influence on the water dynamics in the bulk. Such a condition holds even at concentrations of both kosmotropic (SO_4^{2-}) and chaotropic (ClO_4^-) ions, up to 6 M. The conclusion was that ions do not cause a long-range effect of formation or destruction of water structure [3, 40].

The thermodynamic aspects of the effects of ions on the structure of water were studied by Pielak et al. [41]. The bulk water was considered to be constituted of two species that exchange rapidly from one another – one less dense and more structured, the other denser and less structured. A *structure-making* solute, such as a kosmotropic salt, increases the fraction of the less dense species. This occurs at the expense of the denser species in the hydration of the solute. A *structure-breaking* solute has the opposite effect. Pielak et al. related this change in density to the change in partial molar thermal capacity with pressure at constant temperature. Table 3.1 shows the effect of different solutes on the structure of water, measured in calorimetric

TABLE 3.1
Solute Effects on Protein Stability and $(\partial C_p/\partial P)_T$ at 25°C

Compound	Effect on stability[a]	$(\partial C_p/\partial P)_T$ 10^{-6} at 25°C/J Pa^{-1} K^{-1}[b]
$(NH_4)_2SO_4$	+ +	4.58
NH_4Cl	+	3.67
guanidinium Cl	– –	2.62
guanidinium SCN	– – –	2.64
N-methylglycine (sarcosine)	+ +	1.96
urea	–	2.02
glucose	+	1.08
N-trimethylglycine (betaine)	+ +	0.966
trehalose	+	0.669
sucrose	+	0.710
glycerol	+,–	0.694
stachyose	+	0.537
melezitose	+	0.182
1,3-dimethylurea	–	−0.131
trimethylamine	+ + +	−0.859
N-oxide dihydrate		
1,3-diethylurea	–	−0.595
2-propanol	[c]	−0.268

[a] (+) indicates stabilizing, (–) indicates destabilizing. The number of symbols is related to the magnitude of the effect.
[b] Uncertainty is ±0.02 10^{-6} J Pa^{-1} K^{-1}.
[c] Not applicable because 2-propanol is not commonly used to affect protein stability.
Adapted from J. D. Batchelor, A. Olteanu, A. Tripathy, G. J. Pielak, Impact of Protein Denaturants and Stabilizers on Water Structure, *J. Am. Chem. Soc.* 126(7) (2004) 1958–1961. doi:10.1021/ja039335h.

experiments. Values of $\left(\dfrac{\partial \bar{C}p}{\partial P}\right)_T > 0$ would imply that a solute breaks the structure of bulk water; the opposite sign means that structure is created [41].

In Table 3.1, $\bar{C}p$ is the partial molar thermal capacity at constant pressure. The sign of $\left(\dfrac{\partial \bar{C}p}{\partial P}\right)_T$ is related to the making of water structure by a solute [41]. When a solute creates water structure, $\left(\dfrac{\partial \bar{C}p}{\partial P}\right)_T < 0$, and when $\left(\dfrac{\partial \bar{C}p}{\partial P}\right)_T > 0$ the solute breaks water structure. The resulting sign of $\left(\dfrac{\partial \bar{C}p}{\partial P}\right)_T$ and its dependence on temperature should provide a direct test of the properties of the known effects of chaotropes and kosmotropes. The results obtained for the probe solutes show no obvious correlation between the stabilization effects and the sign of $\left(\dfrac{\partial \bar{C}p}{\partial P}\right)_T$. The conclusion is that the thermodynamic results do not prove any long-term effects of solutes on bulk water [3, 41].

In another set of experiments, Zhang and Cremer [15] followed the influence of Hofmeister anions on the phase transition of a surfactant monolayer. The anions' effects on the structure of the surfactant monolayer were followed by observing the adjacent water structure. The authors observed that the effect of the anions, from most to least ordered monolayer, agreed with the Hofmeister series as: $SO_4^{2-} > Cl^- > NO_3^- > Br^- > I^- > ClO_4^- > SCN^-$. Apparently, this observation is related to the ability of an individual anion to penetrate the polar region of the monolayer, causing hydrocarbon packing to disrupt [15]. These observations shed light on the possible mechanisms of protein folding in HIC. They might also explain the changes in retention observed in IPC due to ion pairing over the structure of bonded stationary phases.

These results (and the distinct effect of each salt) oppose the belief that the salting out effect arises from the effect of kosmotropic ions on the structure of water. These findings were corroborated by Leontidis et al. [42], who highlighted four key questions that still lack answers: 1) Is there a concentration threshold at which specific-ion effects to appear? 2) Are specific-ion effects really an interfacial phenomenon? 3) Are specific-ion effects based on local or collective interactions? 4) Does a unique ion parameter exist to correlate ion effects? [42] The next section outlines some possible mechanisms that propose different answers.

3.4.2 OVERVIEW OF POSSIBLE MECHANISMS FOR THE HOFMEISTER SERIES AND SALTING OUT EFFECTS

Several models have been proposed to explain the Hofmeister effect. One model is known as *water-matching affinities* and was proposed by Collins [43]. It establishes that two oppositely charged ions with similar strengths in their interaction with water can form ion pairs, which dominate the ion-specific interactions. Collins showed that many properties of aqueous ionic solutions are a function of the charge density of the ions [44–46]. An example of these is the strength of water–water interactions in bulk solution. Water–water interactions serve as a critical reference-energy level, and are comparable in strength with ion–water interactions [44].

Another property described by Collins is that chaotropes are monovalent ions of small charge density. This means that chaotropes are able to bind the immediately adjacent water molecules less strongly than water binds itself [43, 46]. Therefore, the polarizability of ions is considered to be important in specific-ion effects. For example, it is manifested through the ion dispersion forces, when the dispersion potential is treated at the same level as the electrostatic forces [36].

The Hofmeister effect has also been explained through the specific interactions between the ions and various surfaces, for instance, hydrophobic solids–water and air–water interfaces, in which the kosmotropes are repelled from the surfaces but the chaotropes are adsorbed thereon [3, 14, 16, 25, 41, 42, 47, 48]. The theoretical approaches that support these possible mechanisms are discussed next.

3.4.3 Theories Proposed to Explain the Hofmeister Series

This following section refers to treatment of the Hofmeister effect based on the approaches of Grover and Ryall [17] and of Lo Nostro and Ninham [26].

3.4.3.1 Hydration Theories

According to hydration theories [49], the salting out of a non-electrolyte by the presence of a salt is attributed to the preferential movement of water molecules to solvate the ions over non-electrolytes. In most cases, cations have a hydration layer greater than that of the anions. This suggests that salting out would be a consequence of the cations' solvation. In the same way, salting in would result from the anions' solvation. The net salting effect of a salt on the non-electrolyte solubility would be the result of these two opposing tendencies [49].

In electrolyte solutions, ion–dipole interactions are favored over dipole–dipole interactions. Because of this, the addition of a salt to the solution of a non-electrolyte causes ions and non-electrolytes to compete for water molecules. The water molecules of the hydration layers around the ions are immobilized. Therefore, the availability of water molecules to solubilize the non-electrolyte causes salting out [49]. This theory does not account for other effects, which are important when water serves as a solvent. Such effects include hydrophobicity, hydrophilicity and polarizability of the non-electrolyte, as well as the breakdown of the water structure.

There are two main drawbacks to the hydration theory. First, the theory implies that the predicted number of water molecules surrounding the ion (the hydration number) is a fixed number. This number is independent of the nature of the non-electrolyte that is being salted out, which contradicts experimental findings [50]. Second, these theories do not explain the dependence of the salting constant k from Equation (3.1) on the size of the non-electrolyte [51]. An additional observation is that hydration theories do not provide an explanation for salting in [27].

3.4.3.2   Dipole Theories of Water

The main idea of these theories is that the solubility of non-electrolytes depends on the variations in the specific effects between the salt ions and the water molecules. Non-electrolytes' solubility arises from the orientation of the dipoles of the water molecules around the ions. Ions of the same sign will orient water dipoles preferentially around polar non-electrolytes, causing salting in. Ions of the opposite sign will cause water dipoles to orient unfavorably, causing salting out [17, 52].

Dipole theories of water give further insight than hydration theories. Dipole theories take into account two conditions: the polarizability of polar non-electrolytes, and the hydrophilic hydration near the ion. Both conditions have been proposed as a primary cause of salt effects. Although these explain the effects on polar solutes, they do not explain the variation in the effects on various non-polar solutes [17, 52].

3.4.3.3   Electrostatic Theories

Electrostatic theories are based on the influence of a solute on the relative permittivity of the solvent. The relative permittivity plays a fundamental role in the salting effects. Some studies describe how salting out occurs in a saturated solution of a non-electrolyte, if its relative permittivity is less than that of pure water. If the relative permittivity of the solution is higher, then the non-electrolyte is salted in [49]. This theory provides an explanation of the interactions and of the extent of salting out and sating in. For example, salting out by sodium chloride and potassium chloride have been described in terms of their salting parameter (k in Equation (3.1)). Theoretical values of k for both salts are in agreement with experimental values when the size and polarizability of the non-electrolyte solute are taken into account [50].

Electrostatic models are unable to explain some deviations from the Hofmeister series. For example, no explanation is provided for salting in caused by large ions, such as tetramethylammonium or naphthene sulfonate or salts of long-chain fatty acids [50]. This is to be expected since the theory considers only the attraction of water dipoles into the electrical field of the ions, and not the structural changes or the displacement of water molecules due to the presence of the ions [50]. A recent review stated that the current knowledge of electrostatic effects in complicated systems, such as protein clusters or protein aggregation, is deficient, and represents a challenge to theorists and experimentalists [1]. There is not a unique electrostatic model that can explain any experimental observation accurately.

3.4.3.4   Internal-Pressure Theories

These theories are based on the observation that the volume of a solution varies when a solute dissolves therein. These changes are related to the presence or absence of the dissolved salts. For example, the volume of water decreases upon dissolution of ethyl acetate. If salts are added, volume contractions are observed to increase in the same way as salting out occurs. The theory introduces the concept of an internal pressure, P_{int}, defined as $P_{int} = \left(\dfrac{\partial U}{\partial V}\right)_T = \left(\dfrac{T\alpha}{\kappa}\right) - P$, where T is the absolute temperature, $\alpha = \left(\dfrac{1}{V}\right)\left(\dfrac{\partial V}{\partial T}\right)_P$ is isobaric thermal expansion coefficient, $\kappa = -\left(\dfrac{1}{V}\right)\left(\partial V / \partial P\right)_T$ is the isothermal compressibility and P is the external pressure. The internal pressure modifies the ion–solvent interactions and can produce a precipitation of a polar solute [26].

A subsequent theory [53] proposed a model that allows an explicit study of salt effects. According to this model, the neutral molecules of a solute merely occupy a volume in the bulk solution. Their presence exerts a pressure on the solvent molecules, which in turn modify the solvent–ion interaction. This change causes the precipitation of a solute. The degree of salting out or salting in of a non-polar solute is accordingly determined by the magnitude of the contraction or expansion of the solvent when ions are present [53]. As the compressibility of the electrolyte solution increases, salting out of the non-electrolyte also increases. The predicted and observed salt effects for non-electrolytes correlate well with the corresponding changes when salts dissolve in water [26, 50, 53]. But this theory is unable to account for the marked variation in the classification of the effects of similar electrolyte salting, and vice versa. According to this theory, the predicted order of salting out for several salts, such as sodium sulfate, sodium chloride, lithium chloride or ammonium nitrate, is almost the same for solutes as different as hydrogen, nitrous oxide and benzene [17, 50]. Practice demonstrates this to be untrue. The greatest deficiency of this theory is that, although it provides an effective explanation for non-polar electrolytes, it provides no explanation for the behavior of polar non-electrolytes [51].

3.4.3.5 Theories Based on van der Waals Forces

These theories are based on the van der Waals forces, which differ from the electrostatic forces mentioned above. Van der Waals forces might be attractive or dispersive [13]. Examples of van der Waals forces include Keesom forces due to the orientation of permanent dipoles. Another type are Debye forces, also known as induction forces because they arise from the interaction between a permanent and a temporary dipole. Apart from Keesom and Debye forces, there are London forces that are quantum-mechanical in nature. London forces consider the interactions between two instantaneous dipoles. The standard description of these interactions is based on perturbation theory for two bodies; however, it is not applicable in condensed media [13]. Although van der Waals forces mostly explain the salting in effect, these theories fail to explain salting out. For example, there is not yet an explanation for the low salting out caused by lithium and hydrogen ions [50].

3.4.3.6 Ion Pairing

Manning proposed that the attractions among opposite point charges, such as counterions, would be manifest to one another along a cylindrical geometry [23]. Counterions in such cylindrical geometry are referred to as "condensed." Ion condensation is a mathematical artefact in Manning's description. Such physical interaction between counterions happens in the local cylindrical geometry, while the relative permittivity of the bulk solvent remains constant [54]. Even though the model does not take into account the existence of repulsive interactions of short range, it has nevertheless proven to be useful because it defines a critical parameter ξ. This parameter corresponds to the minimum charge spacing necessary for counterion attraction, and is dependent on the relative permittivity of the solvent [54]. For water $\xi=1$, the relative permittivity is 78.5, and the distance between two opposite charges is 7.135 Å.

This distance is called the "Bjerrum length" [54, 55], which is the separation at which two elementary charges are attracted to one another by an electrostatic energy comparable in strength to the thermal energy scale [56].

According to this model, the presence of polyions in a solution contributes to deviations in the system from that of the bulk solvent, thus increasing the value of ξ. But the ions tend to condense so as to decrease the charge density in order to maintain $\xi=1$ [23, 54, 55]. In consequence, condensation also refers to a physical mechanism of attraction between ions that leads to aggregation. Manning demonstrated ion condensation by studying the variation of some colligative properties with increasing salt concentration. Condensation occurs between ions to maintain $\xi=1$ through the formation of ion pairs, as colligative experiments show [23].

Recent evidence in ion-pairing interactions confirm the Hofmeister behavior of added salts in RPLC and HIC [57]. However, in IPC there is no appreciable difference between kosmotropic or chaotropic agents. Cecchi emphasized that no existing theory (i.e. stoichiometric or electrostatic or thermodynamic) can explain all experimental observations. To encompass all possible phenomena, Cecchi proposed an extended thermodynamic approach, *extended* meaning that it considers all possible equilibria in the mobile and stationary phases, including electrostatic interactions (see Figure 3.5) [57].

Figure 3.5 has been adapted from reference [57]. It shows all the possible ion pair equilibria occurring in a chromatographic system. On the left side (indicated from *a* to *g*) are chemical equations representing each equilibrium. An ionic analyte

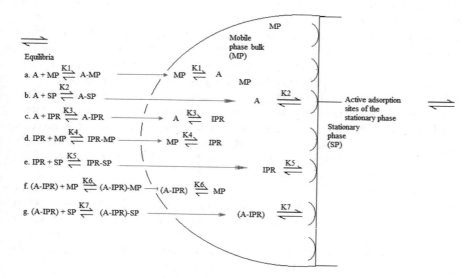

FIGURE 3.5 Equilibrium processes on the addition of an ion-pairing reagent (IPR). (Right) A diagrammatic representation of possible interactions of the analyte (A), the IPR and an analyte-IPR complex (A-IPR) in the mobile phase (MP) and the stationary phase (SP), with the corresponding equilibrium quotients. (Left) Chemical equations representing each equilibrium.

(A) can undergo ion pairing with an ion-pairing reagent (IPR) to form an ion pair moiety determined by the equilibrium constant K_3. Both A and the IPR can undergo, separately, a dynamic equilibrium with the mobile phase (MP). This equilibrium is determined by K_1, K_4 or K_6. Adsorption equilibria with active sites in the stationary phase (SP) are determined by K_2, K_5 or K_7.

Cecchi's model was successfully tested in retention modeling when IPR chaotropes were the chosen additives. Two main mechanistic differences between classic IPC and chaotropic IPC arise from Cecchi's treatment [57]: 1) the affinity of the chaotropic anion and its counterion for the stationary phase (adsorbophilicity) do not differ from one another; 2) the influence of classic IPRs on analyte retention is explained based on the lipophilic portion of the IPR. However, analyte retention with chaotropic salts is explained based on the electrostatic interaction between the chaotropic anion and its counterion [24, 57]. Cecchi's observations agree with Florez and Kazakevich [24], although no similar results are reported for kosmotropes.

According to Cecchi, no new theory has been developed to understand IPC fully, other than that extended thermodynamic approach. Efforts have been undertaken to revisit previous models to test potential new reagents such as chaotropic and kosmotropic salts [57].

3.4.4 A Quantum-Mechanical Approach to the Hofmeister Series

As was summarized in the preceding pages, the Hofmeister series has been known for over 130 years but has never been fully explained. The current explanations of the effects of kosmotropic and chaotropic salts rely on the structure of water and its capability of forming hydrogen bonds. The structure of water in solutions is supposed to be modified by electrostatic interactions with the solutes, including hydrogen bonding and van der Waals forces [12, 13]. Various authors agree that the Hofmeister effect might be explained based on a quantum-mechanical approach [12, 13, 58]. One possible explanation is based on the concept of a quantum vacuum [12].

The quantum vacuum is considered in a description of the Casimir effect. This effect arises from two repelling plates separated 10 pm from one another. The repulsive force drives the formation of a pressure as great as 1 bar, known as a quantum vacuum. Henry proposed that the salting effects and the changes in the structuring of water associated with the Hofmeister series can be explained based on interactions arising in a quantum vacuum [12].

An electrostatic interaction between a salt and a protein might exist because of the presence of opposite charges. From a quantum-mechanical point of view, when an electron of one species is attracted to the nucleus of another species it forms a virtual electron–positron pair. Interaction within this pair results in the disintegration of the approaching electron and the positron, releasing the electron of the pair and leaving a vacuum [12]. This vacuum is so energetic that a new electron fills it, resulting in a new interaction. This reasoning indicates that matter appears and disappears inside a vacuum with a period of attoseconds. Such a short time gives the idea of a continuous electrostatic exchange described as hydrogen bonding or van der Waals forces, which are the main driving forces for chaotropicity and kosmotropicity [12, 13, 58, 59].

3.5 CHROMATOGRAPHIC IMPLICATIONS OF KOSMOTROPIC SALTS AND THE HOFMEISTER SERIES

Apart from the variety of ways in which salts can interact with proteins mediating protein–protein interactions, they also exert an effect on their stability and solubility. These phenomena are due to an increase in the interfacial tension between the proteins and water [25], thus strengthening the hydrophobic interactions of a protein with the stationary phase [47]. These hydrophobic effects induced by the salts are the basis of HIC [48]. The addition of kosmotropic salts has a small effect of increasing resolution, while also improving peak shape and selectivity [5]. Therefore, in HIC the retention is modulated by varying the salt concentration in the mobile phase, in such a way that the protein retains its native structure to a larger extent. There is still no agreement between the theory and experiments of HIC that explains this due to the complexity of the chromatographic system [60].

There is a consensus that the chaotropic or kosmotropic ions from the Hofmeister series affects the solubility and stability of proteins in two possible ways. One, via a change in polarizabilty of the water-mediated ion–protein interaction. Another by direct binding of the ions to the protein [8]. Apart from the various ways in which salts can interact with proteins and mediate protein–protein interactions, salts exert an effect on the stability and solubility of proteins.

Cecchi showed that experimental results and molecular dynamic calculations suggest that chaotropic ions accumulate in two regions: at interface between the mobile phase and the stationary phase; and at the interface between the stationary phase and the protein surface [57, 60]. The surface of the stationary phase is different from the bulk of the mobile phase. These differences arise from the fact that the stationary phase is an inhomogeneous surface. In the stationary phase the electrolyte distribution can induce dipole moments which sum is not zero. Instead, the bulk of the mobile phase is symmetric with respect to electric vector intermolecular forces that sum zero. The experimental results showed that anions and cations partition differently between the surface of the stationary phase and bulk mobile phase. For instance, anions have been observed to accumulate at their interfaces. This tendency to accumulate at the interface follows the Hofmeister series [61]. Non-specific effects are observed for chaotropic salts at a small concentration and are related to chaotropic changes in the water structure and the salting in of solutes, because chaotropic ions are characterized by small electronic densities [57].

3.6 DEVELOPMENT OF THE LIQUID CHROMATOGRAPHIC METHOD FOR HIC PROTEIN SEPARATION

The development of HIC chromatographic methods for protein analyses has been studied by various authors [18, 62, 63, 64, 65]. The use of mobile phases containing kosmotropic salts, such as ammonium sulfate, has been suggested. Separations are performed with an inverse salt gradient, from high to low concentrations in ammonium sulfate. The use of the kosmotropic salt promotes hydrophobic interactions between the proteins and the stationary phases based on the effect predicted by the Hofmeister series [66].

The selection of the mobile phase conditions in HIC is an empirical process that can be simplified by applying RPLC gradient elution relationships. Karger et al. and others showed that retention and separation in the HIC of proteins can be explained by the linear solvent strength (LSS) gradient model for RPLC, shown in Equation (3.3) [67].

$$\log k^* = \log k_w - S\varnothing^* \tag{3.3}$$

In Equation (3.3) k^* is the retention factor in the gradient (the median value of k during gradient elution), $\varnothing^*$ is the median volume fraction of mobile phase B, k_w is the retention in pure water for a particular analyte and S is the negative slope $[-d(\log k)/d\varnothing]$ (the change of $\log k$ with respect to %B). S is generally accepted to be approximately $0.25*(molecular\ weight)^{1/5}$ for molecules with molecular weights over 400. The magnitude of S affects the retention factor k^*. Other values that affect k^* are the gradient steepness, the mobile phase flow rate, the column dead volume and the change in %B during the gradient ($\Delta\varnothing$) [67].

In HIC, the retention k depends on the concentration of kosmotropic ammonium sulfate. This dependence is shown in Equation (3.4) [18, 67].

$$\log k = \log k_0 + A_{HIC}C_{AS} \tag{3.4}$$

In Equation (3.4) C_{AS} is the concentration of ammonium sulfate, k_0 is the value of k when $C_{AS}=0$ and A_{HIC} is the slope $d(\log k)/d(C_{AS})$. As C_{AS} increases, k also increases. C_{AS} must decrease during the gradient; otherwise proteins will remain retained in the column. That is the reason why inverse gradient elution is preferred. Equation (3.4) can be transformed into Equation (3.5) based on ammonium sulfate in order to be analogous to the LSS gradient model as:

$$\log k = \log k_{2.5} - S_{HIC}\varnothing_{HIC} \tag{3.5}$$

$k_{2.5}$ is the value of k for 2.5 M ammonium sulfate, $\varnothing_{HIC}$ is defined as $[-(C_{AS}-2.5)/2.5]$ and S_{HIC} is $-2.5A_{HIC}$. In a linear inverse gradient from 2.5 M to 0 M ammonium sulfate, $\varnothing_{HIC}$ varies from 0 to 1. For example, if C_{AS} gradient values are 2.5 M, 1.25 M and 0.0 M, $\varnothing_{HIC}$ is equal to 0.00, 0.50 and 1.00, respectively. Other kosmotropic salts, such as ammonium acetate and sodium formate, give similar values of A_{HIC} and S_{HIC} at higher concentrations, but are not frequently used [67].

Equation (3.5) for HIC is equivalent to Equation (3.3) for a RPLC gradient, because they follow similar qualitative and quantitative relationships in relation with k. For this reason, RPLC gradient elution rules can be applied directly to HIC method development for protein analysis. For instance, for the analysis of proteins in the range of molecular weight $10^4 \leq M \leq 10^5$, S_{HIC} values range from 4 to 9. In contrast, S values in RPLC for the same analytes are between 25 and 80. These S_{HIC} values are almost an order of magnitude smaller than S values for RPLC. This means that, based on the LSS gradient model, steeper gradients can be used in HIC with shorter gradient times compared to RPLC for the same analytes [67].

Recent interest in the HIC analysis of proteins has sought to provide generic guidelines for method development. This includes the use of other kosmotropic salts besides ammonium sulfate, as well as determining the proper salt concentration, salt gradient and gradient steepness. For instance, Fekete and coworkers studied salt type and concentration to optimize HIC mobile phase for the analysis of monoclonal antibodies [64, 68].

These authors tested the influence of 2 M ammonium sulfate on the selectivity for antibody separation, and used it for comparison with other salts such as sodium chloride, ammonium acetate and ammonium formate. What the authors found is that similar selectivities can be achieved with either salt by adjusting their kosmotropic strength [68]. This is done by determining the kosmotropic salt concentrations at which the selectivity is similar. A summary of the chromatographic results for sodium chloride compared to 2 M ammonium sulfate is shown in Table 3.2.

From Table 3.2 it can be seen that to obtain the same selectivity as 2 M ammonium sulfate it is necessary to have 5 M sodium chloride in the mobile phase. At the same time, selectivity can be manipulated by changing the concentration of Cl^-. Similar selectivities to that observed with 2 M ammonium sulfate can be obtained with sodium formate and sodium acetate in the mobile phase in concentrations between 5.0 and 5.5 M. These results are in line with the Hofmeister series [64].

From a method development point of view, salts should be interchangeable if their positions in the Hofmeister series are close to each other. At the same time their salting out strength should be adjusted for their concentration. The adjustment could be determined using Equation (3.4) by determining A_{HIC} from a plot of $\log k$ vs %B. Similarly S_{HIC} can be determined from Equation (3.5) by plotting $\log k$ vs ϕ. However, the effect of salt type and concentration cannot be determined in advance, and should be determined experimentally [64, 68].

Another alternative for HIC method development has been described by Tyteca et al. The authors separated cytochrome c, ribonuclease A and lysozyme using an ammonium sulfate gradient from 1.5 M to 0.5 M. Their results show that these gradients do not behave linearly with respect to the LSS gradient model proposed by Snyder due to a mixed mode of interaction. At higher salt concentration at the beginning of the gradient, the linearity improves according to the theory. The best results were observed at a concentration of 1.8 M at the beginning of the gradient [65].

TABLE 3.2

Observed Selectivity for Different Concentrations of Sodium Chloride as an Alternative Kosmotropic Salt Compared to 2 M Ammonium Sulfate

Salt	Concentration/(M)	Selectivity factor between two antibodies
NH_4SO_4	2	1.1
NaCl	5	1.1
NaCl	4	1.6
NaCl	3	2.8

Adapted from M. Rodriguez-Aller, D. Guillarme, A. Beck, S. Fekete, Practical Method Development for the Separation of Monoclonal Antibodies and Antibody-Drug-Conjugate Species in Hydrophobic Interaction Chromatography, Part 1: Optimization of the Mobile Phase, *J. Pharm. Biomed. Anal.* 118 (2016) 393–403. doi:10.1016/j.jpba.2015.11.011.

Protein retention in HIC not only depends on the kosmotropic salt concentration, selection of salt type, concentration and gradient program, but also on the stationary phase to be used. Hydrophobic stationary phases, as well as salt concentration, can induce conformational changes of proteins as well as aggregation [62, 63]. Stationary phases in HIC can be either amphipilic [62] or hydrophilic, with hydrophobic ligands linked by spacer arms. HIC stationary phases differ in chemical nature, in surface concentration of the ligand and in the chemistry and particle size of the matrix. A comprehensive description of the main stationary phases used in HIC can be found in [64]. The stationary phase's hydrophobicity can change with the ionic strength of the mobile phase, affecting retention due to conformational changes of proteins [62] as well as aggregation [63].

3.6.1 Optimization of the Separation in HIC

The LSS model is a function of experimental conditions that affect protein separation. These conditions, such as salt gradient, flow rate and column dimensions, have a predictable effect on the separation (at constant temperature and mobile phase pH). The effect of one or more of these conditions can be predicted by the LSS model in order to optimize a method for protein separation. This can be done by computer simulation, performing two gradient runs with different slopes, thus predicting the separation as a function of gradient conditions. For computer simulation, retention times and peak widths are entered into software. The software uses the specified conditions to determine k_w and S for each protein to be separated. Once k_w and S for each component are determined, the software displays its predicted chromatograms, tables or resolution plots. Such results can be obtained using software such as DryLab, ChromSword, ChromSmart, ACD/LC Simulator, Osiris or Preopt-W [67].

For instance, reference [69] describes the use of DryLab 2010 in order to optimize salt gradient steepness in the HIC analysis of antibody-drug conjugates. The computer-assisted optimization led to a better selectivity with a multilinear salt gradient, steeper at the beginning and flatter at the end. Another work is described in reference [70]. The initial calibration runs were performed for the separation of monoclonal antibodies using two isocratic conditions and extrapolating to a proper gradient combined with a suitable stationary phase. A resolution map was constructed indicating the proper chromatographic conditions [70].

Computer optimization also offers the advantage of performing an automatic search of the best combination of initial and final %B in the gradient, and of the time gradient when other conditions, such as temperature, are changed. Reports based on computer simulation include HIC gradients which do not exceed 10–30 min and temperatures of 20°C and 40°C [68]. The experimental retention time, peak widths and peak tailing were imported into DryLab 2010. The software then converts the retention times into retention factors. The final results were a significant decrease in the time of the analysis [64, 68].

Snyder shows that DryLab is also able to perform the selection of stationary phases along with flow rate and pressure. Using the conditions described above allows the practicing analytical chemist to develop a chromatographic method in a very short time [67]. Other discussions on protein separations are available in references [71–73].

3.7 CONCLUSIONS

Kosmotropicity is an important phenomenon in protein separations yet to be elucidated. The main conclusions from this review are as follows:

1. There is no complete theoretical framework that considers all variables involved in the salting effects due to kosmotropicity or chaotropicity.
2. The kosmotropicity and chaotropicity of salts have wide applicability for the optimization of protein separations and of small molecules.
3. The absence of a theoretical framework is no limitation for the development and optimization of new analytical methods, especially in the pharmaceutical industry.

What began as an empirical area of research to improve HPLC in the late 1980s is still an unfinished compendium of theoretical explanations that has found successful continuous applications, especially in protein and biomolecules separations.

REFERENCES

1. H.-X. Zhou, X. Pang, Electrostatic Interactions in Protein Structure, Folding, Binding, and Condensation, *Chem. Rev.* 118(4) (2018) 1691–1741. doi:10.1021/acs.chemrev.7b00305
2. H. Zhao, Are Ionic Liquids Kosmotropic or Chaotropic? An Evaluation of Available Thermodynamic Parameters for Quantifying the Ion Kosmotropicity of Ionic Liquids, *J. Chem. Technol. Biotechnol.* 81(6) (2006) 877–891. doi:10.1002/jctb.1449
3. A. W. Omta, M. F. Kropman, S. Woutersen, H. J. Bakker, Influence of Ions on the Hydrogen-Bond Structure in Liquid Water, *J. Chem. Phys.* 119(23) (2003) 12457–12461. doi:10.1063/1.1623746
4. K. Tsumoto, D. Ejima, A. M. Senczuk, Y. Kita, T. Arakawa, Effects of Salts on Protein–Surface Interactions: Applications for Column Chromatography, *J. Pharm. Sci.* 96(7) (2007) 1677–1690. doi:10.1002/jps.20821
5. Ł. Komsta, R. Skibiński, A. Bojarczuk, M. Radoń, Salting-Out Chromatography – A Practical Review, *Acta Chromatogr.* 23(2) (2011) 191–203. doi:10.1556/AChrom.23.2011.2.1
6. P. Štrop, Hydrophobic Chromatography of Proteins on Semi-Rigid Gels: Effect of Salts and Interferents on the Retention of Proteins by Spheron P 300, *J. Chromatogr. A* 294 (1984) 213–221. doi:10.1016/S0021-9673(01)96128-X
7. E. Loeser, S. Babiak, M. DelaCruz, P. Karpinski, Triflic Acid and Sodium Triflate as Chaotropic Mobile Phase Additives in RP-LC, *J. Chromatogr. Sci.* 49(1) (2011) 57–62. doi:10.1093/chrsci/49.1.57
8. R. Kou, J. Zhang, T. Wang, G. Liu, Interactions Between Polyelectrolyte Brushes and Hofmeister Ions: Chaotropes Versus Kosmotropes, *Langmuir* 31(38) (2015) 10461–10468. doi:10.1021/acs.langmuir.5b02698
9. J. Ståhlberg, B. Jönsson, C. Horváth, Theory for Electrostatic Interaction Chromatography of Proteins, *Anal. Chem.* 63(17) (1991) 1867–1874. doi:10.1021/ac00017a036
10. M. de Souza Gama, M. Simões Santos, E. Rocha de Almeida Lima, F. W. Tavares, A. Gomes Barreto, A Modified Poisson-Boltzmann Equation Applied to Protein Adsorption, *J. Chromatogr. A* 1531 (2017) 74–82. doi:10.1016/j.chroma.2017.11.022
11. L. M. Pegram, M. T. Record, Thermodynamic Origin of Hofmeister Ion Effects, *J. Phys. Chem. B* 112(31) (2008) 9428–9436. doi:10.1021/jp800816a

12. M. Henry, Hofmeister Series: The Quantum Mechanical Viewpoint, *Curr. Opin. Colloid Interface Sci.* 23 (2016) 119–125. doi:10.1016/j.cocis.2016.08.001

13. T. P. Pollard, T. L. Beck, Toward a Quantitative Theory of Hofmeister Phenomena: From Quantum Effects to Thermodynamics, *Curr. Opin. Colloid Interface Sci.* 23 (2016) 110–118. doi:10.1016/j.cocis.2016.06.015

14. W. J. Xie, Y. Q. Gao, A Simple Theory for the Hofmeister Series, *J. Phys. Chem. Lett.* 4(24) (2013) 4247–4252. doi:10.1021/jz402072g

15. Y. Zhang, P. S. Cremer, Interactions Between Macromolecules and Ions: The Hofmeister Series, *Curr. Opin. Chem. Biol.* 10(6) (2006) 658–663. doi:10.1016/j.cbpa.2006.09.020

16. M. C. Gurau, S.-M. Lim, E. T. Castellana, F. Albertorio, S. Kataoka, P. S. Cremer, On the Mechanism of the Hofmeister Effect, *J. Am. Chem. Soc.* 126(34) (2004) 10522–10523. doi:10.1021/ja047715c

17. P. K. Grover, R. L. Ryall, Critical Appraisal of Salting-Out and Its Implications for Chemical and Biological Sciences, *Chem. Rev.* 105(1) (2005) 1–10. doi:10.1021/cr030454p

18. S. L. Wu, B. L. Karger, Hydrophobic Interaction Chromatography of Proteins. In: B. L. Karger, W. S. Hancock (Eds.), *Methods Enzymology vol 270, High Resolution Separation and Analysis of Biological Macromolecules, Part A Fundamentals*, Academic Press Inc., San Diego, CA, 1996, pp. 27–47.

19. J. Ståhlberg, B. Jönsson, C. Horváth, Combined Effect of Coulombic and Van Der Waals Interactions in the Chromatography of Proteins, *Anal. Chem.* 64(24) (1992) 3118–3124. doi:10.1021/ac00048a009

20. A. I. Liapis, B. A. Grimes, The Coupling of the Electrostatic Potential with the Transport and Adsorption Mechanisms in Ion-Exchange Chromatography Systems: Theory and Experiments, *J. Sep. Sci.* 28(15) (2005) 1909–1926. doi:10.1002/jssc.200500240

21. N. K. Boardman, S. M. Partridge, Separation of Neutral Proteins on Ion-Exchange Resins, *Biochem. J.* 59(4) (1955) 543–552. doi:10.1042/bj0590543

22. J. M. Miller, *Chromatography, Concepts and Contrasts*, Wiley-Interscience, NJ, 2005, pp. 207–209.

23. G. S. Manning, Limiting Laws and Counterion Condensation in Polyelectrolyte Solutions I. Colligative Properties, *J. Chem. Phys.* 51(3) (1969) 924–933. doi:10.1063/1.1672157

24. C. Florez, Y. Kazakevich, Influence of Ionic Mobile Phase Additives with Low Charge Delocalization on the Retention of Ionic Analytes in Reversed-Phase HPLC, *J. Liq. Chromatogr. R. T.* 36(8) (2013) 1138–1148. doi:10.1080/10826076.2012.670183

25. H.-X. Zhou, Interactions of Macromolecules with Salt Ions: An Electrostatic Theory for the Hofmeister Effect, *Proteins* 61(1) (2005) 69–78. doi:10.1002/prot.20500

26. P. Lo Nostro, B. W. Ninham, Hofmeister Phenomena: An Update on Ion Specificity in Biology, *Chem. Rev.* 112(4) (2012) 2286–2322. doi:10.1021/cr200271j

27. J. Song, H. Kang, M. W. Kim, S. Han, Ion Specific Effects: Decoupling Ion-Ion and Ion-Water Interactions, *Phys. Chem. Chem. Phys.* 17(13) (2015) 8306–8322. doi:10.1039/C4CP05992A

28. J. Flieger, Application of Ionic Liquids in Liquid Chromatography. In: A. Kokorin (Ed.), *Ionic Liquids. Applications and Perspectives*, InTech, Croatia, 2011, pp. 243–245.

29. A. Kumar, P. Venkatesu, Does the Stability of Proteins in Ionic Liquids Obey the Hofmeister Series?, *Int. J. Biol. Macromol.* 63 (2014) 244–253. doi:10.1016/j.ijbiomac.2013.10.031

30. H. Zhao, Are Ionic Liquids Kosmotropic or Chaotropic? An Evaluation of Available Thermodynamic Parameters for Quantifying the Ion Kosmotropicity of Ionic Liquids, *J. Chem. Technol. Biotechnol.* 81(6) (2006) 877–891. doi:10.1002/jctb.1449

31. M. C. García-Alvarez-Coque, M. J. Ruiz-Angel, A. Berthod, S. Carda-Broch. On the Use of Ionic Liquids as Mobile Phase Additives in High-Performance Liquid Chromatography. A Review, *Anal. Chim. Acta* 883 (2015) 1–21. doi:10.1016/j.aca.2015.03.042

32. R. Kaliszan, M. P. Marszałł, M. J. Markuszewski, T. Baczek, J. Pernak, Suppression of Deleterious Effects of Free Silanols in Liquid Chromatography by Imidazolium Tetrafluoroborate Ionic Liquids, *J. Chromatogr. A* 1030(1–2) (2004) 263–271. doi:10.1016/j.chroma.2003.09.020

33. M. P. Marszałł, T. Baczek, R. Kaliszan, Evaluation of the Silanol-Suppressing Potency of Ionic Liquids, *J. Sep. Sci.* 29(8) (2006) 1138–1145. doi:10.1002/jssc.200500383

34. A. Tot, S. Armaković, S. Armaković, S. Gadžurić, M. Vraneš, Kosmotropism of Newly Synthesized 1-Butyl-3-Methylimidazolium Taurate Ionic Liquid: Experimental and Computational Study, *J. Chem. Thermodyn.* 94 (2016) 85–95. doi:10.1016/j.jct.2015.10.026

35. C. M. S. S. Neves, T. B. V. Dinis, P. J. Carvalho, B. Schröder, L. M. N. B. F. Santos, M. G. Freire, J. A. P. Coutinho, Binary Mixtures of Ionic Liquids in Aqueous Solution: Towards an Understanding of Their Salting-In/Salting-Out Phenomena, *J. Solut. Chem.* 48(7) (2018) 983–991. doi:10.1007/s10953-018-0836-7

36. V. Mazzini, V. S. J. Craig, Specific-Ion Effects in Non-Aqueous Systems, *Curr. Opin. Colloid Interface Sci.* 23 (2016) 82–93. doi:10.1016/j.cocis.2016.06.009

37. Y. J. Zhang, S. Furyk, D. E. Bergbreiter, P. S. Cremer, Specific Ion Effects on the Water Solubility of Macromolecules: PNIPAM and the Hofmeister Series, *J. Am. Chem. Soc.* 127(41) (2005) 14505–14510. doi:10.1021/ja0546424

38. Y. R. Gokarn, R. M. Fesinmeyer, A. Saluja, V. Razinkov, S. F. Chase, T. M. Laue, D. N. Brems, Effective Charge Measurements Reveal Selective and Preferential Accumulation of Anions, but not Cations, at the Protein Surface in Dilute Salt Solutions, *Protein Sci.* 20(3) (2011) 580–587. doi:10.1002/pro.591

39. F. Zhang, M. W. Skoda, R. M. Jacobs, S. Zorn, R. A. Martin, C. M. Martin, G. F. Clark, S. Weggler, A. Hildebrandt, O. Kohlbacher, F. Schreiber, Reentrant Condensation of Proteins in Solution Induced by Multivalent Counterions, *Phys. Rev. Lett.* 101(14) (2008) 148101. doi:10.1103/PhysRevLett.101.148101

40. A. W. Omta, M. F. Kropman, S. Woutersen, H. J. Bakker, Negligible Effect of Ions on the Hydrogen-Bond Structure in Liquid Water, *Science* 301(5631) (2003) 347–349. doi:10.1126/science.1084801

41. J. D. Batchelor, A. Olteanu, A. Tripathy, G. J. Pielak, Impact of Protein Denaturants and Stabilizers on Water Structure, *J. Am. Chem. Soc.* 126(7) (2004) 1958–1961. doi:10.1021/ja039335h

42. A. Aroti, E. Leontidis, E. Maltseva, G. Brezesinski, Effects of Hofmeister Anions on DPPC Langmuir Monolayers at the Air-Water Interface, *J. Phys. Chem. B* 108(39) (2004) 15238–15245. doi:10.1021/jp0481512

43. K. D. Collins, Ion Hydration: Implications for Cellular Function, Polyelectrolytes, and Protein Crystallization, *Biophys. Chem.* 119(3) (2006) 271–281. doi:10.1016/j.bpc.2005.08.010

44. K. D. Collins, Charge Density-Dependent Strength of Hydration and Biological Structure, *Biophys. J.* 72(1) (1997) 65–76. doi:10.1016/S0006-3495(97)78647-8

45. K. D. Collins, Ions from the Hofmeister Series and Osmolytes: Effects on Proteins in Solution and in the Crystallization Process, *Methods* 34(3) (2004) 300–311. doi:10.1016/j.ymeth.2004.03.021

46. K. D. Collins, Sticky Ions in Biological Systems, *PNAS* 92(12) (1995) 5553–5557. doi:10.1073/pnas.92.12.5553

47. R. L. Baldwin, How Hofmeister Ion Interactions Affect Protein Stability, *Biophys. J.* 71(4) (1996) 2056–2063. doi:10.1016/S0006-3495(96)79404-3

48. A. Vailaya, C. Horváth, Retention Thermodynamics in Hydrophobic Interaction Chromatography, *Ind. Eng. Chem. Res.* 35(9) (1996) 2964–2981. doi:10.1021/ie9507437

49. P. M. Gross, The "Salting Out" of Non-Electrolytes from Aqueous Solutions, *Chem. Rev.* 13(1) (1933) 91–101. doi:10.1021/cr60044a007

50. F. A. Long, W. F. McDevit, Activity Coefficients of Nonelectrolyte Solutes in Aqueous Salt Solutions, *Chem. Rev.* 51(1) (1952) 119–169. doi:10.1021/cr60158a004

51. T. J. Morrison, 729The Salting-Out of Non-Electrolytes. Part I. The Effect of Ionic Size, Ionic Charge, and Temperature, *J. Chem. Soc.* 0 (1952) 3814–3818. doi:10.1039/JR9520003814

52. W. Xie, Z. Zheng, M. Tang, D. Li, W.-Y. Shiu, D. Mackay, Solubilities and Activity Coefficients of Chlorobenzenes and Chlorophenols in Aqueous Salt Solutions, *J. Chem. Eng. Data* 39(3) (1994) 568–571. doi:10.1021/je00015a038

53. W. F. McDevit, F. A. Long, The Activity Coefficient of Benzene in Aqueous Salt Solutions, *J. Am. Chem. Soc.* 74(7) (1952) 1773–1777. doi:10.1021/ja01127a048

54. G. S. Manning, Limiting Laws and Counterion Condensation in Polyelectrolyte Solutions. III. An Analysis Based on the Mayer Ionic Solution Theory, *J. Chem. Phys.* 51(8) (1969) 3249–3252. doi:10.1063/1.1672502

55. G. S. Manning, Limiting Laws and Counterion Condensation in Polyelectrolyte Solutions. IV. The Approach to the Limit and the Extraordinary Stability of the Charge Fraction, *Biophys. Chem.* 7(2) (1977) 95–102. doi:10.1016/0301-4622(77)80002-1

56. W. B. Russell, D. A. Saville, W. R. Schowalter, Colloidal Dispersions. In: G. K. Batchelor (Ed.), *Cambridge Monographs on Mechanics and Applied Mathematics*, Cambridge University Press, Cambridge, 1989, p. 475.

57. T. Cecchi, Theoretical Models of Ion Pair Chromatography: A Close Up of Recent Literature Production, *J. Liq. Chromatogr. R. T.* 38 (2014) 404–414. doi:10.1080/10826076.2014.941267

58. A. Berger, G. Ciardi, D. Sidler, P. Hamm, A. Shalit, Impact of Nuclear Quantum Effects on the Structural Inhomogeneity of Liquid Water, *PNAS* 116(7) (2019) 2458–2463. doi:10.1073/pnas.1818182116

59. D. F. Parsons, M. Boström, P. Lo Nostro, B. W. Ninham, Hofmeister Effects: Interplay of Hydration, Nonelectrostatic Potentials, and Ion Size, *Phys. Chem. Chem Phys* 13(27) (2011) 12352–12367. doi:10.1039/C1CP20538B

60. T. Cecchi, Retention Mechanism for Ion-Pair Chromatography with Chaotropic Reagents. From Ion-Pair Chromatography Toward a Unified Salt Chromatography, *Adv. Chromatogr.* 49 (2011) 1–35. doi:10.1201/b10721-2

61. Y. V. Kazakevich, R. LoBrutto, R. Vivilecchia, Reversed-Phase High-Performance Liquid Chromatography Behavior of Chaotropic Counteranions, *J. Chromatogr. A* 1064(1) (2005) 9–18. doi:10.1016/j.chroma.2004.11.104

62. S.-L. Wu, A. Figueroa, B. L. Karger, Protein Conformational Effects in Hydrophobic Interaction Chromatography. Retention Characterization and the Role of Mobile Phase Additives and Stationary Phase Hydrophobicity, *J. Chromatogr. A* 371 (1986) 3–27. doi:10.1016/S0021-9673(01)94689-8

63. N. Grinberg, R. Blanco, D. M. Yarmush, B. L. Karger, Protein Aggregation in High-Performance Liquid Chromatography: Hydrophobic Interaction Chromatography of β-Lactoglobulin A, *Anal. Chem.* 61(6) (1989) 514–520. doi:10.1021/ac00181a003

64. A. Cusumano, D. Guillarme, A. Beck, S. Fekete, Practical Method Development for the Separation of Monoclonal Antibodies and Antibody-Drug-Conjugate Species in Hydrophobic Interaction Chromatography, Part 2: Optimization of the Phase System, *J. Pharm. Biomed. Anal.* 121 (2016) 161–173. doi:10.1016/j.jpba.2016.01.037

65. E. Tyteca, J. De Vos, M. Tassi, K. Cook, L. Xiaodong, E. Kaal, S. Eeltink, Generic Approach to the Development of Intact Protein Separations Using Hydrophobic Interaction Chromatography, *J. Sep. Sci.* 41 (2017) 1017–1024. doi:10.1002/jssc.201701202

66. J. M. Roberts, A. R. Díaz, D. T. Fortin, J. M. Friedle, S. D. Piper, Influence of the Hofmeister Series on the Retention of Amines in Reversed-Phase Liquid Chromatography, *Anal. Chem.* 74(19) (2002) 4927–4932. doi:10.1021/ac0256944

67. L. R. Snyder, J. W. Dolan, *High-Performance Gradient Elution. The Practical Application of the Linear-Solvent-Strength Model*, John Wiley & Sons, Hoboken, NJ, 2007, pp. 228–260.

68. M. Rodriguez-Aller, D. Guillarme, A. Beck, S. Fekete, Practical Method Development for the Separation of Monoclonal Antibodies and Antibody-Drug-Conjugate Species in Hydrophobic Interaction Chromatography, Part 1: Optimization of the Mobile Phase, *J. Pharm. Biomed. Anal.* 118 (2016) 393–403. doi:10.1016/j.jpba.2015.11.011

69. B. Bobály, G. M. Randazzo, S. Rudaz, D. Guillarme, S. Fekete, Optimization of Non-Linear Gradient in Hydrophobic Interaction Chromatography for the Analytical Characterization of Antibody-Drug Conjugates, *J. Chromatogr. A* 1481 (2017) 82–91. doi:10.1016/j.chroma.2016.12.047

70. E. Tyteca, J. L. Veuthey, G. Desmet, D. Guillarme, S. Fekete, Computer Assisted Liquid Chromatographic Method Development for the Separation of Therapeutic Proteins, *Analyst* 141(19) (2016) 5488–5501. doi:10.1039/C6AN01520D

71. M. Baca, J. De Vos, G. Bruylants, K. Bartik, X. Liu, K. Cook, S. Eeltink, A Comprehensive Study to Protein Retention in Hydrophobic Interaction Chromatography, *J. Chromatogr. B Analyt. Technol. Biomed. Life Sci.* 1032 (2016) 182–188. doi:10.1016/j.jchromb.2016.05.012

72. K. O. Eriksson, Hydrophobic Interaction Chromatography. In: G. Jagschies, E. Lindskog, K. Łącki, P. Galliher (Eds.), *Biopharmaceutical Processing*, Elsevier, New York, 2018, pp. 401–407. doi:10.1016/B978-0-08-100623-8.00019-0

73. C. T. Tomaz, Hydrophobic Interaction Chromatography. In: S. Fanali, P. R. Haddad, M.-L. Riekkola, C. F. Poole (Eds.), *Liquid Chromatography. Fundamentals and Instrumentation*, Elsevier, New York, 2017, pp. 171–190. doi:10.1016/B978-0-12-805393-5.00007-5

4 Isotopic Separations Indicating the Efficiency and Selectivity of HPLC Columns and Stationary Phases

Kazuhiro Kimata, Eisuke Kanao, Takuya Kubo,
Kodai Kozuki, Tohru Ikegami, Masahiro Furuno,
Koji Otsuka, and Nobuo Tanaka

CONTENTS

4.1 INTRODUCTION

Increasing the separation efficiency and understanding the interactions associated with the columns and the stationary phases are some of the major objectives of fundamental studies on liquid chromatography. One of the most efficient columns packed with 5-μm octadecylsilylated (C18) silica particles was developed by Lloyd Snyder, Barry Karger, and Roy Eksteen in the late 1970s. The theory of gradient elution and the method of column characterization based on the hydrophobic subtraction model provided later by Snyder, John Dolan, and coworkers taught us how to understand and utilize the characteristics of a variety of high-efficiency columns [1].

The efficiency of a column, or a chromatographic system, can be demonstrated by applications which are commonly believed to be difficult to achieve. Isotopic separation is one such application, with very minor differences in solute structures. Here several examples of separations of isotopic compounds which were barely possible with high-efficiency and/or high-selectivity columns will be described to illustrate the improvement in efficiency and understanding of modern high-performance liquid chromatography (HPLC) columns and stationary phases.

The separation efficiency of HPLC at each stage of the development of columns and instruments can be illustrated as follows. In the first stage of development of modern HPLC columns, H/D isotopic compounds were able to be separated on the basis of secondary isotope effects; later, separations of oxygen ($^{16}O/^{17}O/^{18}O$) and nitrogen ($^{14}N/^{15}N$) isotopic compounds became possible, based on the primary isotope effects on the ionization of carboxylic acids and aromatic amines. In the normal phase mode, the chromatographic discrimination of isotopic chirality created by H/D substitution was achieved.

At the same time, H/D isotopic separations can provide information on the interactions between the solutes and the mobile/stationary phases by showing the difference in behaviors of the H/D isotopic solutes with minimum differences in their structures. Studies on stationary phases with various functionalities using isotopic solutes with different structural features have enabled an understanding of weak molecular interactions involved in the retention and separation of solutes in HPLC, in addition to the hydrophobic interactions and polar interactions that are well accepted. Fundamental information on molecular interactions will be helpful for understanding the retention behaviors of various solutes in combination with diverse stationary phases and mobile phases.

Along with the isotopic separations undertaken in the authors' laboratories, related results including the methods utilized for generating a large number of theoretical plates will be briefly discussed. Increase in column performance has been achieved by reducing the size of packing materials for nearly half a century. The reduction in particle size was accompanied by the development of ultrahigh pressure LC and small-size columns, which enabled faster separations, but which also resulted in a reduction in the maximum number of theoretical plates that could be generated under a certain pressure limit. A long column, or a set of columns with relatively high permeability, has been employed for isotopic separations instead. Because such columns have to be operated under high pressure, the effect of pressure on solute retention and column efficiency in recycle chromatography will also be mentioned.

4.2 SEPARATION OF HYDROGEN, NITROGEN, AND OXYGEN ISOTOPIC COMPOUNDS WITH HIGH-EFFICIENCY C18 COLUMNS

4.2.1 A SIMPLE REVERSED-PHASE SYSTEM FOR THE SEPARATION OF H/D ISOTOPIC COMPOUNDS

Figure 4.1 shows a baseline separation of palmitic acid-h_{31} and -d_{31} by using a μBondapak C18 column (4 mm i.d.; length, $L=30$ cm; generating a number of theoretical plates, $N=6,000-10,000$) packed with the first 10 μm packing material, i.e. irregularly shaped, totally porous C18 silica particles [2]. Using the same column in

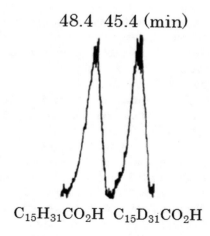

$$48.4 \quad 45.4 \ (\text{min})$$

$$C_{15}H_{31}CO_2H \quad C_{15}D_{31}CO_2H$$

FIGURE 4.1 Separation of H/D isotopic palmitic acids by reversed-phase LC. Column: μBondapak C18 (4 mm i.d., 30 cm). Mobile phase: methanol/water$=80/20$, pH 3.33 with acetic acid. Flow rate: 0.5 mL/min. Solutes: $C_{15}H_{31}COOH$ and $C_{15}D_{31}COOH$. 30°C. (Reprinted with permission from *J. Am. Chem. Soc.* 98 (1976) 1617–1619. Copyright (1976) American Chemical Society.)

40% methanol, a partial separation was achieved for benzene-h_6/d_6 with separation factor $\alpha = 1.043$ and resolution $R_S =$ ca. 0.8, while R_S values of ca. 0.7 and 0.9 were reported for the separation of benzene-h_6/d_6 and toluene-h_8/d_8, respectively, with a Zipax column (1.2 mm i.d., $L = 100$ cm, $N =$ ca. 2,000) packed with spherical, superficially porous 30 μm C18 silica particles in mobile phase of 0–10% methanol in water [3]. The early-middle 1970s was the period of transition from large size silica particles (particle size, $d_P = 30$–50 μm) to 10 μm, then to 5 μm spherical silica particles, while the introduction of the reversed-phase mode started a little earlier.

The separation mechanism of H/D isotopic compounds, or the discrimination between C–H and C–D, was understood on the basis of the difference in the dispersion interactions between the isotopic solutes and the hydrophobic stationary phase, which possesses greater polarizability than the aqueous mobile phase (a methanol–water or acetonitrile–water mixture) [4]. Deuterated compounds possess slightly smaller polarizability, due to their slightly smaller volume, than corresponding H-compounds (a C–D bond is slightly shorter than a C–H bond). The difference in retention factors of isotopic species should be taken into account when deuterated or tritiated compounds are to be used as an internal standard for a LC/MS study [5].

4.2.2 Kinetic Plots Illustrating the Advances in Column Technology

High-efficiency columns developed during 1970s suffered from secondary retention effects caused by unreacted silanols remaining in the bonded phase and metal impurities in the starting silica particles. In the 1980s, ODS silica columns showing much less secondary retention effect were developed. In 1990s and later, technologies leading to higher column efficiency were investigated to reach the limit in N that can be generated per unit time under certain pressure. The limit is shown by the dashed line in Figure 4.2, formed by the minima of the curves (minimum t_0/N^2 values corresponding to the optimum performance, where t_0 is the column dead time, or the retention time, t_R, of a nonretained solute) for each particle size with a 40 MPa pressure limit [6]. Monolithic columns, sub-2 μm fully porous particles, and sub-3 μm core–shell particles realized the possibility of exceeding the limit of performance associated with the 40 MPa pressure limit.

The columns packed with sub-2 μm fully porous particles showed much higher performance than the columns packed with larger sized particles under increased operating pressure as high as 100 MPa in UHPLC [7]. The 2.6–2.7 μm core–shell particles having a 0.35–0.5 μm porous silica layer on the 1.7–1.9 μm solid core provided equivalent column efficiency under lower pressure based on the shorter diffusion path length and the larger particle size [8, 9]. Even smaller 1.3 μm core–shell particles have been supplied to show potentially very high column efficiency under very high pressure [10]. Small-diameter and short columns have been used to avoid a thermal effect and to increase the efficiency of small columns along with instruments with reduced extra-column volume and the sample injection methods that can minimize the extra-column band broadening effects associated with the use of small size columns [11, 12, 13].

As shown in the kinetic plots in Figure 4.2 [14], high-speed separations (the large number of theoretical plates per unit time) have become possible by the use

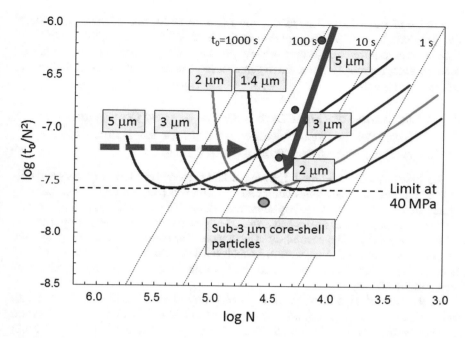

FIGURE 4.2 Kinetic plots illustrating the advances in column technology. The curves describe the limiting performance of columns packed with 1.4–5 μm fully porous particles under the pressure limit of 40 MPa, assuming the following parameters: mobile phase viscosity, $\eta = 0.00046$ Pa s, flow resistance parameter, $\phi = 700$, solute diffusion coefficient, $D_m = 2.22 \times 10^{-9}$ m²/s, and Knox equation, $h = 0.65\nu^{1/3} + 2/\nu + 0.08\nu$. The diagonal dotted lines indicate the t_0 values under each set of conditions. The points for the particle size of 2–5 μm represent the approximate range of typical performance of commercial reversed-phase columns of $L = 15$ cm. The optimum performance for each particle size is shown by the minima of the curves that form the dashed line indicating the limiting performance of particulate columns under the pressure limit of 40 MPa. (Modified with permission from *Anal. Chem.* 78 (2006) 7632–7642. Copyright (2006) American Chemical Society.)

of columns packed with smaller particles, resulting in the increase in N and simultaneous reduction in separation time (t_0), as indicated by the solid arrow along the decreasing particle size. At the same time, however, the maximum number of theoretical plates that can be produced by a column packed with fully porous particles under certain pressure limit has decreased, as shown by the broken arrow directed to the smaller N in Figure 4.2. Figure 4.2 indicates that smaller particles including sub-2 μm fully porous or sub-3 μm core–shell particles are suitable for high-speed separations (small t_0/N), while large particles (5 μm) are suitable for high efficiency (large N) separations.

For a pair of peaks eluting with a retention factor of $k = 4.0$ and separation factor of $\alpha = 1.010$, the plate counts, $N = 250,000$, are required to achieve the resolution of $R_S = 1.0$, according to the relation $R_S = (1/4)(\sqrt{N})(\alpha-1)[k/(k+1)]$. The separation factor for a H/D isotopic solute pair is commonly smaller than 1.010 per one H/D substitution. The separation factor for isotopic compounds to be discussed in Section 4.2.3 is in a range of 1.010–1.012 for $^{16}O/^{18}O$, and 1.010–1.020 for $^{14}N/^{15}N$. The plate counts (N) provided by a C18 silica column available in 1980s were 10,000–25,000, far smaller than required for isotopic separations. Therefore a series of several columns, often in a recycle mode, were used to generate large numbers of theoretical plates for isotopic separations.

4.2.3 SEPARATION OF OXYGEN AND NITROGEN ISOTOPIC COMPOUNDS BASED ON THE PRIMARY ISOTOPE EFFECT IN THE IONIZATION OF CARBOXYLIC ACIDS, PHENOLS, AND AMINES

In the 1980s the chemical modification method was improved to suppress peak tailing for amines and chelating compounds, in order to produce ODS silica packing materials free from the secondary retention effects. The characterization of the secondary retention processes revealed the effects of silanols and metal impurities in silica particles [15]. The secondary effects were eliminated by endcapping and the use of high-purity silica as a starting material [16]. Such ODS stationary phases enabled the separation of $^{16}O/^{18}O$ and $^{14}N/^{15}N$ isotopic compounds by recycle chromatography [17, 18].

In the case of separations with oxygen isotopic benzoic acid (I in Figure 4.3), complete resolution, $R_S = 1.5$, was observed for I-$^{16}O^{16}O$ and I-$^{18}O^{18}O$ with the column recycle method using a 60-cm C18 silica column system (4.6 mm i.d., 15 cm × 2 × 2) 17 times to generate $N = 290,000$ in methanol/0.05 M acetate buffer = 20/80 mobile phase, as shown in Figure 4.3 [19]. The third isotopomer, I-$^{16}O^{18}O$, eluting in between the two peaks was not separated completely. The chromatographic separation factor, $\alpha = {}^{16}k/{}^{18}k = 1.012$, is smaller than the primary isotope effects on the ionization of carboxylic acid ($^{16}K_a/^{18}K_a = 1.020$ for benzoic acid) [17]. The optimum pH for the oxygen isotopic separation is 0.5–1.0 pH units higher than the pK_a of the acid.

For a phenol possessing the pK_a value in a pH range which allows the use of an ODS silica column, oxygen isotopic separation is possible, as for carboxylic acids. The resolution $R_S = 1.5$ was observed for isotopic p-nitrophenol-^{16}OH and p-nitrophenol-^{18}OH in p-nitrophenol-$^{16}O/^{17}O/^{18}O$ mixture with a 12-time column recycle on the 90-cm column system to generate $N = 144,000$ (Figure 4.4) [20]. The p-nitrophenol-^{17}OH was found as an impurity in p-nitrophenol-^{18}OH.

A separation factor of $\alpha = 1.010$ was observed for the PNP-$^{16}OH/^{18}OH$ mixture with a long microbore column packed with ODS silica particles (particle size: $d_P = 10$ μm, 1.5 mm i.d., 50 cm × 12 = 6 m) in 5/95 = methanol/0.02M phosphate buffer (pH = 7.6) [21], while the separation factor was found to be $\alpha(^{16}O)/(^{18}O) = 1.010$ at one cycle on a 60-cm column system (4.6 mm i.d., 15 cm × 2 × 2), $\alpha = 1.015$ at three cycles, and $\alpha = 1.028$ at ten cycles. The separation factor increased with the recycle operation. The reasons for the varying separation factors in recycle chromatography will be discussed later.

Figure 4.5 shows the separation of nitrogen isotopic compounds on a 350-cm column system (4.6 mm i.d., 15 cm × 10 + 25 cm × 8) of C18 silica [20]. Figure 4.5a shows the separation of N,N-dimethylaniline (DMA)-$^{14}N/^{15}N$ standard reagents. A fraction corresponding to the elution range of DMA-^{15}N between the two spikes during the elution of natural DMA (DMA-^{15}N abundance = 0.37%) was collected in

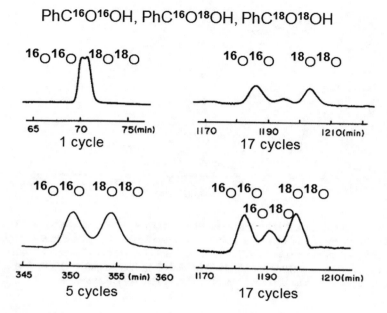

$PhC^{16}O^{16}OH, PhC^{16}O^{18}OH, PhC^{18}O^{18}OH$

FIGURE 4.3 Separation of isotopic benzoic acids in recycle chromatography. Column: Cosmosil 5-C18-P (4.6 mm i.d., 15 cm×2×2). Mobile phase: methanol/0.05 M acetate buffer=20/80 (pH 4.83). Flow rate: 0.8 mL/min. Temperature: 30°C. Cycles: a, 1; b, 5; c, d, 17. 0.5 µg each of I-$^{16}O_2$ and I-$^{18}O_2$ (a-c); 0.25 µg of I-$^{16}O^{18}O$ and 0.5 µg each of I-$^{16}O_2$ and I-$^{18}O_2$ (d). A part of the tail of 1-$^{18}O_2$ was shaved off during the recycle operation in (d). (Reprinted with permission from *J. Chromatogr.* 352 (1986) 307–314. Copyright (1986) Elsevier.)

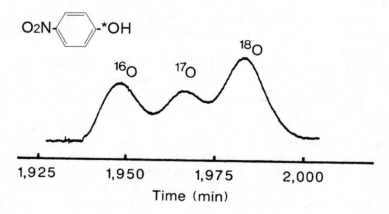

FIGURE 4.4 Separation of isotopic p-nitrophenol-$^{16}O/^{17}O/^{18}O$ by ionization control reversed-phase HPLC. Column: Cosmosil 5C18, 4.6 mm i.d., 90 cm (15 cm×3×2), 12-cycle, mobile phase: CH$_3$OH/0.02 M phosphate buffer=5/95 (pH 7.6), 0.4 mL/min, 40°C. (Reprinted from *Nature,* 341 (1989) 727–728.)

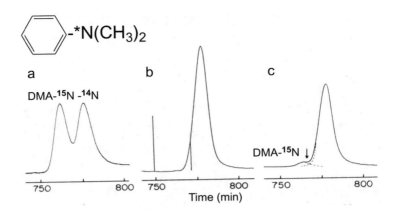

FIGURE 4.5 Separation of N,N-dimethylaniline (DMA)-^{14}N/[15]N. (a) DMA-^{14}N and DMA-^{15}N reagents, (b) enrichment of DMA-^{15}N fraction from DMA of natural abundance isotopic ratio, (c) the reinjection of the enriched fraction in (b). Column: Capcellpak C18-SG, 4.6 mm i.d., 350 cm (10×15 cm+8×25 cm), mobile phase: CH$_3$CN/0.05 M acetate buffer=10/90, 0.01% triethylamine (pH 3.8), flow rate: 0.37 mL/min, 50°C. (Reprinted from *Nature*, 341 (1989) 727–728.)

Figure 4.5b. The concentrated fraction from Figure 4.5b contained ca. 6% DMA-^{15}N, resulting in the chromatogram in Figure 4.5c, which indicated that enrichment of up to 90% isotopic purity of DMA-^{15}N is possible by re-fractionation at the arrow. The isotopic separation method can be applied for various amines in which the nitrogen atom participates in the ionization.

The 350-cm column system provided $\alpha=1.021$ and $R_S=$ca. 0.8 for DMA-^{14}N/^{15}N with $N=36,000$ in 10% acetonitrile at 50°C, as shown in Figure 4.5. A smaller separation factor, $\alpha=1.011$, was observed for DMA-^{14}N/^{15}N with a long microbore column packed with ODS silica particles ($d_P=10$ μm, 1.5 mm i.d., 50 cm×12=6 m) in 20/80=acetonitrile/0.05M acetate buffer (pH=3.8) in the presence of 0.01% triethylamine at 30°C [21]. For aniline-^{14}N/^{15}N, α(^{14}N)/(^{15}N)=1.010 was observed in 5% methanol at 30°C [18].

The oxygen isotopic separations shown in Figures 4.3 and 4.4 were achieved by obtaining the difference in retention as a sum of the retentions of the neutral form and the dissociated form of the acid, reflecting the difference in population of the two forms at a certain pH, based on the difference in pK_a of the isotopic acids, the difference being 0.008–0.009 pH units, as illustrated in Equations (4.1) and (4.2). In Equation (4.1), k is the retention factor at a certain pH, $k_{(AH)}$ is the retention factor of the neutral form, $k_{(A^-)}$ is the retention factor of the dissociated form, and K_a is the dissociation constant. This type of separation is more easily achieved by capillary electrophoresis, where the difference in population of the neutral form and the dissociated form is more directly observed than in ionization-control mode reversed-phase LC [22].

In the case of oxygen isotopic separation with an acid, ^{18}O-acid was retained longer than ^{16}O-acid, while ^{14}N-amine was retained longer than ^{15}N-amine in the nitrogen isotopic separation. This is based on the fact that the covalent bond between the heavier isotope (^{15}N, ^{18}O) and hydrogen is stronger than the bond between the lighter isotope (^{14}N, ^{16}O) and hydrogen; therefore, the population of the neutral form of ^{18}O-acid and the protonated ^{15}N-amine are greater than their counterparts with the lighter isotope, resulting in the longer retention of ^{18}O-acid and the shorter retention of ^{15}N-amine than those with each lighter isotope.

$$k = \frac{k_{(AH)}\left[RCOOH\right]}{\left[RCOOH\right]+\left[RCOO^-\right]} + \frac{k_{(A^-)}\left[RCOO^-\right]}{\left[RCOOH\right]+\left[RCOO^-\right]}$$

$$= \frac{k_{(AH)}}{1+K_a/\left[H_3O^+\right]} + \frac{k_{(A^-)}K_a/\left[H_3O^+\right]}{1+K_a/\left[H_3O^+\right]}$$

(4.1)

$$^{16}K_a / {}^{18}K_a = \frac{k_{(AH)} - {}^{16}k}{{}^{16}k - k_{(A^-)}} \Big/ \frac{k_{(AH)} - {}^{18}k}{{}^{18}k - k_{(A^-)}}$$

(4.2)

A rather similar magnitude of primary isotope effects were observed for oxygen (^{16}O/^{18}O) and nitrogen (^{14}N/^{15}N) isotopic compounds for the difference in the mass number by two and one, respectively. If the relation can be extrapolated to carbon isotopes, a larger isotope effect may be expected for ionization of the isotopic acid (^{12}C/^{13}C), suggesting the possibility of ^{13}C enrichment by the ionization control mode if a suitable working substance containing an ionizable carbon is used.

4.2.4 RECYCLE CHROMATOGRAPHY

Difficult separations have often been carried out by recycle chromatography using a series of several columns to allow the solute mixture band to go through the columns many times before being eluted out. The recycle operation can be performed with a valve to switch the column order of connection to recycle the solute band within the columns only [23] (abbreviated as "column recycle"), or the solute band can be recycled through the entire LC system including the detector, the pump, and the injector after the elution from the column (abbreviated as "system recycle"). On the other hand, monolithic silica columns [24, 25] and a column packed with superficially porous particles for UHPLC [7–9] possessing high permeability can be used as a medium to generate a large number of theoretical plates as a long column [26, 27].

The column recycle method [23] needs an exact timing for switching the column order when the solute band is in the middle of one column to utilize the maximum capacity of the columns until the solute band occupies the whole column on one side. The system recycle method includes the pump, the detector, and the injector (the injector loop is skipped after the first cycle) in the flow path, and is therefore accompanied by the greater extra-column band dispersion, but is easier to perform. In either case, the volume outside the column must be minimized to avoid the additional extra-column band broadening; therefore large size columns are preferred.

During the migration in the columns of the column recycle method, one solute band always exists at the high pressure side, and the other at the low pressure side, for a long time. In the case of oxygen or nitrogen isotopic separations where the dissociating group is different from that of buffering species, the difference in the effect of pressure on the ionization equilibria of the solute and the buffering reagent can cause an increase or a decrease in retention factors and the separation factor of the solutes along with the progress of separation [21, 28]. Also, for a peak of one solute, the leading half and the tailing half always exist under different pressures, resulting in either broadening or compression of the band width compared to normal chromatographic processes. For the examples shown above, increases in band width, retention factors, and separation factors were observed for p-nitrophenol-^{16}O/^{18}O, while decreases in band width, retention factors, and separation factors were observed for DMA-^{14}N/^{15}N in recycle chromatography [21].

The effect of pressure observed in the ionization control mode is one of the pressure effects on solute retention in reversed-phase LC, caused by the difference in distribution equilibrium between the neutral form and the ionized form of the acids and bases under different pressures, which in turn was reflected in the difference in solute retention also affected by the volume change upon transfer of the solute from the mobile phase to the stationary phase [28]. The phenomena are understood based on the relative strength of solvation of the neutral molecule and the ion. Ions are more strongly solvated by the solvent molecules, leading to the reduction of the system volume upon ionization, which is facilitated under high pressure. The effect of pressure appeared as an increase or a decrease in band width and the separation factor for oxygen isotopic p-nitrophenols and nitrogen isotopic DMAs in the examples shown above [21]. In reversed-phase LC, the effect of pressure was also reported for solute retention including dispersion interactions between the aromatic hydrocarbons and the ODS stationary phase [29].

4.2.5 High-Efficiency Separations of H/D Isotopic Compounds by LC Systems Generating Large Numbers of Theoretical Plates

Isotopic separation based on the difference of one deuterium atom on benzene was reported by van der Wal [30, 31]. A partial separation of monodeuterobenzene as a shoulder of benzene peak with a long microbore column (0.32 mm i.d., 225 cm) packed with 5 μm C18 particles in 15% acetonitrile in ca. 700 min was reported. Higher resolution, R_S estimated to be ca. 0.9, was obtained at 19 cycles with a column recycle system with two 4.6 mm i.d., 15-cm columns packed with 3 μm C18 particles in 40% acetonitrile in ca. 450 min [30]. The separation was further improved with the recycle system with two 4.6 mm i.d., 15-cm columns packed with larger particles, $d_P = 5$ μm, to result in the separation of monodeuterobenzene and benzene with resolution R_S=ca. 1.0, after 55 cycles to generate N=ca. 200,000 in about 720 min in acetonitrile/water=20/80 under 38 MPa inlet pressure [31]. The author noted the effect of pressure on the retention factor and the separation factor with the pressure gradient, 38 MPa/30-cm column.

Isotopic separation of benzene-d_6, benzene-1,3,5-d_3, and benzene was achieved by column recycle chromatography using monolithic capillary columns with high permeability. Takeuchi and coworkers utilized two monolithic silica C18 capillary columns (0.1 mm i.d., 45.5 and 44.0 cm) to effect the separation of the three isotopic benzenes with R_S = ca. 2.0 and N=230,000 in ca. 600 min in mobile phase, acetonitrile/water=30/70 with an inlet pressure of 0.8 MPa [32].

Gritti and coworkers reported fast H/D isotopic separation of benzene-d_6, benzene-1,3,5-d_3, and benzene with higher resolution in their study on column recycle chromatography using a small column system with C18 core–shell silica particles (d_P=2.7 µm, 3.0 mm i.d., 15 cm×2) in acetonitrile/water (55/45, v/v). Resolution up to 2.0 was observed for the difference of three D atoms with α=1.023 in a 22-time recycle operation for 4,500 seconds to generate N=275,000 for solutes with k=2.0 [33]. This example is not as high resolution as the one reported by van der Wal [30, 31], but still demonstrates respectable resolutions in a much shorter time indicating the high efficiency of modern columns.

The effect of pressure in recycle chromatography, namely the increase in retention factor and the decrease in column efficiency in the recycle operation, was reported for the behavior of hydrophobic aromatic hydrocarbons, benzoanthracene and chrysene [34], as mentioned earlier, for the ionization-control mode separation of oxygen isotopic p-nitrophenols [21]. Gritti and coworkers successfully predicted the behavior of the aromatic hydrocarbons, taking into account the volume change associated with the transfer of the solute from the mobile phase to the stationary phase, as reported previously [29]. Van der Wal also studied the effect of pressure on retention and column efficiency, and concluded that very high efficiency separations would need high pressure, a column packed with relatively large particles, and/ or long analysis time [31]. Although the importance of increasing the selectivity has been emphasized for difficult separations, the separations based on the difference of 1–3 deuterium atoms in benzene shown above were not optimized in terms of selectivity. This is because these examples were intended to prove the efficiency of the chromatographic systems.

An example showing the resolution with the smallest separation factor (so far as the authors are aware) was presented by system recycle chromatography in reversed-phase mode utilizing an optimized mobile phase for H/D isotopic separations [35]. The separation by the difference of single H/D substitution, with α=1.008, was reported for deuterated benzenes (Figures 4.6a and 4.6b). Even more difficult one was the separation of mono-deuterated toluene isomers, toluene-α-d and toluene-4-d, with a separation factor of 1.0016 in a ternary mobile phase, as shown in Figures 4.6c and 4.6d.

Large columns were employed for the operation of the system recycle chromatography. Recycling 12 times on a C18 column system (6 mm i.d., 15 cm×4) generated ca. 500,000 theoretical plates to achieve separation (R_S> 1.0) of five isotopologues (benzene-d_6, benzene-d_5, benzene-1,3,5-d_3, benzene-d, and benzene) in ca. 460 min based on the difference of one H/D substitution, as shown in Figures 4.6a and 4.6b. The separation between 4-deuterotoluene and α-deuterotoluene (Figures 4.6c and 4.6d) was more difficult to achieve. As shown in Figures 4.6c and 4.6d, recycling on the 60-cm column system for nearly three days to generate ca. 3.5–4 million theoretical plates resulted in a resolution of ca. 0.6 with the separation factor of α=1.0016 between the two mono-deuterated isotopic toluenes.

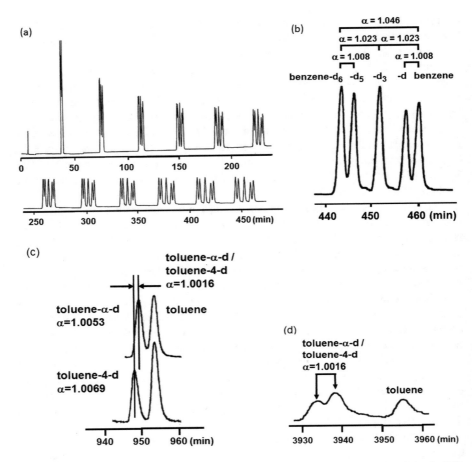

FIGURE 4.6 Separation of benzene and toluene isotopologues on the basis of single H/D substitutions. Column: Consmosil 5C18-MS (6.0 mm i.d., 15cm×4=60 cm). Flow rate: 2.0 mL/min, Detection: UV 254 nm. 30°C. (a, b) Mobile phase: methanol/acetonitrile/water=25/20/55. Sample: 1. benzene-d_6, 2. benzene-d_5, 3. benzene-1,3,5-d_3, 4. benzene-d, 5. benzene. (c,d) Mobile phase: methanol/acetonitrile/water=40/20/40. Sample: toluene-4-d and toluene-α–d. (Reprinted from *LC-GC* Europe, 32, S5 (2019) 14–19.)

Interestingly the use of methanol as a part of the ternary mobile phase produced greater isotopic resolutions, based on the larger separation factors, than the acetonitrile/water mobile phases that are frequently employed for separations of isotopologues. That the isotopic separation factor, α(H/D), is greater in methanol/water than in acetonitrile/water, with a similar hydrophobic separation factor, α(CH$_2$), implies the different mechanisms for CH/CD differentiation from a hydrophobic partition. The effect of organic solvent in the mobile phase on the discrimination between CH and CD will be discussed in Section 4.4.3.

The ternary mobile phases provided plate counts close to those in acetonitrile/water, and were shown to be more advantageous in terms of the column back pressure, the separation factor (α), and the retention factor (k) than binary mobile phases [35]. Utilizing the results obtained in two binary mobile phases, methanol/water and acetonitrile/water (for example, see Section 4.4.3), k and α values in a ternary mobile phase can be interpolated approximately based on the ratio, methanol/acetonitrile, in the mobile phase. The effect of the retention factor, k, on resolution (R_S, Equation (4.3)) in recycle chromatography is different from a single pass operation, because obtainable plate counts in recycle LC, $N_{(obs-recycle)}$, depend on the plate counts per cycle, $N_{(C)}$, the number of recycle operations, t_R/t_C, and in turn k, as indicated in Equations (4.4) and (4.5), where t_C is the time for one cycle and t_0 is the column void time. While $N_{(C)}$ tends to be slightly smaller with the increase in t_R/t_C, smaller k or t_C and a faster flow rate contribute to the larger $N_{(obs-recycle)}$.

$$R_s = (1/4)\left(\sqrt{N}\right)(\alpha - 1)\left[k / (k+1)\right] \tag{4.3}$$

$$N_{(obs-recycle)} = N_{(C)}\left(t_R / t_C\right) \tag{4.4}$$

$$t_C = t_0(1+k) \tag{4.5}$$

As indicated by the example of H/D isotopic separation of benzenes at relatively low pressure [32], monolithic silica columns having high external porosity can potentially exceed the limit of the performance of columns packed with fully porous particles shown in Figure 4.2. Many attempts aiming at 1,000,000 theoretical plates by HPLC have been reported since the 1970s without success for retained solutes. Successful examples were shown for unretained solutes with a series of many columns packed with 5–10 μm particles [36]. An efficiency greater than 1,000,000 theoretical plates was reported in the 1990s by utilizing open-tubular capillary columns of 10–30 μm i.d. for solutes with small retention factors ($k = 0.1$) [37,38].

Hybrid monolithic silica capillary columns prepared from a 3/1 mixture of tetramethoxysilane/methyltrimethoxysilane possess a large external porosity, over 70%. 3–5-m columns providing high permeability, $K = 1.4$–1.5×10^{-13} m², with a plate height of $H = 7$–10 μm, are advantageous for generating a large number of theoretical plates. Three monolithic silica capillary columns (100 μm i.d.) connected to a total length of 12.38 m generated more than one million theoretical plates for aromatic hydrocarbons (k up to 2.4) at 40–50 MPa [39].

The chromatogram in Figure 4.7 demonstrates the high efficiency of monolithic silica C18 capillary columns connected to make $L = 850$ cm, which separated benzene isotopologues based on the difference in one H/D substitution, benzene-d_6, benzene-1,3,5-d_3, benzene-d, and benzene in the ternary mobile phase. The chromatogram implies that the sample contained benzene-d_2, -d_4, and -d_5 as impurities in addition to the four isotopologues indicated, showing the possibility of separation of all benzene H/D isotopologues. A separation factor of 1.008 per one H/D was observed in Figure 4.7 with $N = 240{,}000$ to result in $R_S =$ ca. 0.9.

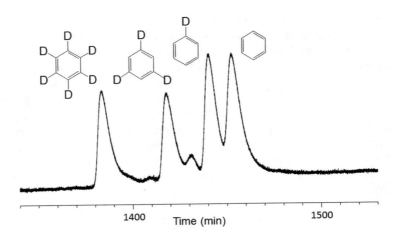

FIGURE 4.7   Separation between benzene isotopologues. Mobile phase: CH₃CN/CH₃OH/ H₂O = 10/5/85. Column: monolithic silica C18, 100 μm I.D., effective length: 850 cm (500 cm + 350 cm). Detection: 210 nm. Temperature: 30°C. u = 1.02 mm/s. Pressure = 34 MPa. Sample: 1. benzene-d₆, 2. 1,3,5-benzene-d₃, 3. benzene-d, 4.benzene. (Reprinted with permission from *Anal. Chem.* 80 (2008) 8741–8750. Copyright (2008) American Chemical Society.)

4.3   H/D ISOTOPIC SEPARATIONS CONTRIBUTING TO THE UNDERSTANDING OF THE SOLUTE–STATIONARY/ MOBILE PHASE INTERACTIONS IN HPLC

4.3.1   Direct Separation of Enantiomers Based on H/D Isotopic Chirality

The separation of isotopic compounds is significant in separation science for how it illustrates the extremely high ability of HPLC systems, and has its own significance, too, as a longtime target of scientific curiosity. Scientists in the 1930s tried to separate enantiomers based on H/D isotopic chirality, soon after the start of production of deuterium oxide. The reportedly successful studies on the separation of enantiomers created by the presence of deuterium atoms [40, 41] were disputed by later studies [42, 43]. One of the examples was the resolution of benzhydrol-d₅ (that is, phenyl(phenyl-d₅)methanol), which was reportedly resolved as its phthalic acid half ester by fractional recrystallization in the presence of brucine [41]. Makino and coworkers synthesized enantiomers with known configurations to argue against the reported resolution of phenyl(phenyl-d₅)methanol enantiomers based on isotopic chirality [43].

Figure 4.8 shows chromatograms illustrating the separation of phenyl(phenyl-d₅) methanol enantiomers based on isotopic chirality by system recycle chromatography using tribenzoylated cellulose-coated silica particles [44]. Figure 4.8a was obtained after 65 cycles on a column system (6.0 mm i.d. × 15 cm × 4 = 60 cm) to generate $N = 350,000$ and resolution R_S = ca. 1.0, with a separation factor of $\alpha = 1.0080$.

The mobile phase was hexane/2-propanol = 95/5 (v/v), and in the stationary phase cellulose tribenzoate was coated onto silica surfaces. The functional groups involved for the interaction with the solute in the stationary phase are the ester group, the

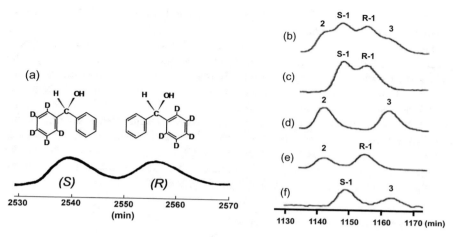

FIGURE 4.8 Separation of enantiomers based on the isotopic chirality created by the presence of H/D substitution. Column: Tribenzoylated cellulose on silica (6.0 mm i.d., 60 cm). Mobile phase: Hexane/2-Propanol = 95/5, flow rate: 2.0 mL/min, temperature: 40°C, detection: UV 220 nm. (a) 65 cycles. Solute: (R)- and (S)-phenyl(phenyl-d_5)-methanol, 20 mg/mL, 0.5 μL was injected. (b–f) 29 cycles. Solute: (S)-phenyl(phenyl-d_5)-methanol (S-1), (R)-phenyl(phenyl-d_5)-methanol (R-1), diphenylmethanol (2), and di(phenyl-d$_5$)methanol (3). (Reprinted with permission from *Anal. Chem.*, 69 (1997) 2610–2612. Copyright (1997) American Chemical Society.)

three phenyl groups per unit structure, and the cellulose backbones of the stationary phase. The mobile phase cannot discriminate enantiomers. The major interaction for the solute binding to the stationary phase is presumably the hydrogen bond between the hydroxyl group of diphenylmethanol and the ester groups of cellulose tribenzoate in the stationary phase.

In this chromatographic system, diphenylmethanol was eluted earlier than di(phenyl-d$_5$)methanol, as shown in Figures 4.8b–f. In other words, the stationary phase retains the phenyl-d$_5$ group preferentially compared to the phenyl group. The results imply that the mobile phase containing hexane as a major component possesses greater polarizability than the binding site of cellulose benzoyl ester in the stationary phase. The ester groups possess very small polarizability, as is discussed later [45].

The separation of enantiomers of phenyl(phenyl-d$_5$)methanol was supposed to be achieved by interactions including the hydrogen bonding between the OH group of the solute and the ester functionality in the stationary phase, as well as by the localized dispersion interactions between the phenyl/phenyl-d$_5$ groups of the solutes with the phenyl groups on the cellulose backbones of the stationary phase, where the phenyl groups were sterically arranged to show the preferential retention of the (R)-isomer of the solute (R-1). The local dispersion interactions could be emphasized in the low polarizability environment of the stationary phase.

This interpretation is supported by the following results. The stationary phase preferentially retains the (S)-form of (4-substituted-phenyl)phenylmethanol with a substituent, Cl, Br, or CH$_3$, on one of the two phenyl groups of diphenylmethanol, as shown in Table 4.1. The (R)-form was preferentially retained in the case of phenyl(phenyl-d$_5$)methanol.

The substituted phenyl group having the Cl, Br, or CH$_3$ substituent possesses greater polarizability than the phenyl group. The phenyl group possesses greater polarizability than the phenyl-d$_5$ group. If the priority of the substituents on the central carbon atom is compared based on the polarizability, (R)-phenyl(phenyl-d$_5$) methanol possesses configuration similar to that of (S)-(4-substituted phenyl) (phenyl)methanol, as shown in Figure 4.9. Therefore the differentiation based on isotopic chirality was assumed to be provided by the difference in dispersion interactions of the phenyl groups of the stationary phase with (R)-configuration of the phenyl and phenyl-d$_5$ groups of the solute; that is, preferred over similar interactions with the (S)-configuration, as shown in Table 4.1 and Figure 4.9.

4.3.2 Separation of Diastereomers Created by H/D Isotopic Chirality by Reversed-Phase LC

Figure 4.10 shows chromatograms illustrating the separation of diastereomers of methyl 3-phenyl-3-phenyl-d$_5$-glycidate (1–4) along with methyl 3,3-di(phenyl-d$_5$)

TABLE 4.1

Retention Factor and Separation Factor Observed for (R)- and (S)-(substituted phenyl)phenylmethanol

Substituent	k(R)	k(S)	α
4-CH$_3$	2.19	*2.95*	1.35
4-Cl	4.25	*6.91*	1.63
4-Br	3.89	*5.76*	1.48
2,3,4,5,6-d$_5$	*4.367*	4.333	1.008

Bold, italic characters indicate the retention factor of the preferred isomer.

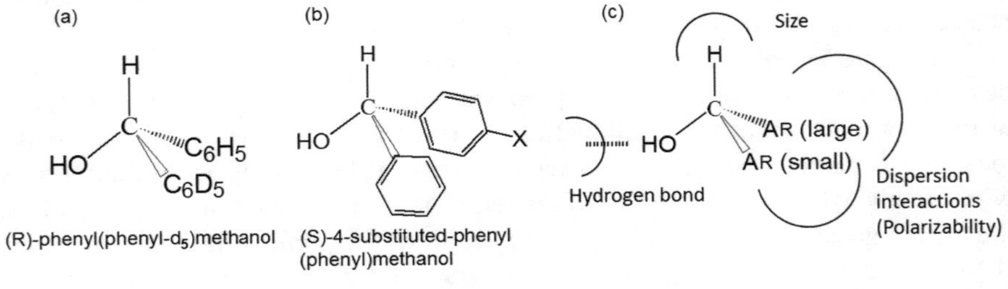

FIGURE 4.9 Stereochemistry of (a) phenyl(phenyl-d$_5$)methanol, (b) (substituted-phenyl) phenylmethanol, and (c) the interactions causing preferential retention of the preferred enantiomer in chiral separation of phenyl(phenyl-d$_5$)methanol on cellulose-ester-coated silica stationary phase.

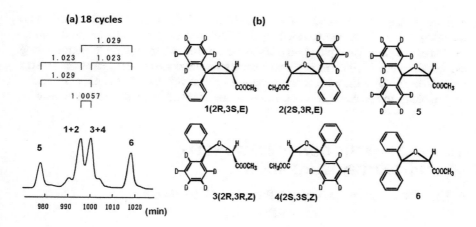

FIGURE 4.10 (a) Separation of diastereomers of methyl 3-phenyl-3-phenyl-d_5-glycidate (1–4) based on isotopic chirality. Compounds 5 and 6 were included in the sample mixture. Column: Cosmosil 5C18-MS, four columns of 6 mm i.d., 15 cm in length. Mobile phase: methanol/water=65/35. Detection: UV 254 nm. Solutes were recycled through the column system, the detector, and the pump. 30°C. (b) Structures of methyl 3-phenyl-3-phenyl-d_5-glycidate (1-4), methyl 3,3-di(phenyl-d_5)glycidate (5), and methyl 3,3-diphenylglycidate (6). (Reprinted with permission from *J. Am. Chem. Soc.* 118 (1996) 759–762. Copyright (1996) American Chemical Society.)

glycidate (5) and 3,3-diphenylglycidate (6) by silica C18 stationary phase in a mobile phase, methanol/water=65/35 [46]. The diastereomers created by the presence of isotopic chirality were separated by 18 cycles in system recycle chromatography using a silica C18 column (6.0 mm i.d., 15 cm×4=60 cm) to afford N=600,000, a separation factor of α=1.0057, and resolution R_S=ca. 0.8.

The separation was achieved by common reversed-phase mode, where the stationary phase and mobile phase cannot differentiate the enantiomers. The four isomers (compound 1–4, Figure 4.10b) were separated into two peaks, or two pairs of enantiomers that are diastereomeric to each other. In Figure 4.10a, methyl 3,3-diphenylglycidate (6) was retained longer than 3,3-di(phenyl-d_5)glycidate (5). In other words, the stationary phase preferentially retains the phenyl group over the phenyl-d_5 group. This corresponds to the greater polarizability of the stationary phase than the mobile phase.

A common observation between the direct separation of the isotopic chirality shown in Figure 4.8 and the separation of diastereomers created by the presence of phenyl and phenyl-d_5 groups (Figure 4.10) is the fact that the phenyl group, having the greater polarizability, was preferentially partitioned into the more hydrophobic environment possessing higher polarizability, and the less polarizable phenyl-d_5 group was preferentially partitioned into the less polarizable polar medium.

Solutes 3 and 4, having the less polarizable phenyl-d_5 group and ester linkage close to each other, exposing the more polarizable phenyl group to the medium, were preferentially retained with the C18 stationary phase with a higher polarizability than the methanol/water mobile phase. In contrast, the greater exposure of the less polarizable phenyl-d_5 group in solutes 1 and 2 provided the preferential partition into the less polarizable, aqueous mobile phase. With this substrate, the eclipsed conformation of the phenyl and phenyl-d_5 groups against the two groups on the neighboring carbon atom provided by the epoxy group emphasized the effect of stereochemistry on polarizability of the molecule, which in turn was reflected in the retention factor of the solutes in reversed-phase LC through the extent of dispersion interactions with the stationary phase.

4.4 MECHANISTIC CONSIDERATIONS ON H/D ISOTOPIC SEPARATION

4.4.1 EFFECT OF THE STATIONARY PHASE ON H/D ISOTOPIC SEPARATIONS IN REVERSED-PHASE LC

The effect of H/D isotopic substitution in alkanes, alkyl alcohols, and aromatic hydrocarbons on retention factors in methanol–water was reported for ten stationary phases including octadecyl (C18), 4,4-di(trifluoromethyl)-5,5,6,6,7,7,7-heptafluoroheptyl (F13C9), and 3-(pentabromobenzyloxy)propyl (PBB) groups bonded to silica particles [45]. A part of the results is shown in Table 4.2, where the total isotope effects on retention (TIE) and those per one H/D substitution (%IE) are listed, as calculated from the separation factor of isotopic compounds, $\%IE = 100[(k_H/k_D)^{1/n} - 1]$.

The isotope effects represent the interaction of the CH/CD bond of solutes with the stationary phase and the mobile phase. In the same mobile phase, the H/D isotope effect was found to be influenced by the stationary phase. The positive isotope effect ($k_H/k_D > 1.0$) indicates that the stationary phase plays an active role in solute binding, providing attractive interactions, and that the CH/CD bond is weakened

---

TABLE 4.2

Hydrogen–Deuterium Isotope Effect on Retention Observed with the Stationary Phases Possessing Different Polarizability

	k_H/k_D, TIE (%IE per D)		
Compound	C18	PBB	F13C9
Naphthalene	1.056 (0.680)	1.056 (0.688)	1.013 (0.168)
Anthracene	1.077 (0.748)	1.078 (0.752)	1.018 (0.178)
Octane	1.064 (0.347)	1.080 (0.427)	0.977 (−0.129)
1-Dodecanol	1.089 (0.343)	1.109 (0.414)	0.971 (−0.118)
Structural unit		α(separation factor)	
α(CH$_2$)	1.680	1.590	1.330
α(C$_4$H$_2$)	2.553	5.240	1.074

Mobile phase: 70% methanol, TIE: total isotope effect = $\alpha(k_H/k_D)$, $\%IE = 100[(k_H/k_D)^{1/n} - 1]$. $\alpha(CH_2) = k(\text{amylbenzene})/k(\text{butylbenzene})$, $\alpha(C_4H_2) = k(\text{naphthalene})/k(\text{benzene})$
(Modified with permission from *J. Am. Chem. Soc.* 125 (2003) 13836–13849. Copyright (2003) American Chemical Society.)

---

in the stationary phase, especially on PBB and C18, compared to the mobile phase. The small, in some cases inverse, isotope effects observed with the F13C9 stationary phase indicate that the stationary phase interacts very weakly, sometimes weaker than the mobile phase, with an alkyl group. These observations are consistent with the retention mechanism dominated by London dispersion forces between the solute and the stationary phase, and with the idea that lipophilic phenomena make an important contribution to the hydrophobic effect. A similar conclusion was provided based on the relation between the retention factors and refractive indices of solutes for these stationary phases (refractive index is related to the polarizability of a molecule) [47]. Valleix and coworkers obtained similar results, and provided similar interpretations for their results, on H/D isotopic effects, the contributions of dispersion interactions (importance of polarizability), and the interactions of CH/CD with both alkyl groups in the stationary phase and the CH part of the mobile phase components [48].

Apparently the aromatic group in the solute and/or the stationary phase contribute to the larger isotope effect, suggesting the contribution of CH/CD–π(aromatic) and Ar-H/D–π(aromatic) interactions to the retention of the alkyl as well as aromatic solutes [45, 48]. A similar effect was observed for aromatic (H/D) hydrocarbons on the C18 stationary phase, indicating that CH–π(ArH/D) interactions also resulted in a large H/D isotope effect. Contribution of interactions in the mobile phase to the hydrophobic effect is also indicated by the dependence of %IE on the methanol content of the mobile phase. The effect is found to be stronger for acetonitrile than for methanol. The total hydrophobic effect is much smaller with the F13C9 phase, as seen with the %IE and $\alpha(CH_2)$ value. The CH bonds existing in the organic solvent molecules in the mobile phase also seem to undergo interactions with the aromatic group of solutes, reducing the %IE for solutes with an aromatic group more than for alkyl type solutes, as indicated by the twice as great sensitivity of %IE to the increase in the organic solvent content of the mobile phase.

4.4.2 Isotope Effects in Relation to Solute–Stationary Phase Steric Complementarity and System Pressure

The effect of pressure on retention and the H/D isotope effect were studied with long alkyl-bonded silica stationary phases, including monomeric C18 (InertSustain C18), polymeric C18 (Inertsil ODS-P), and monomeric C30 stationary phase (Inertsil C30) [49]. The retention factor of deuterated species (k_D), total isotope effect, and single isotope effect (%IE) at low and high pressure are listed for the three alkylsilylated silica stationary phase in Table 4.3. The k_D indicates the retention factor of a deuterated compound in $CH_3CN/H_2O = 80/20$ at 20°C. The k_D(high)/k_D(low) value indicates the ratio of the retention factor at high pressure to that at low pressure. Typically the reproducibility in %IE, calculated from the well-separated peaks in the same run, is within plus or minus 0.01. The differences in %IE observed for the planar solutes, pyrene and triphenylene, on ODS-P and C30 stationary phases between at high and low pressure are outside experimental reproducibility.

Stationary phases consisting of polymeric ODS and C30 alkyl groups were reported to possess ordered structures [50], resulting in the preferential retention of

TABLE 4.3
Effect of Pressure on Hydrogen–Deuterium Isotope Effect for Retention of Aromatic Hydrocarbons with Stationary Phases of Different Alkyl-Bonded Groups

Solute	Stationary phase	Inertsil ODS-P		Inertsil C30		InertSustain C18	
	Pressure (MPa)	4.3	54.0	3.4	54.5	3.1	52.7
Pyrene	k_D	6.40	8.54	6.35	7.72	4.66	5.13
($C_{16}H_{10}/C_{16}D_{10}$)	k_D(high)/k_D(low)	1.334		1.216		1.100	
	TIE	1.088	1.094	1.073	1.076	1.061	1.062
	(%IE)	(0.84)	(0.90)	(0.70)	(0.73)	(0.60)	(0.61)
Ortho-	k_D	3.71	4.25	3.93	4.14	4.44	4.70
terphenyl	k_D(high)/k_D(low)	1.141		1.053		1.059	
($C_{18}H_{14}/C_{18}D_{14}$)	TIE	1.061	1.063	1.059	1.060	1.058	1.060
	(%IE)	(0.43)	(0.44)	(0.41)	(0.42)	(0.41)	(0.41)
Triphenylene	k_D	7.75	11.32	8.10	10.51	5.29	5.96
($C_{18}H_{12}/C_{18}D_{12}$)	k_D(high)/k_D(low)	1.461		1.298		1.126	
	TIE	1.096	1.101	1.080	1.083	1.068	1.069
	(%IE)	(0.77)	(0.81)	(0.64)	(0.67)	(0.55)	(0.56)

Mobile phase: acetonitrile/H_2O = 80/20, temperature: 20°C. The k_D(high)/k_D(low) value indicates the ratio of k_D at high pressure to k_D at low pressure for each set of solute and stationary phase. TIE: total isotope effect = $\alpha(k_H/k_D)$, %IE = $100[(k_H/k_D)^{1/n} - 1]$. (Modified with permission from supplementary materials of *J. Chromatogr. A*, 1339 (2014) 86–95. Copyright (2014) Elsevier.)

planar aromatic hydrocarbons compared to nonplanar solutes [15, 51]. The ODS-P and C30 stationary phases showed about twice as much retention (k_D in Table 4.3) for planar triphenylene as nonplanar ortho-terphenyl, having the same number of carbon atoms and π-bonds. The H/D isotope effects, TIE and %IE, observed for triphenylene are significantly larger than those observed for ortho-terphenyl.

A much larger increase in retention, a large k_D(high)/k_D(low) ratio, was observed by the increase in pressure by about 50 MPa for the planar solute on these stationary phases compared to the nonplanar solute. Significant increase in H/D isotope effect was observed for the planar solutes on polymeric ODS and C30 stationary phases with the increase in pressure, while such an increase in the isotope effect was not observed for nonplanar ortho-terphenyl. The increases in retention and H/D isotope effect caused by the solute planarity and the pressure increase were much smaller on the monomeric C18 stationary phase. Another planar solute, pyrene, provided results similar to triphenylene on the three stationary phases.

The results shown in Table 4.3 are consistent with the proposed mechanism for the preferential retention of planar aromatic hydrocarbons with the ordered long-chain stationary phase with steric complementarity [51]. The planar solutes are supposed to be retained between the long alkyl chains interacting with each other by dispersion interactions and specific CH–π(Ar) interactions resulting in a large isotope effect, especially at high pressure, due to the shorter distance between the solute and the ordered alkyl chains under high pressure. High pressure causes an increase in density of the system components, resulting in a decrease in intermolecular distance and an increase in dispersion interactions.

The results shown in Sections 4.1 and 4.2 suggest that discrimination between CH and CD is possible based on dispersion interactions. In the case of separation based on isotopic chirality, shown in Figure 4.8, the separation was considered to be achieved by a more favorable binding of the R-form of the solute than the S-form to a part of the binding site in the stationary phase having significant polarizability, perhaps sterically arranged phenyl groups of cellulose tribenzoate (Figure 4.9).

Under reversed-phase conditions, the discrimination between CH and CD is influenced by the stationary phase structure, the hybridization state of the carbon atom bearing H or D, the neighboring atom of the carbon (of CH/CD), and other conditions, even in the same mobile phase. Stationary phases and/or solutes with high polarizability tend to provide large isotope effects, or easy discrimination between CH/CD, indicating that the dispersion interactions between the solute and the stationary phase significantly contribute to retention and separation in addition to the hydrophobic interactions in the mobile phase. Therefore a stationary phase having heavy heteroatoms such as the PBB stationary phase provides very large dispersion interactions, showing preferential retention of solutes containing heavy atoms and/or aromatic groups with high polarizability [52]. It is also known that organic solvent molecules distribute between the mobile phase and the stationary phase [53], and that those existing in the stationary phase contribute to selectivity through enhanced polar interactions with the polar functional group of the solute in the nonpolar environment [54]. Polar interactions are emphasized in a medium of low dielectric constant, as are the electrostatic interactions between the opposite charges. The organic solvent molecules existing in the stationary phase also seem to contribute to the isotopic selectivity, α(H/D), as is discussed in the following section.

4.4.3 EFFECT OF THE MOBILE PHASE ON H/D ISOTOPIC SEPARATIONS IN REVERSED-PHASE LC

In Figure 4.11a, the separation factors between toluene and benzene, α(CH$_2$) = ($k_{toluene}/k_{benzene}$) (the right Y-axis), and between benzene and benzene-d$_6$, α(H/D) (the left Y-axis), are plotted against the content of methanol or acetonitrile in the mobile phase. The α(H/D) values are plotted against α(CH$_2$) in Figure 4.11b. The retention factors of benzene-d$_6$ in each mobile phase are attached to the plots in Figure 4.11b. Retention factors (k) and the separation factors, α(H/D) and α(CH$_2$), decrease with the increase in organic solvent content. The retention factors and the separation factors are larger in methanol/water than in acetonitrile/water at the same organic percent. For example, the α(CH$_2$) value in acetonitrile/water=30/70 is similar to that in methanol/water=55/45, and the retention factor of benzene-d$_6$ in acetonitrile/water=35/65 is similar to that observed in methanol/water=45/55 (regarding uracil as a t_0 marker) [35].

The α(H/D) in methanol/water are clearly larger than those in acetonitrile/water when α(H/D) are compared at similar α(CH$_2$) values, as reported by Tchapla, et al. [48] and by Kanao, et al. [55], in spite of the smaller retention factors in

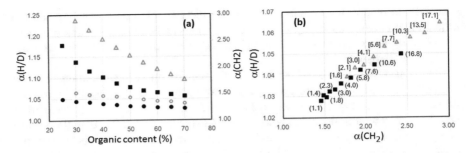

FIGURE 4.11 (a) Plot of the methylene group selectivity, $\alpha(CH_2)$ ($\triangle$, $\blacksquare$, the right Y-axis), between toluene and benzene, and the isotopic separation factor, $\alpha(H/D)$ ($\circ$, $\bullet$, the left Y-axis), between nondeuterated and perdeuterated benzene, against organic solvent content (%) of the mobile phase, in methanol/water (open symbols) and acetonitrile/water (solid symbols). (b) Plot of $\alpha(H/D)$ for benzene against $\alpha(CH_2)$ in methanol/water ($\triangle$, 30–70% methanol) and acetonitrile/water ($\blacksquare$, 25–70% acetonitrile). The retention factor, k, of benzene-d_6 is given beside each plot in methanol/water in brackets and in acetonitrile/water in parentheses. (Reprinted from *LC-GC* Europe, 32, S5 (2019) 14–19.)

methanol/water. The results suggest that the isotopic selectivity, $\alpha(H/D)$, seems to be provided by a mechanism which is different from the hydrophobic interactions that presumably account for the $\alpha(CH_2)$ values. The $\alpha(CH_2)$ values should represent the behavior of a typical hydrophobic group. The results imply that the mechanism differentiating CH from CD results in a greater $\alpha(H/D)$ in methanol/water than in acetonitrile/water, though also in similar $\alpha(CH_2)$ values.

The hydrophobic property or lipophilicity of a stationary phase is affected by the surface density of the alkyl groups, as indicated by the larger $\alpha(CH_2)$ with high-coverage C18 than low-coverage C18 in the same aqueous mobile phase [15]. High-polarizability stationary phases such as PBB and PYE provide larger $\alpha(H/D)$ values for aliphatic CH/CD along with an $\alpha(CH_2)$ smaller than C18 [45]. These results indicate the contribution of attractive interactions between a solute and the stationary phase to retention, namely the dispersion interactions (instantaneous dipole-induced dipole interactions) that play a major role for H/D isotopic separations, because of the greater polarizability of the CH than the CD group [4, 45, 48].

That the $\alpha(H/D)$ values in methanol/water are larger than in acetonitrile/water, under conditions providing the same $\alpha(CH_2)$ values, can be explained by the greater contribution of dispersion interactions in methanol/water. When acetonitrile/water and methanol/water provide similar $\alpha(CH_2)$ values, the free energy changes associated with the transfer of one methylene group (CH$_2$) from the mobile phase to the stationary phase are similar. Acetonitrile is known to be absorbed by the stationary phase more than methanol, at similar organic content of the aqueous mobile phase [53], making the C18 stationary phase less hydrophobic and less polarizable than the C18 phase in methanol/water. In other words, similar $\alpha(CH_2)$, or similar k values, can be obtained with higher water content of acetonitrile/water mobile phase than methanol/water. The intrinsic polarizability of the stationary phase is more strongly reflected in solute retention in methanol/water than in acetonitrile/water, resulting in the larger H/D isotope effects, when the two kinds of mobile phases provide similar hydrophobic retention or hydrophobic selectivity.

The amount of organic solvent extracted into the stationary phase from the mobile phase results in the difference in the phase ratio, also influencing the retention factors. The acetonitrile/water mobile phase provided, compared to methanol/water, a smaller void volume and considerably larger retention factors for a variety of solutes, when the two kinds of mobile phases provided similar $\alpha(CH_2)$ values [54] (also shown in Figure 4.11b). Hydrophobic solutes tend to be associated with organic solvent molecules in aqueous mobile phases which can also contribute to the smaller $\alpha(H/D)$ in a mobile phase containing acetonitrile which is more polarizable than methanol. This way, the generally larger isotope effects observed in methanol/water than in acetonitrile/water can be rationalized.

4.4.4 EFFECTS OF SOLUTE STRUCTURE ON H/D ISOTOPIC SEPARATIONS IN REVERSED-PHASE LC

Chromatograms in Figure 4.12 show that deuterium atoms in different environments provide a different extent of reduction in retention in reversed-phase LC. The chromatograms were obtained using a monolithic silica C18 capillary column in acetonitrile/water mobile phases. The long (450 cm) column generating nearly 300,000 theoretical plates enabled resolution of isotopic peaks with a separation factor as small as 1.01 or even smaller, facilitating the study on the isotope effect.

It is noted that the isotope effect on retention is small for a CH/CD in a terminal methyl group compared to a CH/CD in a methylene group. It is also noted that CH/CD, next to a heteroatom such as nitrogen or oxygen, resulted in large H/D isotope effects, especially when the CH/CD bond existed in between oxygen and carbon. The first two peaks of solute 10 (m-diethoxybenzene) in Figure 4.12b provided the larger separation factor based on the four deuterium substitutions, compared to the separation based on the six deuterium atoms in the two terminal methyl groups for the last two peaks.

Figure 4.13, which is intended to show the magnitude of total isotope effects ($TIE = k_H/k_D$) and the single isotope effects ($\%IE = 100[(k_H/k_D)^{1/n} - 1]$) for various deuterated solutes with 2–10 H/D substitutions simultaneously, illustrates the effects of the number of deuterium atoms and the location of the CH/CD group in a molecule, as well as the neighboring atoms on the isotopic selectivity with respect to a whole molecule (TIE) and a single H/D substitution (%IE). The number of deuterium atoms in a solute molecule was varied from 2 to 10 for the compounds plotted in Figure 4.13.

Larger TIE values (large values along the X-axis) were observed in the isotopologues with the greater number of deuterium substitutions that were expected to provide the greater difference in molecular polarizability [45]. The single isotope effect (%IE) for each H/D substitution (the Y-axis), however, depends on the location, and on the neighboring atoms and groups, of the CH/CD group. A deuterium in the terminal methyl group resulted in a smaller %IE compared to a deuterium in a methylene group. The CH/CD group seemed to show a larger %IE with a neighboring atom in the order N > O > F > S > C. (The %IE for a methylene CH/CD was 0.44, while

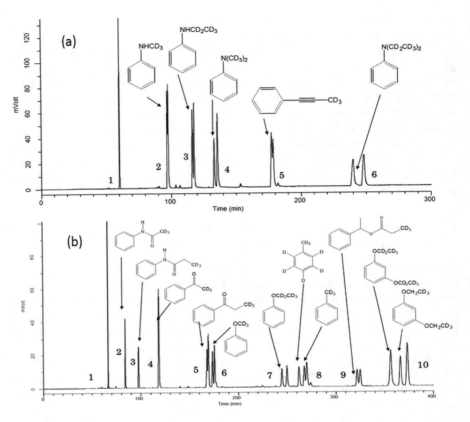

FIGURE 4.12 Chromatograms obtained for various isotopologue pairs on a silica C18 capillary column. Column: monolithic silica C18, 0.1 mm I.D., 450 cm. Mobile phase: (a) acetonitrile/water=50/50, and (b) acetonitrile/water=40/60, provided by a split-flow operation at 41 MPa at 30°C. Detection at 210 nm. Solute structures are indicated over the peak for the deuterated isotopologue which was followed by the nondeuterated one in each pair of peaks. Peak 1 in each chromatogram is the peak of a t_0 marker, uracil.

that in the terminal methyl group was 0.29, calculated from TIE observed for hexane and octane in acetonitrile/water=60/40 with a C18 column, assuming constant %IE for all methylene CH/CD, according to the supplementary information of [45].) Deuterium substitution at the β-position to oxygen resulted in small %IE, while the deuterium on the α-carbon to oxygen provided a large %IE. Thus %IE values seem to increase by the presence of a σ-bond and/or lone pair electrons on the neighboring atom.

The normal isotope effect in retention factors, TIE > 1.0, indicates that the CH/CD bonds are less restricted in the stationary phase than in the mobile phases, corresponding to the weakened CH/CD bonds, weaker in the nonpolar, more polarizable stationary phase relative to the aqueous mobile phase [45]. Such weakening of a CH/CD bond by interactions with another molecule (or the alkyl groups in the stationary phase) was supposed to be influenced by the presence of a σ-bond and/or lone pair

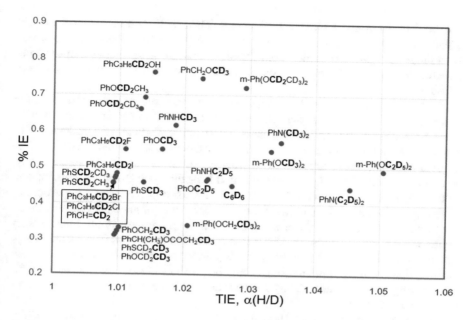

FIGURE 4.13 Plot of single isotope effect (%IE) caused by one H/D substitution against total isotope effect (TIE) which is the chromatographic separation factor between the isotopologues. The location of deuteration is indicated by the bold characters in the solute molecular formula. Mobile phase: acetonitrile/water=50/50. Other conditions are the same as Figure 4.12.

electrons on the neighboring atom. This seems to be reasonable assuming that the solute–stationary phase interactions are dispersion type interactions. In other words, the instantaneous dipole (or the induced dipole) on CH/CD in either orientation will be stabilized by either electron-donating or electron-withdrawing substituents suitably oriented. For a terminal methyl CH/CD, such an effect is missing on one side.

The difference in %IE with a different position of CH/CD in a molecule suggests that the dispersion interactions with binding sites in a stationary phase may be localized to a part of a solute molecule. The separation based on the isotopic chirality created by the presence of deuterated and nondeuterated phenyl groups is supposed to be made possible by such localized dispersion interactions. The variation of the isotope effect by the environment of a CH/CD group may help the understanding of intermolecular interactions in the binding process of biologically active compounds, and will be an interesting subject to study.

4.4.5 H/D ISOTOPIC SEPARATIONS IN NONPOLAR MOBILE PHASES WITH POLAR STATIONARY PHASES

As shown in Figure 4.8d, a polar stationary phase showed preferential retention for di(phenyl-d_5) methanol (3) compared to diphenylmethanol (2) in a nonpolar mobile phase, resulting in a reversed elution order from common reversed-phase systems for the deuterated and nondeuterated solutes. The results can be explained by the greater dispersion interactions of diphenylmethanol than di(phenyl-d_5) methanol with the mobile phase of higher polarizability, containing hexane as a major component, than that at the binding site of cellulose benzoyl ester in the stationary phase.

In order to find other possible interpretations for H/D separations by studying the behavior of solutes with simpler structures in the absence of hydrophobic interactions, a few polar stationary phases, including bare silica (Silica) were evaluated with hexane as a mobile phase using H/D isotopologues of aromatic hydrocarbons. The chromatograms of the isotopologues on Silica are shown in Figure 4.14. Interestingly, in all cases the deuterated compounds eluted later than nondeuterated compounds. In fact, the elution order was reversed from the reversed-phase system. In particular, peak top separations were achieved for the isotopologues of phenanthrene and hexamethylbenzene (HMB). In these cases, the polar silanol groups might interact with aromatic rings in the solutes, the difference in electron density on the aromatic rings by the presence of H or D possibly contributing to the difference in retention. The electron density of the aromatic rings could be greater in D-substituted solutes [56], contributing to the slightly stronger interaction with the silanol groups by π(Ar)–dipole interaction.

According to the α(D/H) values for each solute pair in Figure 4.14, both the number of deuterium atoms and of aromatic rings seemed to affect the magnitude of the isotope effect. Consequently, deuterated phenanthrene and HMB seemed to undergo a slightly stronger π(Ar)–dipole interaction, resulting in the peak top separation.

4.4.6 H/D Isotope Effect Based on Dispersion and CH–π(Ar) Interactions in Nonpolar Mobile Phases

Weak interactions, such as dispersion and CH–π(Ar) interactions, contribute to the H/D separations along with the hydrophobic interaction in the reversed-phase mode. In order to reveal these weak interactions, it is necessary to employ a simple normal phase system, in which the hydrophobic interaction is completely suppressed. Kubo et al. succeeded in the immobilization of fullerenes onto a silica-monolithic capillary to develop a new separation medium, providing the strong π interaction derived from the large aromatic structures of fullerenes [57, 58]. The fullerene-coated silica-monolithic capillaries in the normal phase system showed an effective separation of polycyclic aromatic hydrocarbons (PAHs) based on the dispersion and π(Ar) interactions [59, 60, 61]. Especially, the C_{70}-fullerene (C70)-coated column showed the strong retention for certain PAHs by the π(Ar) interactions. Here, it was expected that the separation of isotopologues based on the dispersion interaction and/or CH/CD–π(Ar) interactions could be achieved with the C70 column. In the evaluation with hexane as a mobile phase, the C70 column provided the stronger dispersion force and/or CH–π(Ar) interaction toward methyl substituted benzenes. The results clearly indicated that the number of methyl substitutions increased the retention factors, especially on the C70 column. Figure 4.15 shows a chromatogram of the isotopologue pairs of HMB and phenanthrene on the C70 column. The peak

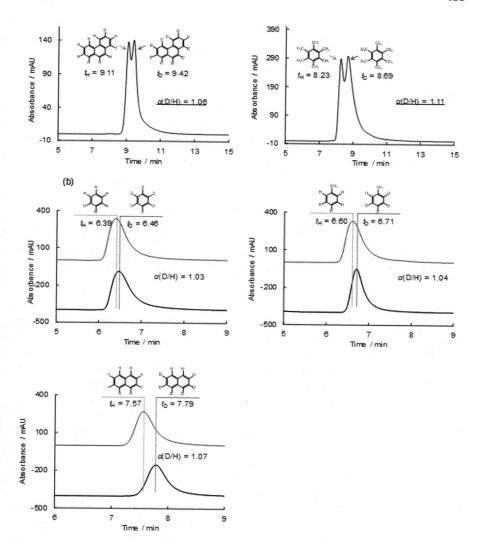

FIGURE 4.14 Chromatograms of isotopologues on the Silica column obtained for mixed isotopologue pairs of phenanthrene and HMB, and for single isotopologues of benzene, toluene, and naphthalene. Conditions: column, silica–150 (TOSOH, 250×4.6 mm i.d.); flow rate, 0.8 mL min^{-1}; mobile phase, hexane; detection, UV 254 nm (for benzene, naphthalene, and phenanthrene), 220 nm (toluene and HMB); temperature, 40°C. (Reprinted with permission from *J. Phys. Chem. C* 122 (2018) 15026–15032. Copyright (2018) American Chemical Society.)

top separation of the isotopologue pairs was confirmed, where the protiated HMB was retained longer than the deuterated one. The results support the interpretation that the CH provided stronger retention due to the dispersion force and/or to the CH–π interaction with C70. On the other hand, despite the longer retention of phenanthrene than that of HMB, the isotopologue separation was not observed as shown

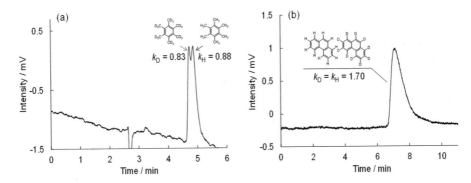

FIGURE 4.15 Chromatograms for the H/D isotopologue pairs of HMB and phenanthrene in hexane with the C70 column. Condition: column, C70-bonded monolithic silica capillary column (32.0 cm × 100 μm i.d.); flow rate, 2.0 μL min^{-1}; mobile phase, hexane; detection, UV 220 nm (HMB), 254 nm (phenanthrene), temperature, 40°C. (Reprinted with permission from *J. Phys. Chem. C* 122 (2018) 15026–15032. Copyright (2018) American Chemical Society.)

in Figure 4.15b. In this case, it seemed that strong π–π interaction occurred, resulting in little differences in retention for H/D isotopologues.

The protiated solutes were more retained on the C70 column, contrary to the results on the columns with polar moieties providing the H/D isotope effect based on the dipolar interaction discussed above. The results showed that the dispersion and/or the CH–π interactions provided the opposite tendency compared to the π(Ar)–dipole interaction. In the latter case, the π–π interaction on HMB was not so effective because of the steric hindrance by methyl groups [62, 63]; then, the dispersion and CH–π interactions were notably observed.

The deuterated solutes may be allowed a stronger dipolar interaction than the protiated analytes because of the higher dipole moment on CD than on CH [64]. On the contrary, the longer bonding distance of CH could provide the preferable interaction and, therefore, the dispersion and π interactions on CH are slightly stronger than those on CD [65]. This way, the magnitude of the H/D isotope effects can be controlled by optimizing the dipolar interaction due to the polarity and dispersion/CH–π interactions with both stationary and mobile phases. A demonstration of the effective isotope separation was carried out by employing the π(Ar)–dipolar interaction with the mobile phase, methanol, and dispersion/CH–π interactions with the stationary phase, the C70 moieties. As shown in Figure 4.16, an almost baseline separation of the H/D isotopologue pair of phenanthrene was successfully achieved. In this case, the complementary isotope effects contributed. Briefly, the greater π(Ar)–dipole interaction with methanol on a deuterated species, and the greater dispersion/CH–π interactions with C70 on a protiated solute, worked cooperatively. Thus, the effective H/D separation can be demonstrated by using methanol as a mobile phase. Controlling the intermolecular interactions by selecting stationary phase and mobile phase will accomplish efficient H/D separations with liquid chromatography.

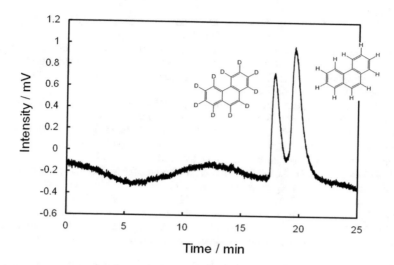

FIGURE 4.16   Separation of H/D isotopologue pair of phenanthrene. Condition: column, C70 column (75.0 cm × 100 μm i.d.); flow rate, 0.8 μL min⁻¹; mobile phase, methanol; detection, UV 254 nm; temperature, 25°C.

CONCLUSIONS

Isotopic separation is an interesting subject for separation science by itself because of the high similarity in the solute structure. It has contributed to the advances of high-efficiency separation systems. The subject has also contributed to the understanding of the mechanism of solute–mobile/stationary phase interactions, providing fundamental information which has in turn contributed to the understanding of the properties of various stationary phases. Selection of a suitable stationary phase and mobile phase will further facilitate the separation of H/D isotopic compounds. Such a fundamental understanding will be important in the future with the increase in diversity of columns and target molecules in the field of separation science.

ACKNOWLEDGMENT

NT's studies on the fundamental aspects of HPLC started with Dr. L. R. Snyder and Prof. B. L. Karger, after an initial study on H/D isotopic separations with Prof. E. R. Thornton. NT, whom Dr. Snyder suggested write an autobiography as a CASSS Award recipient several years ago, would like to dedicate this article to Dr. Snyder with deep appreciation.

REFERENCES

1. L. R. Snyder, J. J. Kirkland, J. W. Dolan, *Introduction to Modern Liquid Chromatography*. Wiley, 2009.
2. N. Tanaka, E. R. Thornton, Isotope effects in hydrophobic binding measured by high-pressure liquid chromatography, *J. Am. Chem. Soc.* 98(6) (1976) 1617–1619.

3. G. P. Cartoni, I. Ferretti, Separation of isotopic molecules by high-performance liquid chromatography, *J. Chromatogr.* 122 (1976) 287–291.

4. N. Tanaka, E. R. Thornton, Structural and isotopic effects in hydrophobic binding measured by high-pressure liquid chromatography, a stable and highly precise model for hydrophobic interactions in biomembranes, *J. Am. Chem. Soc.* 99(22) (1977) 7300–7307.

5. G. K. Worth, R. W. Retallack, Tritium isotope effect in high-pressure liquid chromatography of vitamin D metabolites, *Anal. Biochem.* 174(1) (1988) 137–141.

6. G. Desmet, D. Clicq, P. Gzil, Geometry-independent plate height representation methods for the direct comparison of the kinetic performance of LC supports with a different size or morphology, *Anal. Chem.* 77(13) (2005) 4058–4070.

7. J. R. Mazzeo, U. D. Neue, M. Kele, R. S. Plumb, Advancing LC performance with smaller particles and higher pressure, *Anal. Chem.* 77 (2005) 460A–467A.

8. J. J. Kirkland, F. A. Truszkowski, R. D. Ricker, Atypical silica-based column packings for high-performance liquid chromatography, *J. Chromatogr. A* 965(1–2) (2002) 25–34.

9. F. Gritti, I. Leonardis, D. Shock, P. Stevenson, A. Shalliker, G. Guiochon, Performance of columns packed with the new shell particles, Kinetex-C18, *J. Chromatogr. A* 1217(10) (2010) 1589–1603.

10. S. Fekete, D. Guillarme, Kinetic evaluation of new generation of column packed with 1.3μm core–shell particles, *J. Chromatogr. A* 1308 (2013) 104–113.

11. A. C. Sanchez, J. A. Anspach, T. Farkas, Performance optimizing injection sequence for minimizing injection band broadening contributions in high efficiency liquid chromatographic separations, *J. Chromatogr. A* 1228 (2012) 338–348.

12. S. R. Groskreutz, S. G. Weber, Temperature-assisted on-column solute focusing: A general method to reduce pre-column dispersion in capillary high performance liquid chromatography, *J. Chromatogr. A* 1354 (2014) 65–74.

13. I. Kuroda, H. Uzu, S. Miyazaki, M. Ohira, N. Tanaka, Reduction of the extra-column band dispersion by a slow transport of a sample band from the injector to the column in isocratic reversed-phase liquid chromatography, *J. Chromatogr. A* 1572 (2018) 44–53.

14. T. Hara, H. Kobayashi, T. Ikegami, K. Nakanishi, N. Tanaka, Performance of monolithic silica capillary columns with increased phase ratios and small-sized domains, *Anal. Chem.* 78(22) (2006) 7632–7642.

15. K. Kimata, K. Iwaguchi, S. Onishi, K. Jinno, R. Eksteen, K. Hosoya, M. Araki, N. Tanaka, Chromatographic characterization of silica C18 packing materials. Correlation between a preparation method and retention behavior of stationary phase, *J. Chromatogr. Sci.* 27(12) (1989) 721–728.

16. Y. Ohtsu, Y. Shiojima, T. Okumura, J. Koyama, K. Nakamura, O. Nakata, K. Kimata, N. Tanaka, Performance of polymer-coated silica C18 packing materials prepared from high-purity silica gel. The suppression of undesirable secondary retention processes, *J. Chromatogr.* 481 (1989) 147–157.

17. N. Tanaka, M. Araki, Separation of oxygen isotopic compounds by reversed-phase liquid chromatography on the basis of oxygen isotope effects on the dissociation of carboxylic acids, *J. Am. Chem. Soc.* 107(25) (1985) 7780–7781.

18. N. Tanaka, A. Yamaguchi, M. Araki, K. Kimata, Separation of nitrogen isotopic compounds by reversed-phase liquid chromatography on the basis of nitrogen isotope effects on the dissociation of amines, *J. Am. Chem. Soc.* 107(25) (1985) 7781–7782.

19. N. Tanaka, M. Araki, K. Kimata, Separation of oxygen isotopic compounds by reversed-phase liquid chromatography, *J. Chromatogr.* 352 (1986) 307–314.

20. N. Tanaka, K. Hosoya, K. Nomura, T. Yoshimura, T. Ohki, R. Yamaoka, K. Kimata, M. Araki, Separation of nitrogen and oxygen isotopes by liquid chromatography, *Nature* 341(6244) (1989) 727–728.

21. N. Tanaka, A. Yamaguchi, K. Hashizume, M. Araki, A. Wada, K. Kimata, Separation of isotopic compounds by reversed-phase liquid chromatography. Effect of pressure gradient on isotope separation by ionization control, *J. High Resol. Chromatogr.* 9(11) (1986) 683–687.

22. S. Terabe, T. Yashima, N. Tanaka, M. Araki, Separation of oxygen isotopic benzoic acids by capillary zone electrophoresis based on isotope effects on the dissociation of the carboxyl group, *Anal. Chem.* 60(17) (1988) 1673–1677.

23. R. A. Henry, S. H. Byrne, D. R. Hudson, High speed recycle chromatography using an alternate pumping principle, *J. Chromatogr. Sci.* 12(4) (1974) 197–199.

24. H. Minakuchi, K. Nakanishi, N. Soga, N. Ishizuka, N. Tanaka, Octadecylsilylated porous silica rods as separation media for reversed-phase liquid chromatography, *Anal. Chem.* 68(19) (1996) 3498–3501.

25. K. Cabrera, D. Lubda, H.-M. Eggenweiler, H. Minakuchi, K. Nakanishi, A new monolithic-type HPLC column for fast separations, *J. High Resol. Chromatogr.* 23(1) (2000) 93–99.

26. T. Ikegami, E. Dicks, H. Kobayashi, H. Morisaka, D. Tokuda, K. Cabrera, K. Hosoya, N. Tanaka, How to utilize the true performance of monolithic silica columns?, *J. Sep. Sci.* 27(15–16) (2004) 1292–1302.

27. D. Cabooter, F. Lestremau, F. Lynen, P. Sandra, G. Desmet, Kinetic plot method as a tool to design coupled column systems producing 100,000 theoretical plates in the shortest possible time, *J. Chromatogr. A* 1212(1–2) (2008) 23–34.

28. N. Tanaka, T. Yoshimura, M. Araki, The effect of pressure on solute retention in the ionization control mode in reversed-phase liquid chromatography, *J. Chromatogr.* 406 (1987) 247–256.

29. M. M. Fallas, U. D. Neue, M. R. Hadley, D. V. McCalley, Investigation of the effect of pressure on retention of small molecules using reversed-phase ultra-high-pressure liquid chromatography, *J. Chromatogr. A* 1209(1–2) (2008) 195–205.

30. Sj. van der Wal, Optimization for minimum analysis time of the separation of benzene and deuterated benzene by reverse phase microbore and recycle HPLC, and the utility of D$_2$O, *J. Liq. Chromatogr.* 8(11) (1985) 2003–2016.

31. Sj. van der Wal, A limit to alternate column recycle chromatography, *Chromatographia* 22(1–6) (1986) 81–87.

32. L. Lim, H. Uzu, T. Takeuchi, Separation of benzene and deuterated benzenes by reversed-phase and recycle liquid chromatography using monolithic capillary columns, *J. Sep. Sci.* 27(15–16) (2004) 1339–1344.

33. F. Gritti, S. Cormier, Performance optimization of ultra high-resolution recycling liquid chromatography, *J. Chromatogr. A* 1532 (2018) 74–88. (Some corrections were made by personal communications.)

34. F. Gritti, M. Leal, T. McDonald, M. Gilar, Ideal versus real automated twin column recycling chromatography process, *J. Chromatogr. A* 1508 (2017) 81–94.

35. K. Kimata, T. Hirose, E. Kanao, T. Kubo, K. Otsuka, K. Hosoya, K. Yoshikawa, E. Fukusaki, N. Tanaka, Recycle reversed-phase liquid chromatography achieving the separations based on one H/D substitution on aromatic hydrocarbons, *LC-GC* 32(15) (2019) 14–19. A part of the results was presented at the 72nd Annual Meeting of the Chemical Society of Japan, Tokyo, March, 1997, Abstract I-529.

36. H. G. Menet, P. C. Gareil, R. H. Rosset, Experimental achievement of one million theoretical plates with microbore liquid chromatographic columns, *Anal. Chem.* 56(11) (1984) 1770–1773.

37. R. Swart, J. C. Kraak, H. Poppe, Performance of an ethoxyethylacrylate stationary phase for open-tubular liquid chromatography, *J. Chromatogr. A* 689(2) (1995) 177–187.

38. G. Liu, N. M. Djordjevic, F. Erni, High-temperature open-tubular capillary column liquid chromatography, *J. Chromatogr.* 592(1–2) (1992) 239–247.

39. K. Miyamoto, T. Hara, H. Kobayashi, H. Morisaka, D. Tokuda, K. Horie, K. Koduki, S. Makino, O. Núñez, C. Yang, T. Kawabe, T. Ikegami, H. Takubo, Y. Ishihama, N. Tanaka, High-efficiency liquid chromatographic separation utilizing long monolithic silica capillary columns, *Anal. Chem.* 80(22) (2008) 8741–8750.

40. G. R. Clemo, A. McQuillen, Molecular dissymmetry due to symmetrically placed hydrogen and deuterium. Part I. The resolution of α-pentadeuterophenylbenzylamine, *J. Chem. Soc.* (1936) 808–810.

41. Y. Pocker, The optical resolution of C6H5·CH(OH)·C6D5 and C6H5·CD(OH)·C6D5. The preparation and racemization of (-)-C6H5·CHCl·C6D5 and (-)-C6H5·CDCl·C6D5, *Proc. Chem. Soc.* (1961) 140–141.

42. R. Adams, D. S. Tarbell, The attempted resolution of phenyl-d_5-phenylaminomethane, *J. Am. Chem. Soc.* 60(5) (1938) 1260–1262.

43. T. Makino, M. Orfanopoulos, T.-P. You, B. Wu, C. W. Mosher, H. S. Mosher, Chiral benzhydrol-2,3,4,5,6-d_5, *J. Org. Chem.* 50(25) (1985) 5357–5360.

44. K. Kimata, K. Hosoya, T. Araki, N. Tanaka, Direct chromatographic separation of racemates on the basis of isotopic chirality, *Anal. Chem.* 69(13) (1997) 2610–2612.

45. M. Turowski, N. Yamakawa, J. Meller, K. Kimata, T. Ikegami, K. Hosoya, N. Tanaka, E. R. Thornton, Deuterium isotope effects on hydrophobic interactions: The importance of dispersion interactions in the hydrophobic phase, *J. Am. Chem. Soc.* 125(45) (2003) 13836–13849.

46. K. Kimata, M. Kobayashi, K. Hosoya, T. Araki, N. Tanaka, Chromatographic separation based on isotopic chirality, *J. Am. Chem. Soc.* 118(4) (1996) 759–762.

47. M. Turowski, T. Morimoto, K. Kimata, H. Monde, T. Ikegami, K. Hosoya, N. Tanaka, Selectivity of stationary phases in reversed-phase liquid chromatography based on the dispersion interactions, *J. Chromatogr. A* 911(2) (2001) 177–190.

48. A. Valleix, S. Carrat, C. Caussignac, E. Léonce, A. Tchapla, Secondary isotope effects in liquid chromatography behaviour of 2H and 3H labelled solutes and solvents, *J. Chromatogr. A* 1116(1–2) (2006) 109–126.

49. K. Okusa, Y. Iwasaki, I. Kuroda, S. Miwa, M. Ohira, T. Nagai, H. Mizobe, N. Gotoh, T. Ikegami, D. V. McCalley, N. Tanaka, Effect of pressure on the selectivity of polymeric C18 and C30 stationary phases in reversed-phase liquid chromatography. Increased separation of isomeric fatty acid methyl esters, triacylglycerols, and tocopherols at high pressure, *J. Chromatogr. A* 1339 (2014) 86–95. Supplementary information.

50. G. Srinivasan, C. Meyer, N. Welsch, K. Albert, K. Müller, Influence of synthetic routes on the conformational order and mobility of C18 and C30 stationary phases, *J. Chromatogr. A* 1113(1–2) (2006) 45–54.

51. L. C. Sander, S. A. Wise, Influence of stationary phase chemistry on shape recognition in liquid chromatography, *Anal. Chem.* 67(18) (1995) 3284–3292.

52. K. Kimata, T. Hirose, K. Moriuchi, K. Hosoya, T. Araki, N. Tanaka, High-capacity stationary phases containing heavy atoms for HPLC separation of fullerenes, *Anal. Chem.* 67(15) (1995) 2556–2561.

53. R. M. McCormick, B. L. Karger, Distribution phenomena of mobile-phase components and determination of dead volume in reversed-phase liquid chromatography, *Anal. Chem.* 52(14) (1980) 2249–2257.

54. N. Tanaka, H. Goodell, B. L. Karger, The role of organic modifiers on polar group selectivity in reversed-phase liquid chromatography, *J. Chromatogr. A* 158 (1978) 233–248.

55. E. Kanao, T. Kubo, T. Naito, T. Matsumoto, T. Sano, M. Yan, K. Otsuka, Isotope effects on hydrogen bonding and CH/CD−π interaction, *J. Phys. Chem. C* 122(26) (2018) 15026–15032.

56. C. L. Perrin, Y. Dong, Secondary deuterium isotope effects on the acidity of carboxylic acids and phenols, *J. Am. Chem. Soc.* 129(14) (2007) 4490–4497.

57. T. Kubo, Y. Murakami, Y. Tominaga, T. Naito, K. Sueyoshi, M. Yan, K. Otsuka, Development of a C(60)-fullerene bonded open-tubular capillary using a photo/thermal active agent for liquid chromatographic separations by π–π interactions, *J. Chromatogr. A* 1323 (2014) 174–178.

58. T. Kubo, Y. Murakami, M. Tsuzuki, H. Kobayashi, T. Naito, T. Sano, M. Yan, K. Otsuka, Unique separation behavior of a C60 fullerene-bonded silica monolith prepared by an effective thermal coupling agent, *Chem. Eur. J.* 21(50) (2015) 18095–18098.

59. T. Kubo, E. Kanao, T. Matsumoto, T. Naito, T. Sano, M. Yan, K. Otsuka, Specific inter-molecular interactions by the localized π-electrons in C70-fullerene, *ChemistrySelect* 1(18) (2016) 5900–5904.

60. E. Kanao, T. Naito, T. Kubo, K. Otsuka, Development of a C70-fullerene bonded silica-monolithic capillary and its retention characteristics in liquid chromatography, *Chromatography* 38 (2017) 45–51.

61. E. Kanao, T. Kubo, T. Naito, T. Matsumoto, T. Sano, M. Yan, K. Otsuka, Differentiating π interactions by constructing concave/convex surfaces using a bucky bowl molecule, corannulene in liquid chromatography, *Anal. Chem.* 91(3) (2018) 2439–2446.

62. T. Kawase, Y. Seirai, H. R. Darabi, M. Oda, Y. Sarakai, K. Tashiro, All-hydrocarbon inclusion complexes of carbon nanorings: Cyclic [6]-and [8] paraphenyleneacetylenes, *Angew. Chem. Int. Ed.* 115(14) (2003) 1659–1662.

63. V. G. Saraswatula, D. Sharada, B. K. Saha, Stronger π··· π interaction leads to a smaller thermal expansion in some charge transfer complexes, *Cryst. Growth Des.* 18(1) (2017) 52–56.

64. A. F. C. Arapiraca, J. R. Mohallem, Vibrationally averaged dipole moments of methane and benzene isotopologues, *J. Chem. Phys.* 144(14) (2016) 144301.

65. Y. Harada, T. Tokushima, Y. Horikawa, O. Takahashi, H. Niwa, M. Kobayashi, M. Oshima, Y. Senba, H. Ohashi, K. T. Wikfeldt, A. Nilsson, L. G. Pettersson, S. Shin, Selective probing of the OH or OD stretch vibration in liquid water using resonant inelastic soft-X-ray scattering, *Phys. Rev. Lett.* 111(19) (2013) 193001.

5 Advances in Simulated Moving Bed Technology

Rui P. V. Faria, Jonathan C. Gonçalves,
and Alírio E. Rodrigues

CONTENTS

5.1 INTRODUCTION TO SMB TECHNOLOGY

Chromatographic processes are widely used in the petrochemical, pharmaceutical, and bioprocess industries to partially separate or even obtain pure products. The separation principle is based on the different affinities between the species carried by a mobile phase and a solid stationary phase. Due to different interactions with the packing material (stationary phase), the compounds travel at different velocities through the column leading to concentration gradients, eventually resulting in the separation of the mixture. The species with stronger affinities toward the solid phase exit the column after the others present in the mixture, provided enough residence time is given for the separation to occur. The efficiency of the process can be significantly increased by operating in a continuous countercurrent mode, the most powerful of which, in terms of performance and feasibility, is created by simulated moving bed (SMB) technology. Similar to heat exchangers, the simulated countercurrent contact between both phases maximizes the driving force, resulting in a significant reduction of both phases compared to elution chromatography [1].

The first attempt to conduct chromatography in actual countercurrent mode came in the late 1950s [2]. In the true moving bed (TMB), the feed and desorbent are continuously introduced to the unit through separate inlet streams. Two outlet streams are also continuously collected, one with the more-adsorbed compound, the extract, and one with the less adsorbed compound, the raffinate. The four streams divide the unit into four zones with specific purposes (see Figure 5.1). The more adsorbed product is adsorbed between the feed and the raffinate (zone III), to be transported with the solid to the zone between the desorbent and extract port (zone I), while the less adsorbed is collected in the raffinate. The more adsorbed product is desorbed in zone I, to be transported by the fluid phase to the extract port, while at the same time the solid is regenerated, to be recycled to the zone between the desorbent and the raffinate (zone IV) where the less adsorbed compound can be adsorbed and the desorbent is regenerated. Finally, between the extract and the feed (zone II), the less adsorbed compound

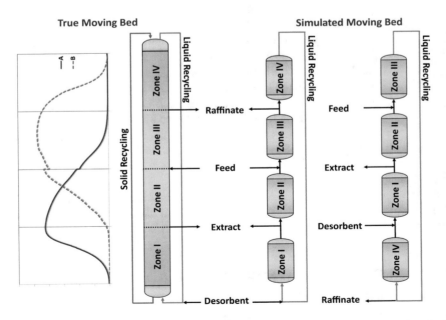

FIGURE 5.1 Simplified scheme of a true moving bed and a simulated moving bed. Compound A is strongly adsorbed and compound B is weakly adsorbed by the solid phase.

is desorbed, to be transported with the mobile phase to zone III to avoid contamination in the extract port. A different convention can be also found in the literature: zone I, between feed and raffinate; zone II, between extract and feed; zone III, between desorbent and extract; and zone IV, between desorbent and raffinate.

The success of the TMB unit was rapidly impeded by the drawbacks of the actual movement of the stationary phase, namely, mechanical stress and back mixing of the solid. The problem was solved in 1961 through the development of SMB technology by Broughton and Gerhold [3] from Universal Oil Products (UOP). The SMB uses a series of short packed bed columns to simulate the countercurrent of both phases by periodically shifting the inlet and outlet ports in the direction of the fluid; the port shifts are conducted at certain time intervals through a rotary valve [4]. The performance of the SMB approaches that of the TMB when a large number of packed bed columns are involved and the time intervals are short [1]. The invention allowed UOP to develop a family of Sorbex processes with applications in the petrochemical and sugar industries, the separation of p-xylene being the most notable application. The first systems operated at low pressure on a very large scale and consisted of a large number of beds. Nonetheless, the lower purity required in the sugar separation allowed an operation with considerably fewer columns [2].

Decades later, the first study of an SMB for pharmaceutical applications was published [5]. In 1997, UCB Pharma installed a large-scale SMB unit for the industrial production of a commercial drug [6]. Later on, the single enantiomer Lexapro (Lundbeck) was the first drug approved by the U.S. Food and Drug Administration to be manufactured using SMB technology [7]. The rise of SMB in pharmaceutical applications was partly due to the generation of chiral stationary phases developed in the 1990s; small particles led to beds as short as 10 cm [2]. Moreover, because small amounts of the chiral drug are normally needed for preliminary biological tests, small-scale SMB units are an excellent tool [8]. Nowadays, well-known pharmaceuticals such as Carbogen–Amcis, Merck, Bayer, GSK, Novartis, and the previously mentioned UCB use SMB for the production of single-enantiomer drugs [1].

Since the revolution in pharmaceutical applications, some enhancements in technology have been achieved by introducing additional degrees of freedom, for instance by varying the concentration or the flow rate of the feed, or by the use of asynchronous port shifts, multiple feeds, or side-streams, among many other techniques. These alternative modes of operation allow the use of fewer columns for maintaining the countercurrent contact between both phases, with the cost of moving away from a truly continuous operation. Applications in biotechnology exploit the best of these so-called unconventional SMBs, as presented in Figure 5.2. As Nicoud [2] states, continuous operation may improve the control and stability of the unit rather than enhancing productivity, yield, or desorbent consumption.

In this work, the most recent advances in SMB technology are addressed, drawing particular attention to the developments reported during the last decade. Due to their historical relevance, the petrochemical, food, and pharmaceutical industries are analyzed in a first instance, describing the enhancements arising from the continuous evolution of this technology even in some of the most well-established SMB industrial processes. Then, the relevance of the emerging bio-based product market is highlighted, as is the vast number of opportunities and challenges within

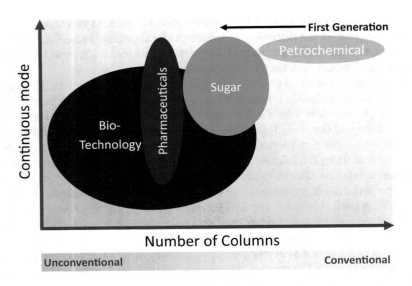

FIGURE 5.2   Qualitative representation of different applications of SMB technology.

it, including new applications for SMB purification processes, alternative configurations and operating modes that improve conventional process performance, and innovative SMB-based technologies. The most relevant technical aspects supporting this progress are discussed.

5.2   THE CONTINUOUS EVOLUTION WITHIN CONVENTIONAL PROCESSES

As stated earlier, SMB technology was initially conceived to perform large-scale industrial separations within the petrochemical industry during the 1960s. The technology was later applied to the food industry with particular success in the separation of sugars; however, before the turn of the century, a new generation of SMB processes and units was developed driven by the demanding separations required by pharmaceutical manufacturing processes. Together, these three industries incorporate the majority of commercial SMB applications. For this reason, the trends and progress observed within these industries associated with SMB technology will be explored and discussed.

5.2.1   Petrochemicals

5.2.1.1   Separation of Xylene Isomers

It may not have been the first SMB unit coming on stream, but separating xylene isomers is definitely the most representative SMB-based separation process in the industry. The most employed technology is UOP's Parex process, the first commercial unit of which came on stream in 1971 and rapidly became the leader licensor in p-xylene separation [9]. Following Parex is IFP's Eluxyl, which was first commercialized in 1995. Recently, BP's crystallization technology started to be licensed by Lummus; according to the licensor, its lower energy consumption is mostly due to the lower energy required to melt p-xylene compared to the vaporization within the distillation columns associated to the SMB unit. One of the disadvantages of the Parex process, pointed out by Colling [10, 11], lies in the stringent specification within the top of the xylene splitter on the number of C_9+ compounds. The author also states that the desorbent-to-feed ratio is already close to the theoretical limit.

UOP recently started to license the LD Parex process, in which toluene is used as a desorbent instead of p-diethylbenzene. The first light desorbent unit was licensed in 1972, but the system based on a heavy desorbent (i.e., p-diethylbenzene) remained preferred since it exhibited a lower desorbent-to-feed ratio [12]. The adsorbent of the new system, ADS-50, is based on ADS-47 adsorbent technology and allows less stringent feed requirements in the SMB unit [13]. When it was introduced in 2011, ADS-47 exhibited 20–25% higher capacity and 32% less desorbent circulation compared to the previous generation [14].

Light desorbent enables an optimized fractionation scheme with 20% less equipment and 15–17% less capital investment [12, 14]. According to Ma [15], the operating temperature is 135°C and the feed requirements on C_9+ was relaxed from 500 to 10,000 ppm. Additionally, the de-heptanizer column was removed, as was the re-run desorbent column since the desorbent is no longer imported; additionally, the thermal integration is based on the raffinate column instead of the xylene splitter. These elements led to a significant reduction in the energy consumed within the complex. Since 2016, seven new LD Parex plants have been licensed [13].

Other modifications have been proposed in order to reduce the energy consumption within the distillation columns. A divided wall column can be used downstream of the gas-phase isomerization reactor to separate toluene (C_7-), xylenes, and heavy aromatics [16]. Another option is to include said type of column to separate the desorbent from two different raffinate streams. The two streams come from different adsorption units handling xylenes with high and low amounts of ethylbenzene. From the top of the raffinate column two streams are obtained, ethylbenzene-rich and ethylbenzene-lean; the desorbent is recovered at the bottom [17].

Leflaive et al. [18] proposed dividing the SMB into two parallel chambers, each having 12 beds. The quantity produced may be made higher by operating the chambers in a non-simultaneous mode. Two rotary valves with staggered switching times have also been proposed [19]. The system allows more flushes and the handling of feeds with different amounts of p-xylene, leading to higher capacity. Lee and Shin [20] suggest injecting the feed with a higher amount of p-xylene (e.g., from selective toluene disproportionation) three beds nearer to the extract node.

A different approach consists of separating a mixture of p-xylene and ethylbenzene from the other isomers. A PSA with a non-acidic medium-pore molecular sieve (e.g., MFI-type) can be used at elevated pressure and temperature to guarantee rapid adsorption and desorption of ethylbenzene and p-xylene, which are then separated in a conventional SMB [21]. The system allows one to conduct isomerization reactions without costly hydrogen. An SMB can also be used to obtain a p-xylene-rich stream containing at least 5 wt% ethylbenzene, so the depleted p-xylene raffinate may be isomerized in the liquid phase without ethylbenzene buildup; p-xylene is then recovered in a second SMB or crystallizer [22].

Other inventions are focused on splitting the isomerization into liquid and gas phases. Two raffinate streams may be obtained from an SMB unit, one desorbent-rich and one desorbent-lean (closer to the feed port); the latter is sent directly to the gas-phase isomerization unit [23]. The desorbent, preferably toluene, from the other raffinate is recovered and the stream is subjected to liquid-phase isomerization. Furthermore, varying feed flow throughout the switching brings several advantages, including two or three raffinate streams from which liquid-phase isomerization can be used in at least one of them due to the small amount of ethylbenzene [24]. Dreux et al. [25] propose two consecutive SMB-isomerization units. The first, using p-diethylbenzene as desorbent, sends the raffinate to a liquid-phase isomerization unit using zeolite ZSM-5 as a catalyst. The near-equilibrium stream is then sent to the second SMB, using toluene as desorbent, from which the raffinate is treated in a gas-phase isomerization unit using zeolite EU-1 with Pt. Even though the isomerization of xylenes in the liquid phase has been studied and proposed, only ExxonMobil [26] has started to offer their LPI process for xylene isomerization in the liquid phase. Low temperature conditions and cheaper catalysts lead to less energy consumption and constitute an alternative to de-bottlenecking existing facilities. Since the conversion per pass of ethylbenzene is very low, only a fraction of the mixed xylenes can be sent to the LPI unit; the rest of the p-xylene-depleted stream must be sent to the gas-phase isomerization unit.

Several works on the simulation of an industrial scale unit, using Ba/K-exchanged faujasite-type zeolite at 177°C and approximately 9 bar with p-diethylbenzene as desorbent, have been recently reported in literature [27–29]. The models included the dead volumes introduced by bed lines, circulation lines, and bed-head dead volumes. Shen et al. [27] determined the adsorption data through batch experiments and lumped mass-transfer coefficients through pulse experiments. The authors used a two-level optimization where productivity was maximized in the first level and the desorbent consumption was minimized in the second level. It was found that the eight-zone SMB exhibits higher productivity than, and similar desorbent consumption to the seven-zone unit. Additionally, the primary flush has greater impact than the secondary flush, and the best location for the tertiary flush (eight-zone SMB) was one column ahead of the feed. Figure 5.3 presents a schematic of the eight-zone SMB and depicts the main difference compared to the seven-zone.

Sutanto et al. [28] used the data reported by Minceva and Rodrigues [30] and also followed a two-level strategy. In the first level, bed configuration and flow rates for the primary and secondary flushing were obtained for a seven-zone SMB. Once the dominant flushing flow rate was determined, the eight-zone SMB was optimized, maximizing recovery, and the bed configuration and the three flushing rates were obtained. One should note that the primary flush in and out have the same flow rate, while the secondary flush in and tertiary flush out are independent. The authors found that the flow rate of the secondary flush was a dominant factor and that the tertiary flush should be located far from the raffinate port but not too close to the feed port. Nonetheless, decoupling of the primary flush in and out has been proposed [31].

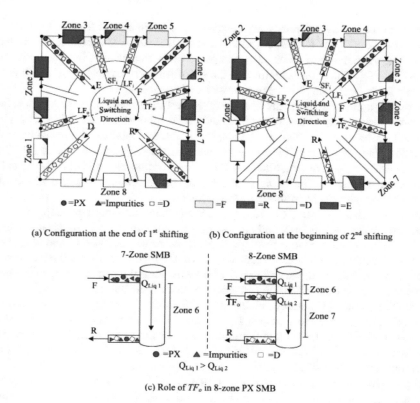

● =PX ▲=Impurities □ =D ▭ =F ▬ =R ▭ =D ▬ =E

(a) Configuration at the end of 1ˢᵗ shifting   (b) Configuration at the beginning of 2ⁿᵈ shifting

● =PX ▲ =Impurities □ =D

$Q_{Liq\,1} > Q_{Liq\,2}$

(c) Role of TF_o in 8-zone PX SMB

FIGURE 5.3 Schematic diagram of eight-zone PX SMB process; D: desorbent, LF_o: line flush outlet, E: extract, SF_i: secondary flush inlet, LF_i: line flush inlet, F: feed, TF_o: tertiary flush outlet, R: raffinate, and Q_{liq}: zone flow rate. (Reprinted from *Separation and Purification Technology*, 96, Sutanto, P.S., Lim, Y.-I. and Lee, J., "Bed-line flushing and optimization in simulated moving-bed recovery of paraxylene," p. 168–181, Copyright (2012), with permission from Elsevier).

Silva et al. [29] used adsorption data from a previous work in batch mode [32]. The authors included the capacity loss of an adsorbent that had been in operation for ten years in the model; the recovery decreased by 3.0% and 9.2% for capacity losses of 10% and 20%, respectively. The addition of aging of the adsorbent improved the fitting between simulation and industrial data, especially in the concentration of *p*-xylene in the zone between the feed and the raffinate. Moreover, the presence of the circulation lines between the two chambers changed the periodicity of the average concentrations from t*-periodic to 12t*-periodic, meaning that the cyclic-steady state can then be defined by 12 switches. If said volume is neglected, the cycle-steady state can be calculated after a single switch using a full-discretization approach [33].

The Sinopec Research Institute of Petroleum Processing developed the RAX series for *p*-xylene adsorption based on faujasite-type zeolite. In 2004, RAX-2000A was applied in the Parex unit at a Sinopec petrochemical facility; the adsorbent showed average p-xylene purity and recovery of 99.81% and 98.6%, respectively. In 2008, RAX-3000 was developed; in 2011, the adsorbent was used in a local plant, exhibiting an increase of about 18%, compared to its predecessor, in unit-production capability [34]. The authors claim that the adsorbent performs very well with unfavorable feeds containing high concentrations of ethylbenzene.

Even though many materials have been studied in the separation of xylenes, faujasite-type zeolite continues to dominate industrial applications. An extensive review focused on the materials that may be employed in the separation of xylenes was presented by Yang et al. [35]. Their review included metal–organic frameworks (MOFs) and molecular sieves, among others, in adsorption, membrane, and chromatographic separation processes. The authors also discussed the main challenges preventing some of the materials from reaching the industry.

Faruque Hasan et al. [36] proposed a multiscale computational approach that combines material screening and process optimization to select appropriate zeolites for the separation of p-xylene through SMB. It was found that MWW and MEL zeolites are excellent candidates for replacing current adsorbents. The authors also state that a layered SMB with multiple zeolites may increase the performance of the unit. Clearly, experimental validation is needed to verify their findings.

As Yang et al. [35] stated, plenty of MOFs have been studied in the adsorption of xylenes; however, only a few MOFs have been used recently in SMB studies. Moreira et al. [37] simulated an SMB for the separation of o-xylene and p-xylene using MIL-53(Al) as adsorbent and m-xylene as desorbent; the adsorption data was determined experimentally. o-Xylene was recovered in the extract and p-xylene in the raffinate at 40°C with a switching time of 115 s. Adsorption in the liquid phase of xylene isomers, including ethylbenzene, was experimentally studied over MIL-125(Ti)_NH2 [38]. Binary and multicomponent experiments were conducted in the liquid phase using n-heptane as eluent at 26, 40, and 70°C; p-xylene was more strongly adsorbed than ethylbenzene at the highest temperature. Ferreira [39] presented separation regions for the separation of p- and m-xylene at the three temperatures over said MOF, and larger areas were observed at higher temperatures.

Finally, process intensification has also been applied to enhance the separation/production of p-xylene, namely through simultaneous adsorption and isomerization of xylenes in the liquid phase. First, a simulated moving bed reactor (SMBR) consisting of reactors inserted between adsorption columns far from the extract point to obtain pure p-xylene was proposed [40–42]. Later, Gonçalves and Rodrigues [43, 44] studied a simpler unit where catalysts and adsorbents were mixed homogeneously within the columns of an existing SMB, but including a crystallization unit to further purify the p-xylene-rich stream from the extract together with another high-p-xylene stream (e.g., from selective toluene disproportionation). A dual-bed configuration, presented in Figure 5.4, proved to be more efficient, allowing to obtain, in the extract, 1.75 of the p-xylene fed to the unit at 200°C [45]. Interesting improvements in the production of p-xylene and benzene were reported [46]; however, an increase in the operation costs, namely due to higher desorbent consumption and to the addition of a crystallization unit, were not reported.

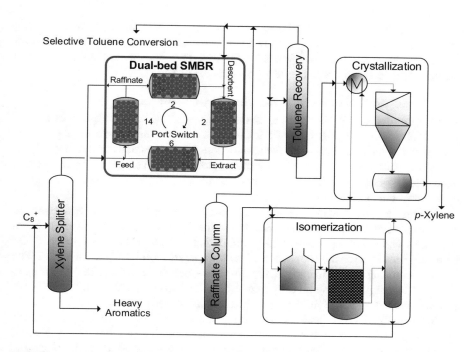

FIGURE 5.4   Simplified scheme of the dual-bed simulated moving bed reactor (catalyst in red and adsorbent in blue) combined with crystallization for the production of *p*-xylene.

A similar approach was proposed by Leinekugel-le-Cocq [47]; however, the feed and desorbent are supplied at different temperatures in order to obtain a temperature profile within the unit. A hotter desorbent and colder feed lead to higher temperatures near the raffinate port and lower temperatures near the extract port. Lower temperatures reduce the conversion of *p*-xylene into the other isomers. In all the SMBR studies previously mentioned, Ba/K-exchanged faujasite-type zeolites were employed as adsorbents and zeolite ZSM-5 as the catalyst. Similar adsorbents but different catalysts at temperatures above 200°C were studied by Shi et al. [48]. The authors considered toluene and benzene to be desorbent and carried out optimizations taking into account energy consumption within the associated distillation columns. The optimum temperature was around 270°C using toluene as desorbent; an increase of about 25% in the recovery of *p*-xylene, compared to the conventional SMB, was obtained. Clearly, a techno-economic analysis of the current and proposed aromatics complex must be conducted to assess the true advantages of including an SMBR in the plant.

In 1995, UOP licensed the first MX Sorbex unit for the production of *m*-xylene by SMB technology [49]. The system is very similar to that of *p*-xylene; indeed, most of the production of *m*-xylene is conducted within complexes where *p*-xylene is also produced [50]. The preferred desorbent is toluene; *m*-xylene is strongly adsorbed in NaY zeolite, although said adsorbent including a small amount of Li has also been proposed [51]. Crystallization can also be combined with SMB [52]. In that system, the SMB can be operated at less severe conditions with an extract purity above 75% in order to avoid the eutectic point in the crystallization unit. The SMB also uses toluene as desorbent; the adsorbent is NaY with a water content less than 1 wt%, and the operating temperature is between 150°C and 170°C.

5.2.1.2 Separation of Normal Paraffins from Naphtha

The first commercial SMB operating in the industry, as part of UOP's Sorbex family, was a Molex unit in 1964 [9]. The Molex process is used to separate normal paraffins from naphtha and, depending on the application, several versions were commercialized with different carbon numbers: Gasoline Molex (C_5-C_6) for octane improvement, Kerosene Molex (C_{10}-C_{13}) for detergent applications, and Heavy Molex (C_{14}-C_{18}) for other surfactant applications [53].

Recently, a more efficient use of naphtha has attracted a lot of attention due mostly to the expansion of crackers and also due to the increase in the production within the catalytic reforming industry [54]. Following the previously mentioned, UOP launched MaxEne for the separation of normal paraffins from C_6-C_{11} naphtha. The first unit came on stream in early 2013 with the support of Sinopec [53]. According to the most recent patent, the preferred conditions are between 170°C and 180°C with enough pressure to maintain the liquid phase using n-dodecane as desorbent [55].

The yield of ethylene and propylene can be increased up to 30% in the cracker by feeding n-paraffins. In the reformer, aromatics can also be increased by feeding the raffinate of the MaxEne unit [53]. Do et al. [56] conducted a techno-economic analysis of a petrochemical complex including a SMB unit for the separation of normal paraffins from C_5-C_{10} naphtha (see Figure 5.5). A simplified model for the SMB was used where the recovery of normal paraffins was 90% with a desorbent-to-feed ratio of 2.5. The desorbent was n-butane and the conditions were 125°C and 25 bar. The dimensions and capital costs were estimated based on those reported for the production of p-xylene. Olefin production increased by 14 wt% and that of the aromatics by 11 wt%; the energy consumption also increased by 25% compared to the conventional petrochemical complex. The authors found the results very promising; however, rigorous calculations of the SMB are required since the recovery of the unit has a major impact on the economics of the complex.

The separation of normal paraffins is normally carried out through size-exclusion adsorption with zeolite 5A. Ragil et al. [57] proposed a mixture of zeolite 5A with silicate or similar. In their invention, zeolite 5A provides a large capacity for straight chain paraffins, which allows efficient separation of mono- from multi-branched paraffins by the other adsorbent with slightly larger pores. MOFs have also been studied in the separation of mono- and di-branched paraffins [58].

Jun et al. [54] conducted an experimental study of an SMB in gas phase for the separation of normal paraffin from naphtha; the normal paraffins were mainly C_5 to C_9. The unit consisted of 16 columns filled with zeolite 5A from UOP and they employed n-pentane as desorbent. They found that the optimal operating temperature was 170°C. The purity of the extract was 98%; in order to increase the purity in the raffinate the authors recycled said stream back into the SMB and, by doing that, the purity of the raffinate reached 92%. The authors also evaluated the experimental performance of steam cracking by feeding the extract from the SMB. The yield of

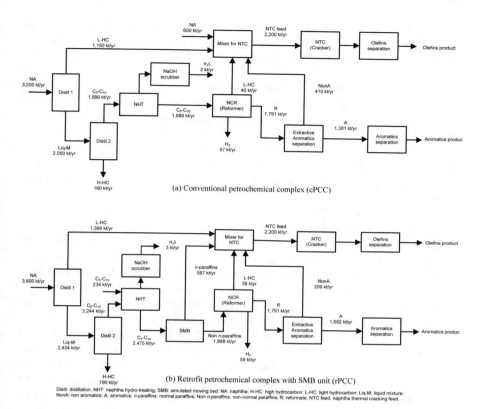

FIGURE 5.5 Conceptual flow diagrams and mass flow rates for the conventional and retrofitted petrochemical complex. (Reprinted from *Chemical Engineering Research and Design*, 106, Do, T.X., Lim, Y.-i., Lee, J. and Lee, W., "Techno-economic analysis of petrochemical complex retrofitted with simulated moving-bed for olefins and aromatics production," p. 222–241, Copyright (2016), with permission from Elsevier).

ethylene increased by 17% and the potential aromatic content of the reformate when feeding the raffinate increased by 10% compared to the conventional naphtha feedstock. Jichang et al. [59] studied the adsorption of naphtha in gas phase over zeolite 5A provided by UOP. They found that the adsorption of long-chain *n*-paraffins was stronger than that of the short-chain.

As previously mentioned, another version of the SMB for separation of normal naphtha is that focused on C₅-C₆ for octane improvement. The principle is the same: the normal paraffins are obtained in the extract to be recycled back into the isomerization reactor; the branched paraffins are used for gasoline production. A recent improvement consists of eliminating the stabilizer column to separate light-ends normally produced within the isomerization. Flash drums upstream of the SMB and within the recycle of the associated distillation columns are used to separate lighter hydrocarbons and hydrogen. The desorbent (e.g., *n*-butane) is obtained as a side stream and the extract and raffinate streams are obtained at the bottom of each column. The SMB also operates at enough pressure to maintain liquid phase at a temperature range from about 100°C to about 180°C [60].

5.2.1.3 Separation of Olefins from Paraffins

There are several processes where the separation of olefins from paraffins is desired: recovery of ethylene from ethane, propylene from propane, C_4 olefins from C_4 hydrocarbon mixtures, and heavier olefins from heavier paraffins for plasticizers and detergent applications. UOP's Olex processes separate olefins from paraffins in liquid phase using SMB technology. The first unit came on stream in 1972; five units were commercialized to process heavy olefin feeds and one to treat a light olefin feed [61].

The heavy version of Olex, also referred to as detergent, is conducted at enough temperature and pressure to overcome diffusion limitations in the liquid phase. The desorbent depends on the carbon number of the feed; normally a mixture of n-hexene and n-hexane is used but heavier compounds such as n-heptane and n-octane have also been used. As the carbon number increases, the separation becomes more difficult [62]. Detergent Olex uses NaX adsorbent to separate C_{10}-C_{14} olefins [63].

The light Olex processes feeds containing mixed C_4 olefins and paraffins to separate either isobutylene, commonly referred to as C_4 Olex, or butane-1, commonly referred to as Sorbutene. Both processes are conducted in the liquid phase; one C4 Olex unit has been commercialized while the Sorbutene process has only been demonstrated in pilot-scale units. The former uses a mixture of n-hexene isomers and n-hexane as desorbent, while the latter uses a mixture of hexene-1 and cyclohexane [62]. Furthermore, an SMB for the separation of propylene from propane in either liquid or gas phase has been proposed; butene-1 or i-butane or a mixture of these two components can be used as desorbent while the adsorbent is zeolite 13X [64]. Strictly speaking, there are no reports of new units coming on stream in recent years; distillation is still the most employed process in spite of the large amount of energy it requires.

Campo et al. [65] obtained experimental adsorption data on propane, propylene, and isobutene over improved zeolite 13X; the data was used for the simulation of an SMB in gas phase with a feed stream of 75% propylene in propane at 100°C and 150 kPa. High purity and recovery (i.e., 99.9%) of propylene was obtained. Later on, Martins et al. [66] developed a bench-scale SMB to conduct experiments in the gas phase. The feed consisted of 67% propylene and 33% propane at 150°C and 1.5 bar, and the adsorbent employed was regular zeolite 13X. Purity of propylene in the extract was 99.93% with a recovery of 99.51%.

Binderless zeolite 13X has also been assessed in the separation of ethylene from ethane. Narin et al. [67] obtained the equilibrium isotherms of propane, propylene, ethane, and ethylene between 50°C and 150°C and pressures up to 5 bar. A mathematical model was validated experimentally and five-step VPSA cycles were successfully designed and tested to produce polymer-grade olefins. A gas-phase SMB comprising eight columns was employed to compare both technologies using the same adsorbent [68]. The SMB cycles were carried out at 100°C and between 120–220 kPa; two ethane/ethylene mixtures were used as feed, 22%/78% and 48%/52%. For similar purities, higher recoveries were obtained in the SMB compared to the VPSA unit. A similar approach was followed to study the separation of ethylene from ethane using Cu-BTC as adsorbent and propane as desorbent [69]. The feed was a mixture containing 39% ethane and 61% ethylene. The SMB was operated in an open loop with three and four zones at 100°C and a pressure ranging from 155–190 kPa. SMB was again superior to the VPSA.

Rive Technology, Inc. was founded in 2006 to develop and commercialize meso-structured zeolites for industrial catalysis and separations. A mesoporous NaX zeolite was developed for propylene/propane separation and compared against conventional NaX adsorbent; the latter exhibited about 50% stronger adsorption but a 260% increase in diffusivity was obtained with the mesostructured adsorbent. The authors estimate between 60% to 80% less energy consumption of a moderate-scale SMB compared to a splitter with the same feed comprising 70% propylene and 30% propane [70].

Even though the difference between the molecular diameters of ethylene and ethane is small, some researchers have focused on finding an adsorbent capable of performing molecular sieving. Ethane exclusion by Ag-exchanged zeolite A led to an enhancement in both ethylene selectivity and capacity. The authors also stated that the selectivity is not affected by the binder material but it is by the silver loading [71, 72].

As opposed to the studies previously mentioned, some researchers have focused on the preferential adsorption of ethane over ethylene. In separation processes such as PSA, the productivity is favored when the target product (i.e., ethylene) is less adsorbed; in the case of SMB it is not straightforward. Lin et al. [73] proposed a simple approach to design an efficient porous MOF for the selective adsorption of ethane from ethylene. A complete separation of ethane from a mixture ethane/ethylene can be realized at ambient conditions by using activated $Cu(Qc)_2$. Wu et al. [74] studied four types of zeolitic imidazole framework (ZIF) with different topologies; the selectivity toward ethane decreases for larger pore diameters at low pressures and increases for larger pore volumes at higher pressures.

Martins et al. [75] studied an SMB in gas phase using ZIF-8 for the separation of ethane from ethylene. The experiments were conducted using two types of desorbent, carbon dioxide (weak desorbent) and propane (strong desorbent). Profiles for both runs are presented in Figure 5.6.

The system with carbon dioxide exhibited higher ethylene purity and recovery in the raffinate for a feed containing about 40% ethane and 60% ethylene at approximately 500 kPa and 50°C. The authors also validated the mathematical model and presented separation regions using both desorbents. The poorer performance of propane as a desorbent was most likely due to strong adsorption and slow diffusion. On the other hand, the recovery of carbon dioxide from ethylene and ethane is yet to be studied. Regarding the recovery of the desorbent, Sivakumar and Rao [76] presented a detailed discussion on different adsorption processes for binary gas mixtures. The authors stated that a SMB with a pressure swing and thermal swing regeneration provide perfect separation and higher productivity.

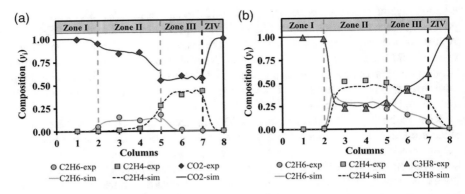

FIGURE 5.6 Internal profile obtained at the middle of the switching time for (a) weak adsorbent and (b) strong adsorbent. (Reprinted from *AIChE* Journal, 65, Martins, V.F.D., Ribeiro, A.M., Kortunov, P., Ferreira, A. and Rodrigues, A.E., "High purity ethane/ethylene separation by gas phase simulated moving bed using ZIF-8 adsorbent," p. e16619, Copyright (2019), with permission from Wiley).

5.2.2 SUGAR SEPARATION

Following the growing need for enriched streams of monosaccharides in the 1970s, UOP developed the Sarex process for the separation of fructose or glucose from mixtures of saccharides in liquid phase through selective adsorption. The products are recovered from the adsorbent by displacement with water which is then removed by evaporation [77]. The Sarex process, part of the Sorbex family, uses SMB technology with X and Y zeolite adsorbents with exchangeable cations; later on, ion-exchange resins, namely polystyrene resins in calcium form, were reported in the patent literature since fructose has a higher affinity for calcium ions [78]. A typical feed to the Sarex process consists of 42% fructose, 53% glucose, and 5% other saccharides; a fructose-enriched corn syrup with 90% is obtained in the extract with a recovery of 90%. The 90% fructose is blended with the feed to obtain a 55% fructose final product (commonly referred to as high fructose corn syrup, HFCS) to be used in the soft drink industry [62]. Maruyama et al. [79] verified that this bypass SMB does not provide any advantages in productivity, but does enhance the flexibility and robustness of the unit.

Even though the separation of fructose can be included in the first generation of SMBs, less conventional modes of operation have been applied to this process as depicted in Figure 5.2. Sreedhar and Kawajiri [80] used the simultaneous optimization-model correction method to optimize several configurations including the conventional, three-zone, and intermittent, as well as a hybrid operation combining an outlet streams swing (OSS) and partial-feed. All of these configurations were experimentally validated using eight columns packed with Dowex Monosphere 99Ca resin with deionized water as the mobile phase at 50°C. At high desorbent-to-feed ratios, the variant of the OSS was superior to the intermittent SMB (ISMB); the opposite occurred at low desorbent-to-feed ratios. Moreover, David et al. [81] investigated the separation of fructose, glucose, and sucrose over a polymeric resin in calcium form by classical SMB, ISMB, and sequential SMB (SSMB) at 60°C. A similar performance was exhibited by ISMB and SSMB comprising six columns; classical SMB required more columns and water (desorbent) to match their performances.

The separation of fructose with and without the recirculation of the fructose-rich stream was studied by Yu et al. [82]; deionized water and Dowex Monosphere 99Ca were used in the SMB at 60°C. The final fructose concentration was 3.7 times higher with recirculation. Tangpromphan et al. [83] proposed a strategy based on port relocation and port closing/opening so that the extract and raffinate were alternately collected with no flow in zone III during raffinate collection. The model was verified experimentally with a three-zone SMB packed with adsorbent resin in calcium form with deionized water as desorbent at 60°C. The proposed strategy resulted in 25% less desorbent consumption compared to the three-zone SMB; additionally, one less pump is needed for that operating mode.

A dual switching strategy was investigated by Vignesh et al. [84]; the operation allowed higher feed flow rates, compared to the conventional separation of fructose and glucose, using exchange resin in calcium form. Simulations suggested that an average extract purity above 99% in the last switch is possible, compared to one obtained by conventional switching (i.e., 96.92%). The switching strategy was further exploited by Yao et al. [85] through the dodecahedron. The authors started with a column configuration of 2/2/2/2 and obtained an overall configuration of 1.94/2.54/2.61/0.91 with four sub-switches; a 15% increase in the feed flow rate, compared to the conventional SMB, was obtained.

Other researchers have used the fructose–glucose system to develop fast computation algorithms, which eventually could be used to control the industrial unit. Sharma et al. [86] proposed a control-relevant multiple linear modeling approach; the experimental unit consisted of one column per zone packed with a strong cation resin of gel-type in calcium form and deionized water as desorbent. Yao et al. [87] employed the conservation element and solution element method combined with continuous prediction technique using data from Beste et al. [88]; compared against the method of lines, the calculation efficiency was improved by 77%. Finally, Vignesh et al. [89] proposed model-based optimization strategies to reduce the off-spec product during transitions between cycle-steady states.

The separation of fructose from glucose was also studied in a thermal SMB [90]. Heat exchangers between the zones provided the required temperature profile within the unit; the optimum temperature for zone I was 70°C while for the rest it was 30°C. Even though the adsorption isotherms have weak temperature dependence, the product concentration was increased by 25% and 75% in the extract and raffinate, respectively; the desorbent-to-feed ratio was decreased by two-fold compared to the isothermal SMB.

The effect of the particle size of gel-type strong cation-exchange resin in calcium form was studied by Heinonen et al. [91]. The authors used 320 and 250 μm resins in an eight-column SMB; the viscosity changed significantly with the pressure drop but no elastic deformation was observed. Water consumption decreased 53% with the smaller particle, leading to 18% less energy consumed in the evaporation step, which outweighed the increase in the energy consumed by the pumps due to the higher pressure drop.

The amount of fructose produced by the enzymatic isomerization is related to the equilibrium, which is about 1:1 at 60°C; based on the aforementioned, some researchers have focused on simulated moving bed reactors to overcome said limitation. Borges da Silva et al. [92] proposed a configuration based on the system from Hashimoto et al. [93] where reactors, inserted between the adsorbers, move along the port, switching so they remain in zone III; however, the feed was only glucose instead of a mixture of glucose and fructose (see Figure 5.7).

The catalyst was a commercial Sweetzyme IT and the adsorbent was a cation-exchange resin in calcium and magnesium forms; the latter must be used since the presence of calcium ions inhibits enzymatic action in the glucose conversion. The unit maximized the conversion of glucose in zone III and the reaction product was obtained with a purity above 90%. Zhang et al. [94] followed a different approach to improve the concentration of glucose at the feed of the SMBR. The authors proposed a four-zone unit where the raffinate passed through a reactor and then mixed with the feed; in another configuration, another reactor was added to zone IV during the initial part of the switch. The data was taken from Hashimoto et al. [93], and higher productivity and purity was obtained with fewer reactors. On the other hand, Toumi and Engell [95] studied a six-column SMBR with both adsorbents and catalysts present in all columns. Pure glucose was injected into the three-zone unit, a mixture of glucose and fructose was withdrawn, and water was used as solvent. The columns were packed with ion-exchange resin Amberlite CR-13Na and immobilized enzyme Sweetzyme T. The authors developed a control scheme that was validated experimentally.

Another growing SMB application is the separation of acids from sugars in the production of ethanol from biomass (see Figure 5.8). Sun et al. [96] studied the separation of sulfuric acid from saccharified liquid from bamboo using an ISMB. The system consisted of six columns packed with Diaion MA03SS anion-exchange resin at 50°C with water as desorbent. The recovery of glucose and xylose was 98.4% and that of acid was 90.5%. The group followed a similar procedure from distilled

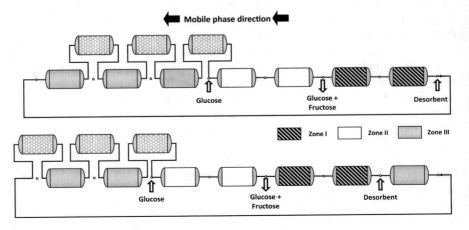

FIGURE 5.7   Schematic representation of an SMBR for glucose isomerization/separation in two consecutive switches.

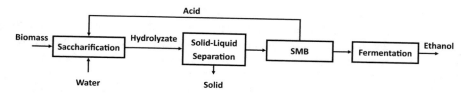

FIGURE 5.8 Simplified scheme for the production of ethanol from biomass.

grain waste from a Chinese spirit-making process; the recovery was slightly lower [97]. The authors claimed that volatile fatty acids were also removed, making the saccharified liquid more feasible to ferment. Additionally, ISMB was more efficient compared to a conventional SMB consisting of more columns. Later on, six commercial resins were evaluated through pulse experiments at 50°C, from which Dowex 1X4 and Dowex 1X8 presented results similar to Diaion MA03SS, which went out of production [98]. Dowex 1X8 exhibited better performance in the SMB separation of sulfuric acid and sugars from acid hydrolysate of bamboo. The authors believed that the higher stability of Dowex 1X8, due to its higher degree of crosslinking, was the main reason for its better performance.

The separation of fucose from other monosugar components by SMB have also been studied with the purpose of developing a process to produce fucose from defatted microalgal biomass [99]. Continuous separation of fucose and 2,3-butanediol was studied over Amberchrom-CG71C resin using two strategies [100]. The first strategy consisted of partial extract collection and partial extract discard, leading to higher 2,3-butanediol concentration. The second included partial extract collection, partial extract recycle, and partial desorbent-port closing, which could lead to 25% less desorbent consumption and higher product concentration. Furthermore, a process for large-scale production of fucose from one of the most abundant seaweeds in South Korea was developed [101]. An open SMB with the flexibility to change port configuration was used with Chromalite-PCG600M resins. Fucose was obtained at nearly 100%.

Two-stage SMBs have been used to separate other, less known sugars; normally, the first stage is used to split mono- from di-saccharides, and then the target product is obtained in the second stage. Choi et al. [102] studied the separation of xylobiose from xylooligosaccharides using Dowex-50WX4 resin in sodium form at 65°C. Xylobiose was obtained with a purity and recovery above 99% and 92%, respectively. Binary separation of xylobiose and xylose was also studied by Lee et al. [103]. On the other hand, separation of trehalose from glucose and maltose was studied by Song et al. [104]. Purolite CGC 100×8 Ca ion-exchange gel-type resin was packed in eight columns and deionized water was used as desorbent. Trehalose was obtained with a 97.6% purity and 95.9% recovery. Other separations of saccharides (mono-, di-, poly-, oligo-) by SMB are discussed by Janakievski et al. [78].

5.2.3 Pharmaceuticals

The introduction of the SMB in the pharmaceutical industry, mainly for the purification of chiral compounds, is seen by some authors as the start of a new era for this technology [2, 105]. Strict constraints were imposed on these separation processes, for instance in terms of target product purity, which drastically increased (values above 99%), and the unit configuration was profoundly changed. While the predecessor SMB processes implemented in the petrochemical industry and sugar separations comprised a large number of chromatographic columns (typically more than eight) of considerable dimensions, the majority of the units used for the purification of pharmaceuticals were constituted by a maximum of six columns with dimensions comprised in the centimeter to meter range. This was a direct consequence of the development of highly specific and efficient stationary phases, which were much smaller than those previously used for other bulk separations (going from the millimeter to the micrometer scale). The elevated costs associated with the stationary phases impelled the search for alternative SMB operating modes that would require a smaller number of chromatographic columns to attain similar performances, in terms of purities and productivities, than the standard operating mode (see Table 5.2 in Section 5.3.2 dedicated to this subject). Among these new operating modes is the Varicol® process.

The Varicol® process was first developed and commercialized by Novasep SAS (France) [106, 107] in around 2000 and is probably the most commonly studied and applied nonconventional SMB operating mode. It consists of the implementation of an asynchronous shift of the inlet and outlet stream positions, which promotes a dynamic variation in the number of columns per zone within each switching interval. By introducing this additional degree of freedom the unit becomes more flexible and the performance can be improved, mostly because this dynamic behavior provides more effective control over the position of the internal concentration bands of the target compounds throughout the unit. In fact, some authors have demonstrated that the performance accomplished in a six-column SMB unit is comparable to the performance of an SMB-Varicol® unit with one column fewer [108, 109].

As previously stated, the development of the Varicol® operating mode is strictly related to the application of SMB technology to chiral separations. For this reason, it is still one of the most studied technologies within pharmaceutical purification processes nowadays. A summary of the main SMB-Varicol® chiral separations reported in open literature within the last decade is provided in Table 5.1.

Apart from the consolidation of Varicol® as one of the most efficient technologies for the purification of chiral compounds, the scientific community has addressed other challenges imposed by pharmaceutical manufacturing processes, such as the separation of more complex mixtures. The advances reported in the purification of one or more compounds from quaternary mixtures of nadolol stereoisomers provides a general idea of the technical solutions that are being implemented.

The simpler approach dealing with the purification of target species from complex multicomponents is to try to reduce these systems to pseudo-binary separation problems; however, this might not always be possible, of course, and in most cases implies an extensive and time-consuming experiment-based mobile and solid stationary phase screening. For the particular case of nadolol isomers separation, the most active stereoisomer (RSR-nadolol) could be purified from the initial quaternary mixture by using Chiralpak® AD as the adsorbent and ethanol: heptane: diethylamine (80: 20: 0.3) as eluent [117]. The SMB process was able to recover 100% of

TABLE 5.1

Literature Review of the Main SMB-Varicol® Chiral Separations Published over the Last Decade

Target product	Stationary phase	Mobile phase	SMB configuration	Performance parameters	Comments	Ref.
R-guaifenesin S-guaifenesin	Chiralcel OD	hexane:ethanol (70:30)	1.0/1.5/1.5/1.0	R-guaifenesin: Purity=99.8% Yield=99.1% Prod.=1.0 g•mL⁻¹•day⁻¹ El. Cons.=0.46 L•g⁻¹ S-guaifenesin: Purity=99.1% Yield=99.8% Prod.=1.1 g•mL⁻¹•day⁻¹ El. Cons.=0.47 L•g⁻¹	Productivity of the five-column Varicol® 26% higher than the six-column SMB. Results equivalent to ModiCon and better than power-feed.	[110]
R-aminoglutethimide S-aminoglutethimide	Chiralcel OD	hexane: ethanol: monoethanolamine (30:70:0.1)	1.0/1.5/1.5/1.0	R-aminoglutethimide: Purity=99.3% Yield=99.4% Prod.=59.1 g•L⁻¹•day⁻¹ El. Cons.=3.0 L•g⁻¹ S-aminoglutethimide: Purity=99.3% Yield=99.1% Prod.=58.0 g•L⁻¹•day⁻¹ El. Cons.=3.0 L•g⁻¹	Productivity of the five-column Varicol® 20% higher than the six-column SMB.	[111]

(Continued)

TABLE 5.1 (CONTINUED)
Literature Review of the Main SMB-Varicol® Chiral Separations Published over the Last Decade

Target product	Mobile phase	Stationary phase	SMB configuration	Performance parameters	Comments	Ref.
Tröger's Base enantiomers	ethanol	Chiralpak AD	1/1.7/2.4/0.9	Prod. = 35.1 g•L^{-1}•day^{-1} El. Cons. = 1.8 L•g^{-1} Extract: Purity = 97.3% Yield = 98.5% Raffinate: Purity = 99.6% Yield = 93.4%	n.a.	[112]
R-(+)-verapamil S-(−)-verapamil	Hexane: isopropanol: ethanol: diethylamine (90:5:5:0.1)	Amylose tris(3,5-dimethylphenylcarbamate)	0.98/2.66/1.40/0.96	R-(+)-verapamil: Purity = 92.0% Yield = 99.4% Prod. = 0.18 g•g^{-1}•day^{-1} El. Cons. = 3.0 L•g^{-1} S-(+)-verapamil: Purity = 93.0% Yield = 99.1% Prod. = 0.20 g•g^{-1}•day^{-1} El. Cons. = 3.0 L•g^{-1}	n.a.	[113]

(*Continued*)

TABLE 5.1 (CONTINUED)

Literature Review of the Main SMB-Varicol® Chiral Separations Published over the Last Decade

Target product	Mobile phase	Stationary phase	SMB configuration	Performance parameters	Comments	Ref.
R-(+)-albendazole sulfoxide S-(−)-albendazole sulfoxide	methanol	Chiralpak AD	0.98/2.38/1.43/1.21	El. Cons. = 1.8 L•g⁻¹ R-(+)-albendazole sulfoxide: Purity = 99.5% Yield = 95.0% Prod. = 60.0 g•kg⁻¹•day⁻¹ S-(−)-albendazole sulfoxide: Purity = 99.0% Yield = 99.0% Prod. = 70.0 g·kg⁻¹·day⁻¹	Long duration run performed. After 55 cycles (11h) it was possible to collect 880 mg of R-(+)-albendazol sulfoxide and 930 mg of S-(−)-albendazol sulfoxide.	[114]
R-(+)-mitotane S-(−)-mitotane	acetonitrile: isopropanol	Amylose tris(3.5-dimethylphenylcarbamate)	0.98/1.46/1.32/1.29	El. Cons. = 0.16 L•g⁻¹ R-(+)-mitotane: Purity = 96.8% Prod. = 0.68 g•g⁻¹•day⁻¹ S-(+)-mitotane: Purity = 97.0%	n.a.	[115]
R-pindolol S-pindolol (target compound)	water: acetonitrile (90:10)	Chiral-AGP	1.5/1.5/1/1	Prod. = 1.1 g·g⁻¹·day⁻¹ R-pindolol: Purity = 83.5% S-pindolol: Purity = 98.0% Yield = 81.5%	n.a.	[116]

the target product in the extract stream with a purity of 100%. As the productivity of the unit was relatively limited (15.6 g of product per liter of bed per day), the authors extended the screening tests and identified Chiralpak® AD and methanol with 0.1% of diethylamine as alternative solid and mobile phases, respectively. The SMB experiments performed demonstrated that the target isomer could be collected in the extract stream with a purity of 99.5% while simultaneously enhancing the productivity by a factor of three [118].

An entirely different approach was followed by Lee and Wankat [119] and by Jermann et al. [120] which resorted to alternative SMB configurations and operating modes to recover intermediately retained nadolol stereoisomers. The first group of authors determined, through an extensive simulation study, that it was possible to recover 99% of the target product (RRS-nadolol) both in an eight-column Pseudo-SMB unit and in a cascade of two eight-column SMB units in series with similar productivities (see Section 5.3.2 for further details regarding these operating modes). The chromatographic columns were considered to be packed with perphenyl car-bamoylated β-cyclodextrin immobilized onto silica gel and the eluent was assumed to be an 80% buffer solution triethylamine (TEA) adjusted with acetic (pH 5.5) and 20% methanol. Although the productivities were similar, the desorbent required for the cascade operation was lower. Jermann et al. [120] optimized and experimentally implemented a cascade of two three-zone closed-loop SMB units operating under ISMB mode to perform this center-cut separation from an equimolar mixture with four stereoisomers (this alternative configuration and operating modes are detailed in Section 5.3.2). Chiralpal® AD was selected as stationary phase and heptane: etha-nol: diethylamine mixtures were used as mobile phase. Through this design meth-odology, the authors demonstrated that the heuristic that proposes that the easier separation should be performed first fails to find the optimal separation sequence for the three-zone ISMB units proposed. Hence, the most retained compounds were separated in the first unit using heptane: ethanol: diethylamine (30: 70: 0.3) as eluent. The outlet stream containing the target product was then mixed with a stream of hep-tane: diethylamine to adjust the eluent composition for the second unit, which was a heptane: ethanol: diethylamine (60: 40: 0.3) mixture. The optimum operating point allowed the purification of the target nadolol stereoisomer with a purity of approxi-mately 98%. The unit productivity was over 50 g of product per liter of adsorbent per day requiring 12 liters of eluent per gram of product. One of the main conclusions of these works is that the use of cascades is generally more advantageous than the use of other alternative operating modes (such as Pseudo-SMB or SMB units with an extended number of zones), since the independent operation of the two units provides additional degrees of freedom, such as the use of different eluent compositions in the work reported by Jermann et al. [120].

The integration of SMB technology with complementary processing units and the application of process intensification strategies to create new SMB-based units has already been proposed by several authors in the context of the petrochemical and sugar industries or in biorefineries; however, only recently have these technical solu-tions started drawing the attention of researchers dedicated to the development of pharmaceutical production processes. Palacios et al. [121] proposed the first SMBR process that performs the racemization reaction and simultaneous enantiomeric separation of chlorthalidone enantiomers (selected as a model system) in a single SMB device with two-zone open-loop or three-zone closed-loop configurations. Side reactors were introduced between the chromatographic columns of zone I in an analogy to the configuration proposed by Hashimoto et al. [93] for four-zone SMBR units. The inexistence of side reactors in the remaining zones is intended to ensure that the desired enantiomer (the least adsorbed one) is not converted near the raffinate collection port which would decrease the overall process performance. The reaction is controlled through a pH gradient imposed by the solvent introduced at the eluent port. The authors demonstrated experimentally that a pure single enantiomer could be produced, achieving 100% conversion and yield values. The performance of the SMBR unit was able to outcome the performance of a process with decoupled reaction and separation units due to the synergetic effects promoted by the combination of these two phenomena simultaneously within the SMBR.

Fuereder et al. [122] proposed a more conservative integration of the enzymatic racemization with SMB separation for the production of enantiopure D-methionine. The process comprised an SMB unit with five stainless steel Chirobiotic TAG columns in a 1-2-1-1 closed-loop configuration dedicated to the separation of the isomers. The feed stream of the SMB was provided by an enzymatic racemization reactor placed before the SMB and by a fresh racemic feed. The target product, D-methionine, was collected through the extract port, since this is the most retained compound, while the raffinate stream, the other enantiomer, was continuously concentrated through a nanofiltration module and recycled back into the reactor. In the experiments performed aiming at the demonstration of the feasibility of the process, the final product purity attained was 98%, with an overall yield of 93.5%. A more detailed economic analyses of the process would have to be performed to assess its viability but the results obtained so far are encouraging.

The search for suitable process intensification strategies to be applied in the pharmaceutical manufacturing process is expected to intensify in the coming years. Nevertheless, Varicol® technology will most likely keep its prominent position within this field.

Finally, it is important to highlight that the advances observed in the development of new chiral stationary phases, especially core-shell particles and monoliths [123], which are already being applied in high-throughput and high-efficiency analytical separations, might be exploited for the development of continuous preparative processes as well within the near future.

5.3 TECHNOLOGICAL IMPROVEMENTS AND NEW APPLICATIONS

5.3.1 NEW APPLICATIONS

Nowadays, the SMB is widely recognized as a well-established technology that is able to perform demanding separations, either in terms of target purities or required productivities, with remarkable efficiency under mild operating conditions, low energy and materials consumption, and high reliability. The long-term industrial applications within the petrochemical, food, and pharmaceutical industries, and their latest advances (some of them discussed previously in Section 5.2), have significantly contributed to this fact. However, with the continuous progress of manufacturing processes, new and challenging separations must be efficiently carried out to ensure the feasibility and sustainability of such processes, and the SMB might represent a technical solution in many cases. In fact, there are countless examples of the application of these chromatographic processes across a large number of industries.

Biochemicals and biopharmaceuticals are currently among the top growing industries. Since many of their processes require the separation of quite similar molecules at high purity levels, or involve heat-sensitive species, for instance, the relevance of SMB technology within these industries has been increasing at an interesting rate supported by several successful applications. In this context, an assessment of some of the most promising applications reported in open literature will be provided together with an analysis of the envisioned advantages and limitations of SMB in the biochemical and biopharmaceutical industries.

5.3.1.1 Biorefineries

The transition from petro-based to bio-based chemicals is a trend observed across the majority of the chemical industries. In this context, the concept of a biorefinery, and especially an integrated biorefinery, is becoming the focus of the R&D activities of an increasingly large number of research institutions and companies; however, this is an extremely broad concept which encompasses hundreds of possible products and production processes. In fact, the sugar purification processes previously described in Section 5.2.2 could be considered within the biorefineries topic as well, for instance. For this reason, the research reported in the present work will be restricted to a few production processes of industrial relevance that have been using SMB technology consistently over the last few years, and will analyze the major advances accomplished.

The separation of carboxylic acids is one of the major applications of SMBs within the biorefineries context. For instance, the bio-based succinic acid production through the fermentation of glucose can be considered as a case study of the commercial implementation of biorefineries [124]. With an estimated global production capacity of approximately 30,000 tons/year (regardless of the production route), companies like BASF, BioAmber, Myriant, and Reverdia have attempted to transition their processes from demonstration scale to a fully commercial scale. The process proposed by Myriant [125], for instance, includes an SMB purification step among the complex sequence of downstream separation stages. The SMB is responsible for the separation of succinic acid from ammonia-salts, residual sugars, and other contaminants (e.g., acetic acid and other organic acids) from a previously filtered and acidified fermentation broth (see Figure 5.9). The process comprises an eight-column closed-loop SMB unit packed with an ion-exclusion resin and uses evaporation and crystallization units to recover the purified succinic acid in the extract stream.

In an attempt to assess the efficiency of the implementation of alternative operating modes and configurations, some authors have resorted to diluted model mixtures of succinic and lactic acids [126–128]. The performance of three different SMB units was compared: conventional three-zone open-loop (without zone I), conventional three-zone open-loop with partial-feed, and unconventional three-zone open-loop (without zone I). Further details regarding the underlying principles and advantages

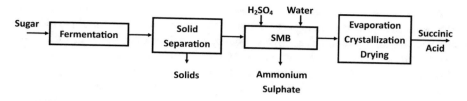

FIGURE 5.9 Simplified scheme for the production of succinic acid.

of each operating mode will be provided later (Section 5.3.2). The three-zone open-loop SMB with an unconventional port arrangement was the most efficient solution for the purification of succinic acid. A succinic acid purity of 98% was achieved by all the processes but the throughput of the SMB without zone I was approximately 40% higher than the three-zone unit to which the partial-feed was applied and over 65% higher than that of the conventional three-zone unit (the only unit for which an experimental validation of the results was performed [126]).

A succinic acid purification method was also proposed by Choi et al. [129] but this time to remove acetic and formic acids from a feed mixture produced by fermentation. An unconventional open-loop SMB without zone I was packed with Amberchrom-CG300C for that purpose and water was used as eluent. The authors performed an optimization of the most relevant operating variables by setting a minimum succinic acid purity of 99%. It was demonstrated that a maximum feed concentration of 18 g/L could be used without compromising the process yield (nearly 100%). Conversely, by setting the yield to 95% the SMB productivity in terms of succinic acid was estimated to be approximately 30 g per liter of adsorbent per day. The performance of a conventional three-zone open-loop SMB (without zone IV) was compared to the previously reported results. Under the same purity and yield constraints, the conventional three-zone open-loop unit was three times less productive. Although this study provided a clear insight into the advantages of implementing an unconventional SMB operating mode to purify succinic acid, all the conclusions were based on numerical simulation results (since the model validation was performed in a previous work [130]).

The same system was studied by Park et al. [130] in a previous work, with the main goal, however, of recovering formic acid. By applying an SMB with the same three-zone open-loop alternative port arrangement, at 40°C, a formic acid yield of 98% was achieved with a purity of 96%. The feasibility of the process was experimentally confirmed.

The production of biodiesel through transesterification reactions leads to the formation of considerable amounts of glycerol, with this by-product representing approximately 10% of the overall product mass. However, at the end of the biodiesel production process glycerol is obtained in a mixture with other alcohols (used as reactants in the previous steps), water, and organic and inorganic salts (due to the use of homogeneous catalysts), in what is usually called crude glycerol. Rohm and Haas (a subsidiary of Dow Chemical) and Novasep developed a commercial glycerol purification process, taking advantage of SMB technology, for the valorization of this by-product [131]. The Ambersep BD50 ion resin is used as stationary phase, water as mobile phase, and glycerol is collected in the extract stream. Through a series of complementary steps (heating, evaporation, splitting, decantation, SMB purification, and vacuum distillation) a final glycerol purity of 99.5% can be reached. Companies like Lanxess Sybron Chemicals Inc., EET Corporation, and the SRS Engineering Company also provide glycerol purification solutions.

Recently, new strategies have been proposed for the valorization of the waste glycerol from biodiesel plants. After its purification through the aforementioned processes, for instance, refined glycerol can be reacted to generate chemicals with high added value; however, the reactions involved in this processes typically generate a complex mixture of products and unconverted glycerol that requires further purification. In this context, Coelho et al. [132, 133] proposed the implementation of an SMB process for the separation of glyceric and tartronic acids obtained through the aerobic oxidation of glycerol in aqueous media. Starting from a model mixture containing the two carboxylic acids, the authors demonstrated the feasibility of the process by performing the separation in a lab-scale unit comprising six columns packed with Dowex 50WX2 and using a solution of sulfuric acid at 4 mM as eluent. Under optimal conditions, the unit was able to achieve a productivity of 86 g of tartronic acid and 176 g of glyceric acid per liter of adsorbent per day, considering the typical commercial purity value of 97% of these species. The estimated eluent consumption was 0.30 liters per gram of products.

Other polyols with interesting commercial value can also be produced in integrated biorefineries. Companies like Archer Daniels Midland Co [134] or LanzaTech [135] have patented SMB purification processes for this particular purpose. In a recent publication [136], the implementation of a three-zone open-loop SMB process with an additional independent zone designed to carry out a regeneration step is described. Butanediol and propanediol were separated by means of a Mitsubishi SP70 resin. Although 100% pure butanediol was collected in the extract stream, the maximum propanediol purity obtained during the experiments performed was 92.6%. The process productivity was 101 g of product per kilogram of adsorbent per day.

Bio-based chemical production processes comprising SMB units cannot be dissociated from SMB reactor (SMBR) technology, a hybrid unit that can simultaneously perform reaction and chromatographic purification steps in the same device, conceived through the implementation of the most advanced process intensification principles. In the SMBR, a reactive mixture is introduced in the unit through the feed port and, in its most common implementation, this mixture is simultaneously converted and separated (according to the SMB principles) by a solid stationary phase with both catalytic and adsorptive properties while one of the reactants is used as eluent [137, 138]. Alternative packed bed configurations and eluents can also be used. In fact, most of the reported applications of this multifunctional reactor were related to the production of bio-based chemicals, with particular focus on the production of bio-derived fuels [139] and oxygenated additives [140]. Examples include the use of bio-derivable reactants such as ethanol to produce 1,1-diethoxyethane [141] and 1,1-diethoxybutane [142], and the valorization of glycerol, obtained as a by-product from the biodiesel manufacture, through the production of glycerol ethyl acetal [140], among others [143, 144]. The target purities obtained through this processes ranged from 90% to 99% with the reactor being able to convert almost 100% of the limiting reactants. High productivities were reported (5–55 kg of product per liter of adsorbent per day) with reasonably low eluent consumptions (below 8 L of eluent per kilogram of target product).

Although some successful applications of SMB technology can already be found within biorefineries, it is important to ensure that upcoming research activities will address the main issues related to these processes. For instance, many studies are still carried out assuming model systems; however, real industrial streams from biorefineries are known to present a rather complex nature that can ultimately invalidate a purification process that has been designed without having accounted for it. Scale issues and long-term SMB operation must also be evaluated during the development of new processes.

5.3.1.2 Biopharmaceuticals

The global biopharmaceuticals market was valued at approximately 188,000 USD in 2017, with cumulative sales between 2014 and 2017 reaching over 650,000 USD [145] and some projections forecasting a compound annual growth rate (CAGR) of almost 14% until 2025 [146]. Monoclonal antibodies are the dominant class of biopharmaceuticals, representing approximately 65% of the overall sales value in 2017. Another relevant aspect is that recombinant proteins represent over 85% of the genuinely new biopharmaceutical active ingredients that reached the market between 2014 and 2018 [145]. Apart from these general market trends, several other biopharmaceuticals have been produced through a variety of manufacturing processes in which, in many cases, downstream separation is known to account for more than 80% of the overall capital costs [147, 148]. Considering this, the prominence of separation technologies within this industry becomes clear. As previously stated, the SMB has been suggested as a potential solution to many of these challenges, mainly due to its ability to achieve high product resolution under moderate operating conditions; however, the complexity of the mixtures to be separated has also prompted significant advances in terms of practical implementation. Some of these issues will be discussed using a few illustrative examples reported over the last decade.

As far as our knowledge goes, the first application of SMB to protein desalting was described by Hashimoto et al., who performed the separation of bovine serum albumin (BSA) from ammonium sulfate using Sephadex G-25Fine [149]. The experimental runs were carried out in a conventional four-column SMB and in an SMB without zone IV, allowing the recovery of approximately 70–80% of BSA, with an outlet concentration of half of the feed value; however, simulations demonstrated that the yield could be increased up to 98% and the outlet stream could reach 90% of the feed value.

A desalting step is also required to remove zinc chloride during the production of insulin before the crystallization unit, for instance. An SMB process based on size exclusion chromatography (SEC) has been proposed and experimentally validated for conducting the zinc chloride removal step simultaneously with the removal of contaminant high-molecular-weight proteins [150]; however, this comprises a complex center-cut separation, since insulin is the intermediately retained species. To overcome this issue, a cascade of SMB-SEC units was used in which the first unit was responsible for the separation of the high-molecular-weight proteins (early eluting species) from insulin, without imposing any constraints on the zinc chloride concentration bands, while the second unit was responsible for the final insulin purification through the removal of zinc chloride. Both insulin purity and yield reached 99%. The advantages of the SMB over the batch system become clear in this case since the overall yield attained by the former did not surpass 90%, even when operating at one third of the SMB productivity. Recently, this separation process was the focus of a deep optimization study which significantly improved the performance of the SMB-SEC process [150]. While keeping the same yield as the tandem SMB system initially proposed, it was possible to achieve a productivity ten times higher and to reduce the overall production costs by 15%, by adjusting the main process design variables. These improvements were even more significant when compared to the traditional batch systems, since the insulin overall production costs dropped to one third.

The combination of SMB with SEC, as in the previously reported work, has been an increasing trend in biopharmaceutical processes recently. The typically large dimensions of biomolecules can be exploited to simplify more complex systems into pseudo-binary separation systems or even to reduce the buffer consumption [151, 152]. The integration of recombinant proteins that refold with purification using SMB-SEC for the development of continuous production processes are among the most interesting applications. To perform a feasibility study of this new process design concept, Wellhoefer et al. [153] selected two model proteins: N^{pro} autoprotease fusion peptides EDDIE-pep6His and EDDIE-MCP-1. Firstly, a continuous dissolution reactor was designed to promote an efficient dissolution of inclusion bodies from N^{pro} fusion peptides by NaOH, instead of the commonly used chaotropes, and subsequent reduction with 1-thioglycerol. The outlet of the dissolution reactor was then mixed with a concentrated stream of refolding buffer and fed to the SMB-SEC unit packed with Sephacryl S-300. One of the main advantages of the proposed process is that it can be operated isocratically by using the refolding buffer as desorbent, which allows the implementation of a classical four-zone closed-loop SMB operation. The refolded or partly refolded target proteins were collected through the extract port and fed to a continuously stirred tank reactor responsible for carrying out a final off-column refolding stage. Inside the SMB-SEC unit, the aggregates' formation is potentially reduced through the spatial separation of the folding intermediate species due to their differences in size; in this way, higher yields can be obtained. An increase of 10% was obtained for both proteins; the refolding equilibrium is shifted toward the formation of refolded proteins through this new process in comparison with conventional batch ones. The aggregates, on the other hand, were collected through the raffinate port and the refolding buffer/desorbent was recovered through a tangential flow filtration unit and recycled back into the process. The SMB closed-loop operation and the buffer recycling strategy allow for a reduction of the overall buffer consumption of over 98%, which constitutes an important advantage since buffer consumption represents a significant cost in downstream processing for these kinds of biopharmaceuticals.

An adaptation to the previously described process was proposed and experimentally validated using the same model system [154]. An additional isocratic four-column closed-loop SMB-SEC unit (packed with Sephacryl S-100 and using the same buffer solution as eluent) was added to the outlet of the off-column refolding reactor to recover the cleaved target peptide (extract stream) by separating it from the residual proteins (raffinate stream). The buffer solution from the raffinate stream of the second SMB is recovered and recycled with a similar procedure as the one reported for the raffinate stream of the first SMB. Outstanding results were reported for the integrated process, including a target product purity and yield of nearly 100%, together with a productivity 180 times higher and a buffer consumption 28 times lower, compared to conventional batch processing.

More recent studies regarding the oxidative refolding of lysozyme in a SMB-SEC unit have been published based on comprehensive mathematical models and complemented with single-column validation experiments [155]. It was concluded that the aggregation process occurs when the local concentration exceeds a critical value corresponding to the solubility of the early intermediates. Moreover, the authors support the idea that the refolding reaction can be carried out off-column after the SMB-SEC, as observed in the works of Wellhoefer et al. [153, 154]; this was verified by the absence of urea (initially present in the buffers) after the native protein was recovered in the raffinate stream (for this reason the authors suggest considering the total soluble protein rather than only the native refolded protein). Although the results look promising, the process still lacks experimental validation.

Ion-exchange chromatography also plays an important role in biopharmaceutical manufacturing processes based on SMB technology. The salt gradient ion-exchange SMB has been applied for the separation and purification of proteins with considerable success by exploiting the advantages of the additional degrees of freedom provided by the step salt gradients [156]. Nevertheless, the system complexity is naturally increased for this particular operating mode and, for that reason, the available tools must be used to select the most effective salt gradient strategy [157]. Some of the most successful applications comprise the separation of BSA from myoglobin [156], the separation of bovine IgG from lysozyme [158], the separation of the dimeric form of the bone morphogenetic protein-2 from a renaturation solution, which also contained its inactive monomeric form among other impurities [158], and the separation of β-lactoglobulins A from β-lactoglobulins B [159]. It is worth mentioning that, in many of these processes, open-loop units were often more suitable than closed-loop ones since they allow more accurate control of the salt concentration throughout the unit.

In a work published in 2018, the purification of human influenza A virus A/PR/8/34 (H1N1) was accomplished by capturing it from a clarified harvest cell culture fluid, through an SMB process relying on ion exchange [160]. This study comprised two distinctive technical details. First, the stationary phase used consisted of a commercial quaternary amine anion-exchange monolithic material, designated as CIM® QA, since monoliths are known to provide high adsorption capacities and to allow high flow throughputs under manageable pressure drops, as typically required by capture steps in biopharmaceutical processes. Second, the SMB unit was run in a three-zone open-loop configuration with an additional detached zone included for regeneration purposes using a high salt concentration step gradient (2M NaCl). The implementation of such a complex configuration was vital to making it possible to perform a ternary center-cut separation in which the contaminant host cell proteins were eliminated through the raffinate port, the purified fraction of the target product (influenza A virus) was collected through the extract port, and the strongly bounded DNA molecules were eliminated during the regeneration step gradient procedure. By loading the monoliths up to 10% of their dynamic binding capacity, a virus yield of approximately 90% was achieved. Moreover, the amount of contaminant proteins was reduced by more than 50% compared to the feed stream concentration values, while DNA concentration was reduced to 1% of its initial value, meeting the stringent regulatory requirements of human influenza vaccines prepared in cell cultures. Due to the low values defined for the dynamic binding capacity, the productivity of the SMB was relatively low, approximately half that of the corresponding batch system, while the buffer consumption was slightly lower. By setting a higher dynamic binding capacity (50%), the product yield dropped to less than 50% and the DNA contaminations surpassed the imposed limits, which required an additional step (Benzonase® treatment) to complete the purification process. Therefore, the authors suggest that the process performance parameters can be improved, particularly productivity, by optimizing the operating conditions. Other influenza virus purification processes were reported by the same research group; however, SMB-SEC was the separation technique adopted [161].

The purification of valine, a branched-chain amino acid of industrial interest, through the removal of isoleucine and other contaminants within the feed mixture, using reversed-phase chromatography SMB processes, has been thoroughly investigated by Park et al. [162–164]. In an initial approach, the binary separation between valine and isoleucine was carried out in a three-zone open-loop SMB combined with a partial-discard strategy, according to which the first fraction of the raffinate stream obtained during the switching interval would be discarded while the second fraction containing purified valine would be collected. On the other hand, isoleucine was continuously withdrawn through the extract port. The separation was performed based in an SMB with three columns packed with Amberchrom-CG161C. After a preliminary optimization study, it was determined that only half of the raffinate volume would be collected during the switching interval. Under these conditions, it was possible to achieve a valine purity and yield of over 98%. Moreover, by rejecting 50% of the raffinate stream volume the concentration of valine in this outlet stream doubled compared to the process without the discard step [162]. In another publication, the same research group proposed an alternative three-zone open-loop SMB configuration with a modified port location. In this new configuration, zone I is bounded by the desorbent and the feed streams, zone II by the feed and the raffinate stream, and zone III by the raffinate and the extract [163]. The feasibility of this operating mode was confirmed and the mathematical model developed to describe the dynamic behavior of the unit was validated with the experimental data collected. The improvements resulting from this minor adaptation were remarkable. The unit throughput increased by 65% and the valine concentration in the raffinate stream was 1.6 times higher than the expected value for the conventional three-zone open-loop SMB unit.

Finally, the same authors developed a continuous SMB process for the purification of the valine present in a model mixture obtained from a fermentation reactor used in an industrial valine production process [164]. For that purpose, a cascade of two consecutive three-zone open-loop SMB units packed with Chromalite-PCG600C was designed and optimized to perform the valine center-cut separation in an attempt to replace the two separation units present in the conventional process, namely, an ion-exchange chromatography stage for the removal of salts (ammonium sulfate) and a crystallization stage to remove the remaining amino acids (mainly leucine and alanine). The authors followed the empirical rule of performing the easier separation first; therefore, the first SMB unit, which adopted a modified port location, was dedicated to the removal of leucine through the extract port and to the recovery of valine through the raffinate port. The port arrangement was defined in order to maximize the valine concentration by attributing to zone III the role of enriching the raffinate stream. The raffinate stream was then fed to the second SMB unit which consisted of a three-zone open-loop SMB; however, in this case, as valine was to be collected through the extract port, a conventional port arrangement was adopted in order to use zone II as a valine enrichment zone. Leucine and alanine, the least retained species, were eliminated through the raffinate port. The feasibility of the process was demonstrated and a valine purity higher than 99% and a yield of approximately 93% were attained. Although two SMB units were used, the concentration of valine after the final purification stage was still higher than 50% of the feed concentration. The previously reported study clearly evidenced the challenges associated with the design and optimization of the separation of amino acids by SMB. For that reason, this topic has been investigated by several research groups [165, 166].

An experimental and simulation study of the purification of single-chain fragment variable (scFv) antibodies in a three-zone open-loop SMB with a pH gradient, containing three columns for separation purposes and an additional column in the cleaning and equilibration zone, is used as an example of an SMB process based on affinity chromatography for the purification of biomolecules or, more specifically, immobilized metal ion affinity chromatography [167]. In this work, the pH gradient was imposed by the different pH values of the desorbent (20 mM sodium phosphate, 0.5 M NaCl, pH 4.0) and feed streams (clarified cell culture supernatant diluted in 20 mM sodium phosphate, 0.5 M NaC, pH 6.5) while the open-loop configuration was used to avoid cross-contamination of the target product by buffer recycling. Water was used for the cleaning step and a sodium phosphate and sodium chloride solution, pH 5.3, was used as an equilibration buffer. The enriched target product was collected with a yield of 91% through the extract port, approximately 2.5 times more concentrated than the target product obtained through a conventional batch reactor. The SMB also presented a productivity value 11 times higher and a desorbent consumption almost six times lower; however, both processes must still be optimized.

Successful examples of application of SMB technology have been presented above. The complexity of biopharmaceutical systems has tested the limits of this technology, which has been able to accomplish rather demanding separations that would have been impossible to perform in conventional SMB units, including ternary center-cut separations, through the implementation of alternative configurations and operating modes (Section 5.3.2 will be entirely dedicated to this topic given its relevance). Other important advances include the use of innovative stationary phases (as monoliths), the implementation of processes that take advantage of the different types of interaction between liquid mobile phases and solid stationary phases, and the implementation of step gradients. Nevertheless, some limitations still persist and the solution for some of those might comprise the development and implementation of completely new SMB-based technologies, as will be discussed in Section 5.3.3.

5.3.2 ALTERNATIVE OPERATING MODES

The conventional SMB concept and operating principle have been carefully addressed in Section 5.1 of the present work. Most of the SMB processes developed and implemented in the industry follow those same characteristics, namely, the existence of two inlet streams, two outlet streams, an internal recycling stream, four sections, and the synchronous shift of the stream locations; however, as one can easily conclude, the number of decision variables that can be defined for such complex multicolumn units can be largely increased by altering some of these features. Throughout the years, several adaptations and alternative operating modes have been proposed to either enhance the performance of this technology or expand its application field.

Commercial examples of these advancements comprise, for instance, the CSEP and ISEP systems commercialized by Calgon Carbon Co. (United States of America), which represent pioneering solutions for water treatment and for further purifying complex mixtures obtained through biomass and sugar processing. The systems enabled the cyclical performance of adsorption, regeneration, and recycling steps [168, 169], by introducing additional inlet and/or outlet streams and independent zone operation. Another example is the Varicol® process [106], which was developed by Novasep SAS (France) to reduce the costs associated with the specialized stationary phases used in chiral separation by promoting an asynchronous shift of the SMB inlet and outlet streams. The Varicol® process is probably the most commonly applied unconventional SMB operating mode; however, other alternative operating modes have become widely accepted, finding several applications and being the focus of a large number of research studies. These alternative operating modes, which can be classified according to the major changes that are promoted in the unit configuration or operating mode when compared to conventional SMB technology, are summarized in Table 5.2.

Previous reviews [168, 169, 200] have already described the majority of the alternative operating modes presented in Table 5.2. Nevertheless, despite the large number of solutions already proposed, new alternative operating modes have been developed by different companies and research groups throughout the world over the past few years. The main motivating forces directing this research line comprise the need to perform separations that were not possible with prior techniques and the need to reduce the capital and operating costs associated with the equipment and the stationary phases.

5.3.2.1 Dynamic Configuration Variations

The improved-SMB® (ISMB) and the sequential SMB (SSMB) dynamic operating modes (described in Table 5.2) are among the most modern multicolumn countercurrent chromatographic processes applied to the industrial-scale production of fructose. Among the main advantages of these technologies is the ability to achieve the

TABLE 5.2

Main Features of the Most Relevant Alternative SMB Operating Modes

Classification	Designation	Main Modifications	Main Advantages	Ref.
Dynamic configuration variations	Varicol®	Asynchronous shift of the inlet and outlet ports	Reduces the number of chromatography columns required	[106, 107]
	Pseudo-SMB	1st step: open-loop operation, ternary mixture fed at the beginning of zone III, eluent fed at the beginning of zone I, intermediately adsorbed compound collected in the raffinate 2nd step: conventional SMB operation for a cycle, feed flow rate suppressed	Allows ternary separations	[119]
	Improved-SMB®	1st step: conventional SMB operation, no flow in section IV 2nd step: inlet and outlet streams closed, recycling re-established	Allows ternary separations; reduces the number of chromatography columns required	[170–172]
	Sequential-SMB	1st step: inlet and outlet streams in the conventional SMB arrangement, no flow in zone II, no flow in zone IV 2nd step: inlet and outlet streams closed, recycling re-established 3rd step: eluent fed at the beginning of zone I and raffinate collected at the end of zone III, feed and extract streams suppressed, no flow in zone IV	Allows ternary separations; reduces the number of chromatography columns required (particularly in zones I and IV)	[81]
Flow rate modulation	Power-feed	Volumetric zone flows changed within a switching interval	Lower desorbent consumption or higher purity	[173–175]
	Partial-feed	Discontinuous introduction of the feed within a switching interval, sectional flow rate maintained by changing raffinate flow rate	Reduces the number of chromatography columns required	[176]
	Partial-discard and partial-withdrawal	Discontinuous collection of the products in the outlet streams within a switching interval	Higher purity	[177]
	Outlet streams swing	Dynamic variation of the extract, raffinate, and eluent flow rates within a switching interval	Higher purity; lower desorbent consumption	[178, 179]

(Continued)

TABLE 5.2 (CONTINUED)
Main Features of the Most Relevant Alternative SMB Operating Modes

Classification	Designation	Main Modifications	Main Advantages	Ref.
Concentration modulation	ModiCon	Dynamic variation of the feed concentration within a switching interval	Higher purity	[180, 181]
	Enriched-extract SMB	Extract collected at the outlet of zone I and partially reinjected at the beginning of zone II after an (external) concentration step	Higher recovery of the more adsorbed compound; higher productivity	[182, 183]
Gradient operation	Solvent gradient	Modulation of the eluent concentration throughout the unit	Lower desorbent consumption; increases product selectivity	[184–187]
	Temperature gradient	Modulation of the temperature throughout the unit	Lower desorbent consumption; increases product selectivity	[188, 189]
Alternative configurations	Reduced number of zones	Elimination of a conventional SMB zone (most commonly the elimination of zone IV leading to an open-loop operation)	Reduces the capital costs; easier control; increases extract purity	[190, 191]
	Extended number of sections	Increase in the number of zones (additional outlet streams, recycling of streams within the unit, etc.)	Allows ternary separations	[192–199]
	Tandem SMB/SMB cascades	Several SMB units placed in series	Allows ternary separations	[192, 193, 197]

same performance as a conventional SMB process, but with fewer columns (typically four to six), thus requiring approximately 10% less eluent [2]. A recent study
in which a direct comparison between ISMB and SSMB was carried out revealed
that both performed quite similarly for monosaccharide separation and corroborated
one another's superior efficiency, compared to the conventional four-zone SMB [81].

Although the ISMB was developed in the 1990s by the Mitsubishi Chemical
Corporation (Japan) [201], recently, two adaptations to this alternative operating mode
were proposed that allow the separation of ternary mixtures [202]. In its initial concept, the ISMB time switch was sub-divided into two steps. A first step, in which a
three-zone open-loop configuration was adopted while zone IV was idle, was followed by a second step, in which all the inlet and outlet streams were eliminated and
the fluid was recycled in the unit. A simple modification consisting of performing the
second step in an open-loop configuration through the introduction of an additional
extract stream at the outlet of zone I (suppressing the connection to zone II) allows
the collection of a third fraction containing the more retained species. The authors
denominated this operating mode 3S-ISMB. Another approach for separating ternary mixtures, designated 3W-ISMB, requires deeper adjustments. In the first step,
the feed stream is still introduced between zones II and III but the least retained
compound is no longer collected at the outlet of zone III during this step. Instead,
this stream is eliminated and all the flow is diverted to the following columns. The
most retained compound is then collected at the outlet of zone I, and zone II is independently operated so that the intermediately eluting species can be collected at its
outlet. In the second step of this operating mode, zone I is idle, eluent is fed to the
column immediately after this zone, and the least retained compound is collected at
the other extreme of the unit. The two steps, performed within each switching time of
the ISMB, 3S-ISMB, and 3W-ISMB, are presented in Figure 5.10.

For the 3S-ISMB to be able to perform ternary separation, the interstitial flow
rate ratios between the mobile and stationary phases must be set in such a way that
the separation between the two least retained species can be carried out as in the
standard ISMB, while the most retained compound is confined to the column to
which the mixture is fed and then carried with the solid phase in the opposite direction of the fluid flow so it can be collected in the extract port during the second step.

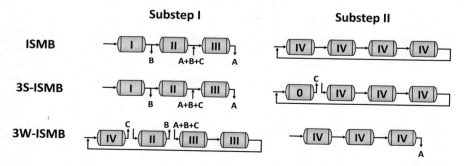

FIGURE 5.10 Schematic representation of the ISMB, 3S-ISMB, and 3W-ISMB operating
modes. A, B, C represent the compounds in ascending order of affinity toward the solid.

The main parameter governing the feasibility of these separations is the selectivity between the most retained species and the remainder.

As previously stated, in the first step of the switching time of a 3W-ISMB, a purified fraction of the two more retained compounds are collected at the outlet of zones I and II, which are disconnected from the subsequent zones where the feed mixture is separated. Therefore, to accomplish a ternary separation through this operating mode, one must ensure that, at the end of the first step, zone I is completely regenerated. For that reason, zone I is not used in the second step of the switching time, during which the remaining columns are connected in series and fed with an eluent stream to collect the least retained compound at the other end of the unit. The eluent also allows for the readjusting of the relative position of the concentration bands of the two more retained compounds throughout the unit, in preparation for the next switching period. The selectivity between the weak and the intermediately adsorbed compounds is the key parameter that defines the size of the separation region.

The new ISMB operating modes described above were applied in the purification of pharmaceuticals (simultaneously separating the two target enantiomers and the impurities present) [202] and in the purification of building blocks obtained through biotechnological processes, such as lactic acid [203], providing a solid proof-of-concept and demonstrating its potential for ternary separations in several industries. Further adaptations and improvements have been proposed for the ISMB operating mode, such as the reduction of the number of zones within the unit; however, as this is the result of the combination of an alternative configuration with an unconventional operating mode, it will be addressed in a later section of this work dedicated to this specific issue (Section 5.3.2.6).

5.3.2.2 Flow Fate Modulation

Several adaptations of the conventional SMB operating mode have been proposed based on flow rate modulation strategies such as the power-feed [173–175], the partial-feed [176], the partial-discard [177], or the OSS-SMB [178, 179]. In most cases, the target compounds were obtained with higher purity, and a few studies also report benefits in terms of eluent consumption [175]. The majority of the most recent flow rate modulation strategies have been implemented in conjunction with other modifications to develop alternative operating modes that can overcome the performance of state-of-the-art solutions, as will be discussed in Section 5.3.2.6.

The power-partial-discard SMB (PPD-SMB) [204] consists of adding a partial-discard feature to the power-feed SMB. In practical terms, this means that during the first fraction of the switching interval the extract flow rate is at a specific set point, the stream is being discarded (to eliminate the impurities), and, after a predefined amount of time, the product is collected at a different flow rate until the end of the switching period. Similarly to the partial-discard SMB, the raffinate stream is collected during the initial moments of the switching time and discarded at the final stage (to remove late eluting contaminant species). Each of these steps are carried out at different flow rates in this new operating mode. The average flow rates can be determined through the equilibrium theory and two additional variables are introduced in comparison with the conventional SMB: the discard duration, which is the ratio between the discard period and the switching time; and the discard amount, which is the ratio between the discard volume and the total volume of fluid collected within a switching interval. By selecting the adequate value for these two additional degrees of freedom, it is possible to overcome the purity obtained in the classical SMB and even in the partial-discard SMB. Moreover, the recovery of the PPD-SMB largely surpasses the value obtained in the partial-discard SMB; however, this value is still far from the value obtained in the conventional SMB (considering the differences in the final product purities) [204].

The partial-feed and the OSS flow rate modulation strategies have also been applied simultaneously in a single SMB device [80, 205, 206]. This operating mode, designated SimCon, is accomplished by dividing the switching interval into three fractions: a first period in which the raffinate is the only active product withdrawal port, or in which the extract stream flow rate is at its lowest level while the feed stream can be active or suppressed (with the latter being the most common status); an intermediate period in which the unit is operated as a conventional SMB; and a final period in which the extract is the only active product withdrawal port, or in which the raffinate stream flow rate is at its lowest level while the feed stream can be active (most common status) or suppressed. Alternatively, the elimination of the intermediate period has also been proposed [205]. The synergetic effects between the partial-feed and the OSS allows the SimCon SMB to exceed the performance of the conventional SMB in terms of product purity, recovery, productivity, and desorbent consumption. The expansion and contraction of the concentration bands imposed by the flow rate modulation and/or partial port closing strategy adopted retards the migration of the extract front and accelerates the migration of the raffinate tail, enhancing separation efficiency in this way. The performance of SimCon is similar to that of the standard OSS operation; however, by adding the partial-feed it is able to achieve better control of the maximum pressure drop through the unit. Some authors have also concluded that the SimCon SMB outperforms the ISMB at moderate eluent consumptions but is less efficient than the three-zone open-loop SMB if high eluent consumptions can be afforded [80]. Although this technology results from the combination of two flow rate modulation strategies, under certain conditions it can also be considered a dynamic modification of the SMB configuration, since the activation and deactivation of the feed and the product withdrawal ports promote a variation in the number of zones in the SMB throughout the switching interval.

5.3.2.3 Concentration Modulation

The concentration modulation (ModiCon) [180] operating mode is known to improve the classical SMB separation performance by promoting minor adjustments on the migrating concentration band velocities in non-linear systems through the dynamic variation of the feed concentration throughout the switching period. More recently, this concentration modulation strategy was combined with the partial-discard technique, originating similar alternative SMB operating modes, designated fractionation and feedback SMB (FF-SMB) [207–209], recycling partial-discard SMB (RPD-SMB) [210], and backfill SMB (BF-SMB) [211]. The basic concept behind these operating modes is the momentary replacement of the feed stream by a recycled portion of the raffinate and extract streams during the initial and final stages, respectively, of the switching interval. These dynamic variations generate a profile in the composition fed to the unit, comprising a primary enriched fraction of the less retained compound followed by the original feed composition and then by an enriched fraction of the more retained compound, leading to an increase in the concentration of the target products near the corresponding collection ports. Therefore, the main advantage that these technologies can provide over the conventional SMB will be an increase in product purity. Moreover, compared to the partial-discard technique, these alternative operating modes are able to achieve much higher recoveries.

Although the FF-SMB, RPD-SMB, and BF-SMB follow the same general concept, some differences can also be highlighted between them. For instance, in the original FF-SMB [207], only one of the outlet streams was partially recycled through the feed port; the RPD-SMB, on the other hand, may also encompass a partial-discard strategy [210], while the BF-SMB is more flexible in terms of how the recycled streams are reinjected into the SMB. In fact, the authors suggest four recycling strategies, including the possibility of recycling the raffinate or the extract streams or both into the corresponding intermediate nodes (namely, into columns between the raffinate and the feed ports and between the extract and feed ports, respectively, in a similar approach to that of the enriched-extract SMB [182]), allowing the modulation of the internal concentration profiles together with or instead of the feed concentration modulation [211]. To some extent, the latter approach can also be considered to be an SMB with an increased number of zones.

Some of the partial-discard and partial recycling techniques presented previously have also been applied to SMB units with a reduced number of zones (the combination of these alternative operating modes and configurations will be analyzed in Section 5.3.2.6).

In another approach that follows a feed concentration modulation strategy, the FeedCol SMB, a mixture is fed to the system by pulse injections, synchronized with the SMB switch; however, instead of entering directly into the SMB, the mixture is pre-separated in an external chromatographic column packed with the same stationary phase as the SMB [212]. In this way, an enriched fraction of the less retained compound enters the SMB at the beginning of the switching period and, conversely, an enriched fraction of the more retained one is fed to the unit at the final stage of the switch. As a consequence, this concentration modulation strategy reduces the impurity concentration near specific collection ports, enhancing the purities of the outlet streams and allowing the separation of binary mixtures composed of species that present considerably low selectivities (a value of 1.10 in the reported case) [212]. The aforementioned strategy simultaneously achieves slightly higher recovery values and productivities when compared to the conventional SMB. Moreover, it was reported that the FeedCol SMB is more effective at accommodating the negative impact of the heterogeneity between the chromatographic columns (uneven aging or packing) within the SMB train, keeping the target product within the specifications and with minor adjustments in the operating conditions [213]. Nevertheless, one must not forget that the capital costs of this hybrid system will be somewhat higher.

Although following a completely different strategy, the bypass SMB (BP-SMB) [79, 214] might also be considered a concentration modulation strategy. However, instead of adjusting the inlet feed concentration, the SMB is operated as a classical

SMB under relatively strict purity constraints and the outlet stream composition is adjusted (raffinate, extract, or both at the same time) by mixing it with the feed stream (bypass stream) in processes with an overall low purity requirement. The main advantages of this operating mode are increased process robustness and the flexibility provided by these additional degrees of freedom compared to a process in which these low purity streams are directly withdrawn from the SMB. The process becomes easier to control since, in the presence of a disturbance, the final product purity can be achieved by adjusting the feed bypass flow rate instead of adjusting the internal SMB flow rate ratios. The SMB units within this process design strategy become smaller and the eluent evaporation costs are also reduced. This alternative operating mode has demonstrated its potential in the industrial production of HFCS with a fructose content of 55% [214], as is discussed in further detail in Section 5.2.2.

5.3.2.4 Gradient Operation

The advantages provided by applying concentration [215], pH [216], or temperature [188] profiles, for instance, in an SMB separation process have been exploited by several research groups for more than 20 years now. Solvent gradients, in particular, are the most common type of gradient operation and their relevance in recent times has considerably increased due to the potential they have demonstrated in biotechnological applications, especially when combined with other complex alternative operating modes. The relevance of these new SMB-based technologies is such that this topic will be discussed in detail in Section 5.3.3.

SMB units that use supercritical fluids as eluent (SF-SMB) represent one of the first practical implementations of solvent gradient operation, in which the elution strength was adjusted by controlling the pressure in each zone of the unit. Although several successful applications have been reported for the purification of compounds of high added value, including fine chemicals and pharmaceuticals, this technology has not found any relevant industrial application up to this moment, as far as our knowledge goes. According to some authors, the main reasons for this are the elevated costs associated with the equipment, the difficulty in controlling all the process variables, the low packing reproducibility, the low long-term stability of the expensive stationary phases, and the complexity inherent to the physical–chemical phenomena going on inside the unit and the lack of solid theoretical and practical knowledge of such phenomena [217, 218]. In order to reduce the required equipment and the complexity of the unit, recently, the elimination of the fourth zone of the SF-SMB has been proposed. In fact, as carbon dioxide can be vaporized and easily separated from the target products of the raffinate and the extract streams, zone IV, conventionally used for eluent regeneration purposes, is no longer strictly required under this operating mode (further advantages of SMB units with reduced number of zones will be disclosed in Section 5.3.2.5; however, those are not specific to SF-SMBs). The three-zone open-loop SF-SMB control system is simpler and this new configuration has been mainly applied to the purification of sesamin, sesamolin [219], and triterpenoids [220], among other fine chemicals [221, 222], typically obtained from crude extracts of natural products as plants or fungi.

The use of temperature gradients within the SMB has been reported before, but recent studies provide a deeper insight into the process [90, 223]. In its most common implementation, the temperature decreases from the first to the fourth zone, which can be accomplished by feeding the eluent at a high temperature and the target product mixture at a lower temperature, or by changing the chromatographic column temperature, or by adding heat exchangers between zones or columns. Nevertheless, it is worth mentioning that the practical implementation of this operating mode is not straightforward since it is rather difficult to have an accurate control of the temperature throughout the unit. The main feature previously associated with this operating mode was the reduction of the eluent consumption, as a direct consequence of the exothermic character of the adsorption equilibrium, since less eluent would be required to regenerate the stationary phase within zone I [188]. This fact was also confirmed by these new studies that concluded that under specific temperature profiles, the temperature gradient SMB can inclusively generate a stream of pure eluent, as long as the adsorption selectivity between the most retained compound in zone I (at the highest temperature within the unit) and the least retained compound in zone IV (at the lowest temperature within the unit) is less than one, because this implies that the flow rate required to regenerate zone I is lower than the actual flow rate leaving zone IV [223]. Moreover, it was verified that enriched fractions of the extract and raffinate products could be obtained with high purity levels through an efficient modulation of the temperature difference between zones I and II and the temperature difference between zones III and IV, considered the key control variables for that purpose. In fact, the reported increase in the outlet stream concentration ranges from two to ten times. As one can easily conclude, all these advantages are more easily exploited in dilute systems with linear adsorption isotherms (since the selectivity for non-linear systems decreases considerably at high temperatures), particularly those that strongly depend on temperature [90]. Finally, the analysis of the dynamic behavior of the temperature gradient SMB revealed that temperature equilibration typically occurs in less than half of a switching period.

Altogether, these results prompted the development of alternative technologies, in parallel, such as the two-zone thermal simulated moving bed concentrator [224, 225]; however, as the name specifies, these units were conceived to enrich and concentrate a target product in a dilute solution and partially to recover its solvent instead of performing conventional product separation. For this reason, no further details will be provided regarding this technology.

5.3.2.5 Alternative Configurations

SMB units with a reduced or enlarged number of zones cannot be considered to be extremely innovative approaches. In fact, they are among the first reported adaptations to SMB technology. For instance, Baker and coworkers in the 1970s developed a three-zone SMB in which zone I was disconnected from the remainder zones and zone IV was eliminated. This operating mode was applied to gas phase separations [226] but was particularly successful for liquid-phase separations, namely for fructose–glucose mixtures [227]. At the turn of the millennium, several studies were reported suggesting the use of five- to twelve-zone SMB units and SMB cascades for the separation of complex mixtures [192–199]. Nevertheless, these alternative SMB configurations have been drawing attention over the last decade since they represent a valid technical solution, addressing the main issues associated with the conventional SMB. Units with a reduced number of zones decrease the technology costs and require less complex control systems while units with an enlarged number of sections and a cascade operation are able to perform ternary separations, a key aspect in chromatographic separation technologies nowadays.

The elimination of zone IV can be considered the most straightforward approach for reducing the number of sections and, expectedly, the number of chromatographic columns used in an SMB. It enables an increase in productivity per unit of volume of adsorbent at the expense of a higher raffinate dilution and an increase in the desorbent consumption, since the regeneration and recycling of the desorbent is no longer possible due to the elimination of zone IV, which exists in conventional SMB units for those specific purposes [190, 191]. More recently, an alternative to this strategy has been proposed that consists of the elimination of zone I [228] or, as other authors describe, of a rearrangement of the port location in a three-zone open-loop SMB [129, 163]. Figure 5.11 provides a comparison between the two three-zone SMB configurations previously mentioned.

The redesigned three-zone SMB configuration has been applied for the purification of amino acids, such as valine [162–164], and for the separation of specialty chemicals obtained in bio-based industries [100, 130], with special focus on succinic acid production [127–129]. The results obtained in these research studies demonstrated that the port rearrangement in the three-zone open-loop SMB allows these units to work with much higher feed concentrations than the three-zone open-loop SMBs deprived of zone IV, and without compromising the target product recovery. This is due to the ability of the redesigned three-zone SMB to confine the heavier compound concentration front to zone II during the entire switch interval, unlike the first three-zone SMBs, for which the heavier compound concentration front propagates through zones II and III during the switching period, leading to a decrease in the recovery for the latter. Hence, it can be concluded that the new configuration allows a more efficient utilization of the adsorbent in terms of the desorption of the compounds with higher affinity toward the stationary phase, by performing this task in both zones I and III of the SMB in opposition to what happens when only zone IV is removed, and zone I becomes the only zone of the SMB responsible for the desorption of the heavier compounds [127, 129]. This also helps to explain why it is particularly more effective to suppress zone I instead of zone IV for systems presenting anti-Langmuir isotherms [228]. Nevertheless, the compromise between recovery and productivity is the same as that observed for any SMB, as to increase the first will consequently decrease the second, and this decrease will be increasingly significant for higher recovery set points (>97%).

The concept of SMB units with an enlarged number of sections has been exploited by several authors but most of the research is dedicated to simulation-based studies [192, 193, 197]. Only a few works report the experimental results obtained with SMB units with more than four zones, including a five-zone SMB for the separation of dimethyl, dibutyl, and dioctyl phthalate [195], a five-zone SMB to recover sugars from corn-stover hydrolysate [196], and a nine-zone SMB process for purifying glucose and xylose from biomass hydrolysate [194]. Apart from those studies, recent publications have described the application of an eight-zone SMB unit with internal recycle (8Z-SMB-IR) to carry out ternary center-cut separations [229, 230],

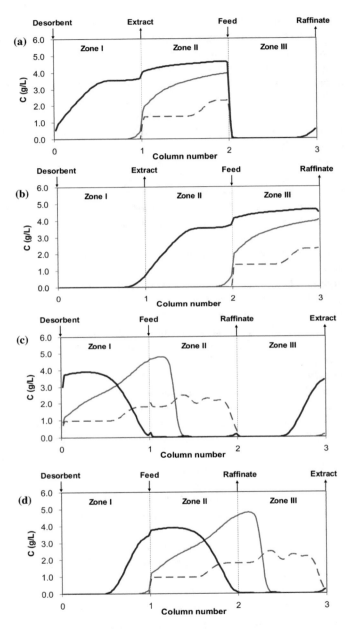

FIGURE 5.11 Comparison between three-zone SMB units with conventional port arrangements at the beginning (a) and the end of the switch (b) and with unconventional port rearrangement at the beginning (c) and at the end of the switch (d). The black, grey and dashed lines represent the succinic acid, acetic acid and formic acid concentration bands, respectively. (Reprinted from *Journal of Industrial and Engineering Chemistry*, 58, Choi, J.-H., Nam, H.-G. and Mun, S., "Enhancement of yield and productivity in the 3-zone non-linear SMB for succinic-acid separation under overloaded conditions," p. 222–228, Copyright (2018), with permission from Elsevier).

for which the experimental technology demonstration has already been performed as well [231]. This was accomplished by separating cyclohexanone (intermediately adsorbed compound) from a stream that also contained cyclopentanone and cyclo-heptanone, with a minimum experimental purity of 98%, using LiChroprep® RP-18 as stationary phase and water-methanol mixtures as mobile phase.

The 8Z-SMB-IR concept consists of combining a cascade of two conventional four-zone SMBs in a single unit, sharing the same separation principles. In practical terms, either the raffinate or the extract stream is recycled three zones forwards or backward, relative to the fluid flow direction, respectively, generating two sub-units within the SMB with four zones each. These internally recycled streams contain the intermediately adsorbed components and either the least or the more retained one, depending if it is the raffinate or the extract that is being recycled. Furthermore, these units comprise three outlet ports that allow the collection of purified fractions of each of the species initially present in the ternary mixture fed, and two eluent ports. An example of an 8Z-SMB-IR configuration is depicted in Figure 5.12.

Compared to the conventional four-zone SMB, the main advantage of the 8Z-SMB-IR is the ability to perform ternary separations, including center-cut sepa-rations. When compared to a cascade of two four-zone SMB units, this configuration presents some limitations such as the existence of a single switching time, operating temperature, and mobile and stationary phases unlike when two independent SMB units are used [232]; however, this loss in terms of degrees of freedom is compen-sated, in a first instance, by the reduction of the capital costs and the complexity of the equipment (namely, reducing the number of required valves and pumps), but, more importantly, by the enhancement of the separation efficiency, since the binary mixture internally recycled is pre-separated, which leads to a decrease in the overall eluent consumption (this is not verified for the tandem SMB configuration because, typically, intermediate tanks must be used to accommodate flow rate fluctuations and to ensure a continuous feed to the second unit) [230, 233].

The latest advances in tandem SMB result from the combination of SMB units with alternative configurations in the cascade, particularly the development of sepa-ration processes based on consecutive three-zone open-loop units [101, 164]. One of the most interesting works in this field is the experimental and simulation study of the separation of valine (the intermediately adsorbed component) from leucine, alanine, and ammonium sulfate in a cascade of three-zone open-loop SMBs with a

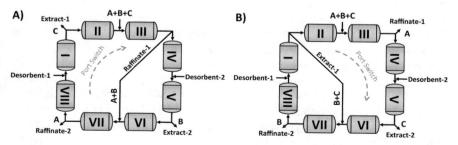

FIGURE 5.12 8Z-SMB-IR scheme considering the raffinate (A) or extract (B) recycling. A, B, C represent the compounds in ascending order of affinity toward the solid.

different port configuration [164]. The separation strategy and the port configuration played a crucial role in the process performance, having a significant effect on valine concentration. Considering this fact, the optimum process comprised a first SMB unit without zone I, in which valine was collected in the raffinate stream, and a second SMB unit without zone IV, in which valine was collected in the extract stream. Taking this into account for three-zone SMB unit separation principles, one can conclude that these configurations generate a valine enrichment zone in zone III of the first unit and in zone I of the second unit, near the respective raffinate and extract collection ports. In this way, apart from the expected reduction in capital costs associated with this SMB configuration in comparison with other chromatographic center-cut separation processes, the authors of this study state that the developed SMB-based process outdoes the current industrial valine purification process, comprising batch ion-exchange and crystallization stages, in terms of valine productivity, purity (99.3%), and recovery (94.4%).

5.3.2.6 Combination of Multiple Alternative Configurations and/or Operating Modes

As the understanding of the physical–chemical phenomena and process dynamics of SMBs grows, and as simulation tools become more precise and comprehensive, the scientific community is able to take more risks and be more innovative in the development of new SMB operating modes. For that reason, several adaptations have been recently implemented in this technology through the combination of multiple alternative configurations and/or dynamic modulation strategies. The most common combinations comprise the use of SMB units with a reduced number of columns with either dynamic configuration variations or flow rate modulation strategies.

The three-zone intermittent SMB or three-column intermittent SMB (3C-ISMB), as originally proposed, is the practical implementation of the ISMB operating mode in a unit comprising only three zones [234, 235]. As previously stated, the switching interval in the ISMB is divided into two stages, an initial stage in which it is operated as an open-loop three-zone SMB without a flow through zone IV, and a second stage in which zone IV is re-integrated in the column train, the recycling is re-established, and all the inlet and outlet ports are closed. A deep analysis of the ISMB operating mode and the internal concentration profiles reveals that nearly one fourth of the stationary phase is not being used at all for separation purposes over the entire switching interval. This is a direct consequence of the absence of zone IV in the first stage but it also arises from the fact that zone I is almost completely regenerated at the end of the first stage of the switching period. To overcome this issue and improve stationary phase usage, Jermann et al. suggested the elimination of zone IV, keeping the dynamic configuration variations unchanged, which originated 3C-ISMB technology [234]. Figure 5.13 presents the internal concentration profiles for the ISMB and the 3C-ISMB, allowing a better understanding of the previously described system.

Similarly to the three-zone SMB with partial recycle [236], the regeneration of the solid and liquid phases typically carried out in zones I and IV of conventional SMB units is performed in zone I of the 3C-ISMB, exclusively, by recycling the less retained compound. This is only possible because, at the beginning of the second stage of the switching period, zone I is already almost completely regenerated.

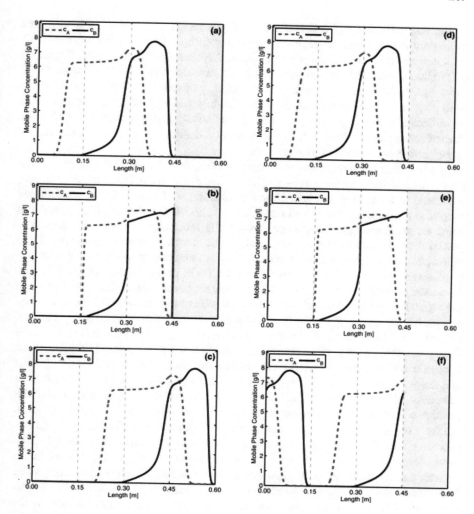

FIGURE 5.13 Comparison between the ISMB and 3C-ISMB internal concentration profiles: ISMB at the beginning of the first stage of the switching interval (a), ISMB at the end of the first stage of the switching interval (b), ISMB at the end of the second stage of the switching interval (c), 3C-ISMB at the beginning of the first stage of the switching interval (d), 3C-ISMB at the end of the first stage of the switching interval (e), and 3C-ISMB at the end of the second stage of the switching interval (f). (Reprinted from *Journal of Chromatography A*, 1361, Jermann, S. and Mazzotti, M., "Three column intermittent simulated moving bed chromatography: 1. Process description and comparative assessment," p. 125–138, Copyright (2014), with permission from Elsevier.)

In this way, a more extensive use of the stationary phase can be accomplished, leading to a considerable enhancement of the process productivity. In fact, it was experimentally demonstrated that it is possible to implement a 3C-ISMB process with one column per zone (packed with Chiralpak AD™) that was able to overcome the ISMB productivity by 80% for the separation of Tröger's base enantiomers [235].

Subsequently, the application of the 3C-ISMB was extended to other systems and more complex configurations. A center-cut separation of nadolol enantiomers from a quaternary mixture in a heptane: ethanol: diethylamine solvent using Chiralpak AD™ as stationary phase was performed in a cascade of two 3C-ISMB units [120]. Once again, the advantages in terms of flexibility of cascades of SMBs over SMBs with an increased number of sections were highlighted and the target product could be collected with a high purity (97.8%) at a relatively high productivity, 2.10 g per liter of adsorbent per hour. Moreover, a general design methodology based on tri-angle theory was proposed by this study.

A comprehensive study was published by Mun, focusing on the application of a partial-feed strategy to SMB units with three zones only [128]. In an initial approach, a theoretical assessment of the expected performance of four different systems was carried out, based on the equilibrium theory. The SMB configurations considered were: a conventional three-zone open-loop SMB, a three-zone open-loop SMB with partial-feed, a three-zone open-loop SMB with an alternative configuration of the outlet ports (eluent-feed-raffinate-extract), and a three-zone open-loop SMB with partial-feed and an alternative configuration of the outlet ports. It was concluded that, as discussed in Section 5.3.2.5, the outlet port rearrangement can considerably enhance the separation performance of conventional three-zone open-loop SMB units. When comparing the effects of partial-feed and port rearrangement independently, the conclusion was that the latter is the more effective strategy. These results are in accordance with the results published in another study of the same author [127]. However, it was verified that the synergetic effect between the simultaneous port rearrangement and the partial-feed strategies applied would allow for an even further increase in separation efficiency. The key factor contributing to that is the superior control of the position of both the rear of the less adsorbed compound concentration band, as well as the front of the most adsorbed compound concentration band, allowed by the simultaneous combination of these adaptations. After the evaluation of the potential of the newly developed technique, the study proceeded to a second stage in which validation of these premises was made by considering the separation of lactic and succinic acids in the four three-zone open-loop SMB units previously described as making up the case study. The conclusions were the same as those obtained during the primary theoretical analysis; the three-zone open-loop SMB with partial-feed and port relocation demonstrated an ability to attain the highest throughput for the same product purity requirements, outperforming the remainder separation units by a factor of 1.2–2.0. Even when compared to the throughput of a conventional four-zone SMB, this new technology was able to provide an improvement of over 15% (besides the reduction in the associated investment and maintenance costs).

Several other flow rate modulation strategies have already been combined with three-zone SMBs in order to develop new SMB processes. For instance, the separation of fucose and 2,3-butanediol was successfully accomplished by applying partial-extract-discard and partial-extract-recycling procedures to a three-zone open-loop SMB with an unconventional port configuration (eluent-feed-raffinate-extract) [100]. As previously addressed (Section 5.3.2.5), to recover high purity products in an SMB with this port configuration, one must provide some safety margin between the front of the concentration band of the lighter compound collected in the raffinate and the rear of the concentration band of the heavier compound collected in the extract port (placed immediately after the raffinate collection port in this configuration). While conventional partial-discard techniques are typically applied to avoid contamination of the target product with other species, the technique proposed in this work aims to improve the product concentration by discarding the desorbent volume corresponding to the safety gap between the two compound concentration bands. The discard/recycling period must be carefully determined in order to maximize product concentration and yield without compromising the required purity constraint. For the partial-extract-discard strategy, an increase of 33% of the 2,3-butanediol concentration was achieved with almost no change in the yield and purity of both products. Recycling the extract stream instead of discarding it improves, even further, the desorbent consumption, the 2,3-butanediol concentration, and the fucose concentration, compared to the original three-zone open-loop SMB with alternative port rearrangement (25%, 33%, and 7%, respectively). Nevertheless, the control system and equipment necessary to implement the partial-extract-recycling operating mode is considerably more complex than for the other approaches (note that, for instance, the flow rate in section III must be constant; therefore, when the extract stream is suppressed, the flow rate in this section must be adjusted).

An adaptation of the OSS operating mode was also applied to a three-zone open-loop SMB with an unconventional port configuration. Basically, in the first moments of the switching interval, the raffinate is collected and the extract port is closed while in the second fraction of the switching time the extract becomes the active collection port, the raffinate is suppressed, and the eluent stream moves one zone forwards. As only three streams were active at each moment, this technology was designated three-port operation in three-zone SMB or simply TT-SMB [237]. The TT-SMB was able to improve both the raffinate and the extract product purity and recovery because it led to a more efficient use of the stationary phase by allowing the resolution of concentration bands overlapping near the product collection ports during its idle periods. Additionally, in terms of equipment, the required number of pumps of the TT-SMB is lower than in conventional three-zone SMBs (since it is possible to replace them with valves), which can decrease the capital costs even further and simplify the unit control system.

One of the solutions proposed to address the challenges of ternary separations by continuous chromatography was a five-zone SMB [195, 199]; however, in its standard implementation, the purity of the intermediately adsorbed species was severely limited. In this context, a modified five-zone SMB unit was designed by combining it with a partial-collection strategy [238]. While the least and most retained species were continuously collected, the intermediately adsorbed one was only collected during the first fraction of the switching interval. When the overlapping concentration bands of the late eluting species reach this point, the port is closed and the stream is eliminated. These concentration bands are allowed to percolate a small distance within zone III so that they can be separated once again in zone II during the subsequent switching period. The most outstanding improvement provided by this combination of an alternative configuration and an alternative operating mode is the ability to recover all three products simultaneously with extremely high purities.

When compared to the standard five-zone SMB, this modified version was able to largely overcome the throughput of the former with the same purity requirements.

Despite the interesting results reported, it is worth mentioning that most of these strategies are very recent and the number of successful applications is rather low, suggesting the potential of the technology but not indicating its robustness or flexibility in terms of the different separation problems faced by current industrial processes.

5.3.3   Emerging SMB-Based Technologies

Since it was first developed, the SMB has proven to be an extremely efficient and competitive separation technology, as evidenced by the large number of commercial applications found over the years within different industries. This technology has been able to provide products with exceedingly high purity levels, typically mandatory for the pharmaceutical and the fine chemical industries, but has also been able to reach the high productivity levels required by bulk chemical production processes, as described in previous sections of this work. This success is in part due to the notorious flexibility of this separation technique and to the advances accomplished through the various adaptations promoted in its configuration and operating mode. However, new challenges imposed by emerging industries have been leading to the development of new SMB-based technologies that fall within a broader concept of continuous multicolumn countercurrent chromatographic separations. One of the main drivers for this paradigm change has undoubtedly been the biopharmaceutical industry, from which monoclonal antibodies account for a considerable market share.

Downstream processing of biopharmaceuticals, in its most conventional implementation, relies on three chromatographic purification steps [148, 152, 239]: first, a capture step, accomplished by affinity chromatography and step elution, which is focused on the removal of non-product-related impurities and on target product concentration enrichment, leading to a significant reduction of the amount of material to be processed in subsequent stages; then, a first polishing step, carried out in a bind-elute mode using gradient elution and cation-exchange resins, to remove product-related impurities; and finally a second polishing step, carried out in a flow-through mode using anion-exchange resins, to remove strongly binding residual impurities. The previously reported stages may involve the purification of intermediately adsorbed species from ternary or pseudo-ternary mixtures, solvent gradients, and complex series of loading, washing, elution, and equilibration steps typically performed in columns momentarily isolated from the column train; these purification processes encompass some of the most widely acknowledged limitations of conventional SMB technology. Furthermore, biopharmaceutical manufacturing processes were conventionally carried out in a batchwise mode to avoid failure and disturbance propagation within the expensive production processes, which could jeopardize product quality requirements. Nowadays, these processes are experiencing a transition into more integrated and continuous operating modes, which has increased the importance of chromatographic separation. All these facts together prompted the development and implementation of continuous multicolumn countercurrent chromatographic separations.

The relevance of the biopharmaceutical market in the evolution of these technologies is also evidenced by the large number of technical solutions that have been developed and commercialized by both emerging and established companies, ranging from the life sciences field to equipment manufacturers. Some of the most relevant examples include the XPure from Xendo Holding B.V. (Netherlands; now part of the ProPharma Group); the Octave® Chromatography System from Semba Biosciences, Inc. (United States of America); the Cadence™ BioSMB PD System from the Pall Corporation (United States of America); the Contichrom® CUBE from ChromaCon AG (Switzerland), which recently became a part of YMC Co., Ltd. (Japan); the EcoPrime Twin from YMC Process Technologies (Japan); the ÄKTA PCC from GE Healthcare (United States of America); and the BioSC® – Sequential Multicolumn Chromatography from Novasep (France). To provide a deeper insight into the underlying principles associated with this wide variety of technical solutions, three illustrative examples will be addressed in further detail: sequential multicolumn chromatography (SMCC); three-column periodic countercurrent chromatography (3C-PCC) with interconnected wash, which will offer a comparison between two different capture step technologies; and the multicolumn solvent gradient process (MCSGP) applied to a polishing step.

As previously stated, the main goal of the capture step is to remove non-product-related impurities (e.g., host cell proteins) and concentrate the target product. It is one of the first downstream separation units and therefore it must be able to process large volumes of harvested cell culture fluid [148, 239]. For that reason, the ideal stationary phase should have a high capacity for the target species and a relatively large average particle size, to maximize the throughput and minimize the pressure drop. For single-column processes, the capture step was accomplished by loading the column up to a maximum value of 1% of its dynamic binding capacity (DBC), to which, in many cases, an additional safety factor of 10% to 20% was applied to avoid product losses. Then, the feed was stopped, and a series of washing, elution (through which the target product is collected), regeneration, and equilibration steps were sequentially carried out. After this procedure, the column was ready for a new cycle. It is clear that this technique results in poor utilization of the stationary phase (which represents a considerable fraction of the overall process costs) as well as low productivity values. The SMCC technique proposed by Novasep (France) [240] is able to overcome these issues by dividing the single chromatographic column considered for batch operation into two to six chromatographic columns of smaller dimension connected in series. Figure 5.14 presents one of the possible operating modes of a three-column SMCC unit as an example.

In the simplest implementation of this technology [241], during the loading stage the feed stream is introduced in column one, which is interconnected with column two. This allows the product that breaks through column one to be captured in column two, largely increasing the solid stationary phase utilization for the first column. Values above 90% of the static binding capacity (SBC) have been reported [242] and unit productivity can be increased by four to five times compared to the analogous batch process [147]. During this period, the third column can be independently eluted, regenerated, and equilibrated, typically through step gradients. Then, the washing is carried out to recover the unbounded target product present in the

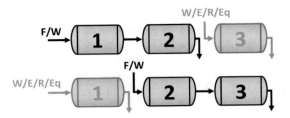

FIGURE 5.14   Three-column SMCC operating mode (F=Feed, W=Washing, E=Elution, R=Regeneration, Eq=Equilibration).

interparticle voids, increasing the product yield. After a specific period of time, the feed is introduced in column two, which is interconnected with column three, while column one is operated independently, similarly to the first step described, only this time all the streams are shifted one column in the direction of the fluid flow. Hence, this operating mode can be repeated cyclically. It is important to note that, while for single-column batch processes the effective column loading is limited by the DBC, in the new multicolumn chromatographic processes the SBC becomes the most relevant parameter [243].

Different operating modes can be considered for SMCC technology [240, 244, 245], increasing the potential for biopharmaceutical downstream processing. Girard et al. [147] reported a four-column SMCC process to capture an IgG1 subtype monoclonal antibody from a clarified and filtered crude obtained from the batch reactor cultures of a stable CHO cell line using a commercially available protein-A adsorbent (not disclosed). In this process, a 10 mM Tris at pH 8.0 buffer was used for the washing buffer and a glycine buffer at pH 3.7 for the elution step. The experimental performance of the four-column SMCC was compared to a single-column batch process and the results clearly demonstrated the superiority of the continuous separation technique. For a minimum IgG purity of 96%, the productivity of the SMCC was approximately 4.5 times higher and the buffer consumption was 5% lower, for a total volume of adsorbent four times lower than the batch process. This study also reports an additional ion-exchange step (SMCC) and an additional polishing step (using chromatographic membranes) through which it was possible to achieve a final IgG purity above 98% with an overall recovery of approximately 75%. Simultaneously, some important guidelines for developing a downstream processing mAb platform were provided.

Another alternative technology widely used for protein capture is the periodic countercurrent chromatography (PCC) developed by GE Healthcare (United States of America) [246]. This technology is typically implemented in three- or four-column units, designated 3C-PCC [247] and 4C-PCC [248], respectively. One of the major differences between the PCC and the SMCC is that the former allows continuous feeding or loading steps while the latter does not. Nevertheless, all the remainder streams are intermittent. A representative example of the 3C-PCC operating mode is provided in Figure 5.15 [247].

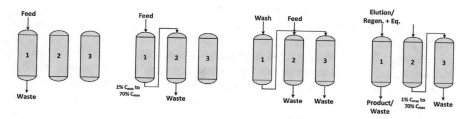

FIGURE 5.15 3C-PCC operating mode (F = Feed, W = Washing, E = Elution, R = Regeneration, Eq = Equilibration).

As shown in Figure 5.15, the 3C-PCC is initiated by feeding the harvested cell culture fluid to column one, previously regenerated and equilibrated, while its outlet is discarded. When the traces of the target product are detected at its outlet (typically 1% of the feed concentration), column one is connected in a series with column two, which will allow an increase in the loading capacity of column one to values that can reach almost 100% of its SBC [248], similarly to that described for the SMCC process. Then, the feed stream is moved to the adjacent pre-loaded column (column two) while the washing buffer is fed to column one; however, the strategy for recovering the unbounded protein eluted during the washing step in the 3C-PCC is conceptually different from that proposed in the SMCC, since the outlet of column one is connected to column three instead of being connected to the following column, as in the latter technology. Afterward, the target product can be eluted from column one, which is subsequently regenerated and equilibrated. In parallel, the outlet of column two will start presenting traces of the target compound and this stream will be fed to column three. This sequence of steps is repeated cyclically from this moment on, in an analogous manner. PCC processes with a larger number of columns and slightly different operating modes have also been proposed, including clean-in-place (CIP) steps [248–250].

The majority of the successful applications of PCC technology reported in open literature were aimed at the capture of monoclonal antibodies [242, 248–250]. In an extensive experimental study, Mahajan et al. [247] demonstrated the economic potential of 3C-PCC-based processes compared to single-column discontinuous ones, using two different test systems. A low-titer mAb (1 g/L) was captured using MabSelect SuRe resin and a high-titer mAb (4 g/L) was captured using ProSep® vA resin. The elution, washing, and equilibration buffers were the same for both resins (25 mM Tris/25 mM sodium chloride, 0.4 M phosphoric acid, and 0.1 N acetic acid); however, a 0.1 N sodium hydroxide buffer was used to regenerate MabSelect SuRe while a 0.1 M phosphoric acid was used to regenerate ProSep® vA. The main conclusion of this work was that the 3C-PCC was able to reduce the resin volume, buffer consumption, and processing time by approximately 40% in comparison with the batch elution process, attaining a yield above 98%. More recently, the dynamic ability to control the protein loading has been improved even in the presence of high concentration cell cultures feeds (over 30 g/L) [250].

The challenges imposed by the polishing step are somewhat different from those previously addressed for the capture step. As stated earlier, the polishing step is designed to eliminate product-related impurities, which might comprise protein fragments and aggregates, among other species. Hence, it can be easily concluded that the target products and the impurities might have similar adsorptive properties, and consequently that separation at high purity levels becomes harder to achieve; however, multicolumn countercurrent chromatographic processes in general, and the SMB in particular, are widely known for their ability to overcome the yield–productivity trade-off limitations, even for systems with extremely low selectivities, by internally recycling the overlapping target product concentration bands. A few purification processes have been proposed within the biopharmaceutical field based on SMB technology, with size exclusion chromatography playing a crucial role, as was reported in Section 5.3.1. Still, the limitations previously highlighted regarding the implementation of linear gradients and center-cut separations persist, which hinders its application to a large number of processes. In this scenario, the MCSGP represents one of the most suitable alternatives for biopharmaceutical polishing steps [152, 239]. The MCSGP was originally designed as a six-unit column with continuous feed and product withdrawal [251, 252], but intensive research led to the development of units with a lower number of columns [253, 254], which in turn gave way to its final version, a two-column semi-continuous process [243, 255]. Its operating mode is schematically represented in Figure 5.16, assuming a ternary separation between A, B, and C with A being the least adsorbed species and C the most adsorbed one.

As is depicted in Figure 5.16, the MCSGP operating cycle is sub-divided into four different stages. In the first stage, column one and column two are interconnected in order to allow the recycling of the overlapping concentration band containing the target product (B) and the least retained compound (A), after mixing it with the eluent (which may be a buffer or a solvent), so that the target product leaving column two can be adsorbed within column one, avoiding product losses. During the second stage, the connection between the two columns is suppressed. A purified fraction of the target product can be collected at the outlet of column two if the most retained compound (C) does not start eluting from the column. In this way, a fraction of B corresponding to the overlapping concentration bands of B and C is also retained inside column two. Simultaneously, column one is fed with the ternary mixture to be separated. Afterward, the connection between column two and column one is restored and the overlapping concentration bands of B and C previously retained in column two are mixed with the eluent (which may be a buffer or a solvent), until

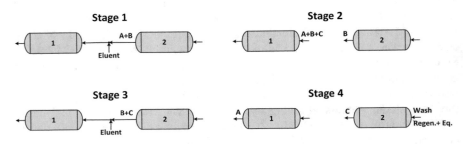

FIGURE 5.16 Multicolumn solvent gradient process operating mode. A, B, C represent the compounds in ascending order of affinity towards the solid.

component B is completely eluted from column two. At the same time, one must guarantee that all of B is re-adsorbed in column one to avoid target product losses. In the final stage, the two columns are independently operated and the least retained compound is collected at the outlet of column one, ensuring that the overlapping concentration bands of A and B are retained, while the more adsorbed compound is collected at the outlet of column two, followed by a series of washing, regeneration, and equilibration procedures conducted in the latter column.

Although this operating mode is much more complex than the conventional single-column batch process, it is able to overcome most of the limitations imposed by the compromise that must be established between yield and productivity, and it allows the implementation of complex solvent gradients. Moreover, by using only two columns, it minimizes the dimensions of the required units, which might imply considerable savings since the cost of downstream processing equipment can represent up to 80% of the total capital investment required in a biopharmaceutical plant, and the highly specialized stationary phases used are extremely expensive [147, 148]. To mitigate eventual product losses and disturbance that might occur during the complex series of discrete steps performed cyclically during the MCSGP operation, researchers have dedicated particular attention to the design and control of the main process variables [255–258].

Several successful applications of the MCSGP (with two to six columns) have been reported within the biopharmaceutical field, including peptide (calcitonin) separation from industrial streams [252], separation of monoclonal antibodies from fragments and aggregates [255], sequential capture and polishing of monoclonal antibodies (IgG_2) from a clarified cell culture supernatant [259], light-chain bispecific antibody separation [260], PEGylated protein separation (focused on the mono-PEGylated α-lactalbumin) [243], and isolation of the main charge isoform of a monoclonal antibody [258], among others. The work published by Krättli et al. [255] can be used as an example to illustrate the potential of this technology. To purify a monoclonal antibody from a clarified cell supernatant with a concentration of 1 g/L, a Fractogel EMD SO3 resin (Merck, Darmstadt, Germany) was used as stationary phase in a twin column MCSGP. Three different buffers were used: a 25 mM phosphate solution with a pH of 6 as adsorbing buffer, a 25 mM phosphate solution with a linear gradient from 0 to 1 M sodium chloride as desorbing buffer, and a solution of 0.5 M sodium hydroxide for the cleaning in place. The authors optimized the process parameters through the design strategies reported in literature [252] and applied an innovative online control concept based on two independent PID controllers that continuously adjusted both the starting point and the ending point of the product elution time window. The new multicolumn chromatographic process was able to achieve a monoclonal antibody purity of 90% with a yield of 80%, while in an equivalent single-column batch process a yield as low as 30% would be attained to meet the same purity requirements.

The downstream purification processes currently integrated in the biopharmaceutical industries have been responsible for a paradigm change in multicolumn countercurrent chromatographic separation pushing these technologies to the limit. A large number of SMB-based units have been developed to meet the demanding needs of this growing industry and some authors suggest that this trend has created a whole new generation of multicolumn countercurrent chromatographic processes [2, 169]; however, although industrial acceptance for these technologies is growing at an interesting pace, the next few years will be crucial to test their reliability.

5.3.4 TECHNICAL INNOVATIONS AND ADAPTATIONS

Despite the specificities of the operating mode or application field, all SMB units share a common concept: potentiating the chromatographic separation by imposing a countercurrent movement between a solid stationary phase and a liquid mobile phase, which is achieved by changing the position of all the inlet and outlet streams at predefined time intervals, in a continuous and cyclic process. Considering this, the development of the pioneer rotary valves by UOP, Inc. (United States of America) [3, 4, 261] can be recognized as one of the major milestones for the implementation and dissemination of SMB technology. As expected, the number of technical solutions proposed for the practical realization of the countercurrent operating mode has largely increased over the years, as a result of extensive work done by both academics and companies. SMB equipment design and instrumental aspects have been extensively reviewed by previous authors [168, 169], with particular attention to valve design, considered a key feature for unit construction. In general, according to the type of valves used and the design strategy adopted, two main groups of systems can be defined: central valves and distributed valve systems.

Each of these groups can be further sub-divided into additional categories. For instance, for distributed valve systems, different types of valves can be used (as two-position or multi-position valves) and these valves can either be associated with the SMB streams or with the SMB columns. Hence, several design strategies can be adopted: the two-position valves-to-column design [262, 263], in which each external stream is connected to a manifold that distributes it to valves that allow or hinder the introduction of these streams into the transfer lines between every two adjacent chromatographic columns; the multi-position valves-to-stream design [264–266], in which one or more multi-position valves are responsible for directing the external streams toward the transfer lines between every two adjacent chromatographic columns which they will be added to or collected from; or the multi-position valves-to-column design [168, 267], in which complex multi-position valves are directly connected to a chromatographic column and to the internal transfer lines allowing the introduction or withdrawal of streams in the columns or even the stoppage of the flux between columns. All of these general types of designs have variations and adaptations depending on the process specifications, without experiencing any groundbreaking advancement in recent years. Nevertheless, several SMB units have been installed in industrial plants following this design strategy, mainly due to its flexibility, the possibility of implementing most of the unconventional operating modes, and the advanced stage of development of the technology, regardless of the large number of valves typically necessary for these units, the control problems associated with it, and the existence of considerable dead volumes.

As previously stated, the UOP's rotary valve was the first engineering solution for the implementation of SMB technology. These large complex valves comprised a lower static disc with several concentric channels as internal fluid transfer lines to which the external streams were connected; several ports in the periphery that were connected to the chromatographic columns; an upper rotating disc with several transfer lines that directed the fluid from the internal channels to the valve ports; and a sealing system, all assembled together inside a housing [4]. With this device it was possible to change position of each inlet and outlet streams simultaneously while the chromatographic columns were static. The design and functionalities of these valves were improved over time, enhancing the sealing system and mechanical resistance while decreasing the cost; it represents nowadays one of the most reliable solutions in terms of valve design, and has been installed in several large industrial plants worldwide by different companies [268–270]. In opposition to central rotary valves with static columns, some companies have developed SMB units comprising central valves with rotating columns, like the ISEP system developed by Advanced Separation Technology, Inc. (United States of America) [271]. In this case, the valves are also composed of two discs, an upper static disc containing the transfer lines that allow the connection between two adjacent columns and the inlet and outlet ports for the external streams, and a lower rotating disc, to which all the column inlets and outlets are connected, and which is attached to a rotor where all the chromatographic columns are installed in a carrousel-like system [271, 272]. Although more flexible than central rotary valve with static columns, these systems still present some relevant limitations such as the inability of implementing the Varicol® and Pseudo-SMB operating modes.

Although the central valve systems previously described have long been applied to a large number of industrial SMB separation processes, this is the valve design strategy that has experienced the most marked evolution in the last decade, through the development of an innovative concept, designated central valve blocks. Central valve blocks are basically a group of modular elements that form a matrix of independent two-position valves, typically pneumatic actuated, assembled in a single device comprising a complex network of channels that enables the implementation of virtually any SMB operating mode. This concept, exploited and applied in units such as the XPure, developed by Xendo Holding B.V. (Netherlands; now part of the ProPharma Group) [273], the Octave® Chromatography System developed, by Semba Biosciences, Inc. (United States of America) [274], or the Cadence™ BioSMB PD System, developed by Tarpon Biosystems Inc. (now part of the Pall Corporation, United States of America) [275], has emerged as a direct solution for the new problems imposed by the dissemination of the use of SMB technology for downstream separations in the biotechnology and biopharmaceutical industries, namely the need of smaller and more flexible units or even the need for single-use or disposable flow paths.

Even though this type of valve falls within the central valve group, it has integrated some of the features of distributed valves systems into a single modular device. A thorough analysis of the technical details of such devices, taking as an example the solution proposed by Xendo [273], allows for a better understanding of its operating principles. The central valve blocks developed by this company consist of a set of manifolds assembled together. Each manifold has a central channel and several branch channels connected to it, all of them possessing a two-position valve (including the central channel). Typically, the central channel is responsible for connecting two consecutive chromatographic columns, and some of the branch channels are connected to the external streams while others provide an internal connection between the inlets and outlets of non-adjacent chromatographic columns. The position of each of the independently actuated valves defines the active fluid transfer line. In terms of their construction, the manifolds are composed of three plates, containing the connection ports and transfer lines, a flexible membrane, and the valve actuators. The central channel is longitudinally engraved in the top plate and in the upper side of the central plate, while the branch channels are transversally engraved in the base plate and in the lower side of the central plate. Additionally, the top plate and the flexible membrane placed over it form chambers that, depending on the valve position, allow or hinder the passage of the fluid from the central channel through several holes that cross the top and central plates before it converges into the branch channels in the base plate. The valves act directly in the flexible membrane (using pneumatic or electric actuators). As previously suggested, these manifolds can then be stacked together to form a single central valve block.

With the design strategy adopted in these central valves, independent port switching, zone bypassing, and other dynamic configuration variations (such as the momentary suppression of an outlet stream, for instance) can be performed, enabling the implementation of virtually any operating mode desired, which also includes the possibility of performing complex sequences of steps in isolated columns, as is often necessary in biopharmaceuticals. Moreover, the fluid paths inside the manifolds can be custom designed to meet the specifications of any particular application, and these components can be easily disposed of and replaced, given the reduced contact area between the mobile phase and the valve block elements. Added to this, the modular nature of its construction enables the increase or reduction of the number of columns within the system as desired.

Considering all of this, it can be easily concluded that central valve blocks represent a highly flexible solution particularly suited to downstream separations in biotechnology industries, as has been already demonstrated, for instance, for the case of protein processing [152, 239]. Nevertheless, it is important to mention the production scale limitations (which falls in the range of milligrams-to-grams) inherent to this design strategy and the complexity of its control system.

5.4 CONCLUDING REMARKS

With over half a century of successful applications over a wide range of industries, SMB technology is nowadays generally recognized as an efficient, flexible, and reliable separation technique. The evolution of this technology cannot be dissociated from the trends observed in the petrochemical, food, or pharmaceutical industries since those were the first to integrate SMB units within their manufacturing processes and still account for the majority of the reported commercial applications. For these reasons, the aforementioned industries were the primary focus of this study.

It is possible to conclude that within the petrochemical field, recent advances rely on the optimization of the process integration strategy rather than on the adsorption process itself. In other words, the improvements in SMB-associated equipment have lowered operating costs and therefore improved the efficiency of the processes. Nevertheless, there have been strong efforts to develop new separation processes, such as the separation of light gases through an SMB, in which several challenges exist like the technical adaptations required for equipment that was conventionally designed to process liquid-phase systems.

The classic fructose–glucose separation is still the main focus of R&D within the sugar industry. The latest progresses include not only the development of more efficient separation modes and improved control strategies but also the implementation of process integration concepts. Additionally, the increasing relevance of biorefineries and other bio-based production processes represents the main factor contributing to the application of the SMB chromatographic separation to new sugars.

The pharmaceutical industry, on the other hand, has intensively explored the advantages of the Varicol® operating mode and the development of new stationary phases, to improve the efficiency of its SMB processes.

SMB technology has also been able to follow some of the most prominent global manufacturing process trends such as the transition from petro-based to bio-based chemicals and the pursuit of an increasingly large number of specialized biopharmaceuticals, finding interesting applications within these emerging markets. However, the technical specificities of such processes have imposed major challenges to SMB technology, such as the limitations in accomplishing ternary center-cut separations or performing elution gradients. Some of these limitations could be overcome through the introduction of minor modifications on the unit configuration and operating mode, while others require the development of innovative SMB-based units exploiting new concepts and separation strategies. This paradigm change has intensified the development of new technical solutions and equipment to carry out multicolumn countercurrent chromatographic separations and has prompted the implementation of alternative design, optimization, and control methodologies.

Finally, in the forthcoming years, SMB technology is expected to experience considerable growth, mainly due to the integration of SMB units in the manufacturing processes of emerging bio-based industries. This fact is evidenced by the increasing number of biotechnology and related equipment manufacturing companies that have been created over the last few years as well as by the investments made by large biotechnology companies and the projected growth of these markets. In the course of the research undertaken during this work, no significant threats were identified regarding the most mature implementations of SMB technology in petrochemical, food, and pharmaceutical processes. Hence, this technology is expected to keep its prominent position within these industries. Finally, it is worth mentioning that, although the progress of some SMB applications was relatively limited during the last decade, such as the supercritical fluid SMB or hybrid processes like the SMB reactor, progress might well speed up in the upcoming years due to the knowledge that has accumulated in those specific research topics.

5.5 ABBREVIATIONS

3C-ISMB	three-column intermittent simulated moving bed
3C-PCC	three-column periodic countercurrent chromatography
3S-ISMB	three-column intermittent simulated moving bed for the strongly adsorbed species
3W-ISMB	three-column intermittent simulated moving bed for the weakly adsorbed species
4C-PCC	four-column periodic countercurrent chromatography
8Z-SMB-IR	eight-zone simulated moving bed with internal recycle
BF-SMB	backfill simulated moving bed
BP	British Petroleum
BP-SMB	bypass simulated moving bed
BSA	bovine serum albumin
CAGR	compound annual growth rate
CIP	clean-in-place
DBC	dynamic binding capacity
FF-SMB	fractionation and feedback simulated moving bed
HFCS	high fructose corn syrup
IFP	Institut Français du Pétrole (French Institute of Petroleum)
ISMB	intermittent/improved simulated moving bed
MCSGP	multicolumn solvent gradient process
ModiCon	concentration modulation
MOF	metal–organic framework
OSS	outlet streams swing
PCC	periodic countercurrent chromatography
PPD-SMB	power-partial-discard simulated moving bed
RPD-SMB	recycling partial-discard simulated moving bed
SBC	static binding capacity
SEC	size exclusion chromatography
SF-SMB	supercritical fluid eluent simulated moving bed
SimCon	simultaneous partial-feed and outlet streams swing
SMB	simulated moving bed
SMBR	simulated moving bed reactor
SMCC	sequential multicolumn chromatography
SSMB	sequential simulated moving bed
TMB	true moving bed
TT-SMB	three-port operation in three-zone simulated moving bed
UOP	Universal Oil Products
ZIF	zeolitic imidazole framework

5.6 ACKNOWLEDGMENTS

This work is a result of: Project "AIProcMat@N2020 – Advanced Industrial Processes and Materials for a Sustainable Northern Region of Portugal 2020," with the reference NORTE-01-0145-FEDER-000006, supported by Norte

Portugal Regional Operational Programme (NORTE 2020) under the Portugal 2020 Partnership Agreement, through the European Regional Development Fund (ERDF); Base Funding – UIDB/50020/2020 of the Associate Laboratory LSRE-LCM – funded by national funds through FCT/MCTES (PIDDAC); Project PTDC/QEQ-ERQ/2698/2014 – POCI-01-0145-FEDER-016866 – funded by FEDER funds through COMPETE2020 – Programa Operacional Competitividade e Internacionalização (POCI) and by national funds through FCT – Fundação para a Ciência e a Tecnologia, I.P.

REFERENCES

1. Rodrigues, A.E., Pereira, C., Minceva, M., Pais, L.S., Ribeiro, A.M., Ribeiro, A., Silva, M., Graça, N. and Santos, J.C., Principles of simulated moving bed, in *Simulated Moving Bed Technology*, A.E. Rodrigues, C. Pereira, M. Minceva, L.S. Pais, A.M. Ribeiro, A. Ribeiro, M. Silva, N. Graça, and J.C. Santos, Editors. 2015, Butterworth-Heinemann: Oxford. p. 1–30.
2. Nicoud, R.-M., The amazing ability of continuous chromatography to adapt to a moving environment. *Industrial & Engineering Chemistry Research*, 2014. **53**(10): p. 3755–3765.
3. Broughton, D.B. and Gerhold, C.G., Continuous sorption process employing fixed bed of sorbent and moving inlets and outlets, US 2,985,589, 1961.
4. Carson, D.B. and Purse, F., Rotary valve, US 3040777 A, 1962.
5. Nicoud, R.-M., Fuchs, G., Adam, P., Bailly, M., Küsters, E., Antia, F.D., Reuille, R. and Schmid, E., Preparative scale enantioseparation of a chiral epoxide: Comparison of liquid chromatography and simulated moving bed adsorption technology. *Chirality*, 1993. **5**(4): p. 267–271.
6. McCoy, M., SMB emerges as chiral technique. *Chemical & Engineering News*, 2000. **78**(25): p. 17–19.
7. Anon, Chiral separations are enduring items in the toolbox. *Chemical & Engineering News*, 2003. **83**: p. 18.
8. Francotte, E.R., Enantioselective chromatography as a powerful alternative for the preparation of drug enantiomers. *Journal of Chromatography A*, 2001. **906**(1): p. 379–397.
9. Johnson, J.A., UOP sorbex family of technologies, in *Handbook of Petroleum Refining Processes*. 2003 Editor by Robert A. Meyers, McGraw-Hill, New York. p. 10.29–10.35.
10. Colling, C., Maximizing energy efficiency in paraxylene production–Part 1. *Hydrocarbon Processing, Process Optimization*, 2017. p. 57–60.
11. Colling, C., Maximizing energy efficiency in paraxylene production–Part 2. *Hydrocarbon Processing, Process Optimization*, 2018. p. 43–47.
12. Robertson, J., LD ParexTM aromatics complex: Lowest cost of production for PX technology, in *5th IndianOil Petrochemical Conclave*. 2016: Mumbai, India.
13. O'Neil, K., Advanced technology in aromatics: The success of LD Parex, in *Honeywell UOP ME-TECH Seminar*. 2018: Dubai, UAE.
14. Bhattacharya, N., Advances in aromatics technology: Lowers paraxylene cost of production, in *Honeywell Technology Summit*. 2017: Kuwait.
15. Ma, D., Design of *para*-xylene complex using LD Parex technology. *Petroleum Processing and Petrochemicals*, 2018. **49**: p. 35–38.
16. Molinier, M., Knob, K.J., Stanley, D.J., Pol, T.S.V., Cao, C., Zheng, X., LeFlour, T., Rault, J., Claudel, S., Prevost, I., Pigourier, J. and Fernandez, C., Aromatics production process, US 9,708,233, 2017.
17. Tinger, R.G., Pilliod, D.L. and Molinier, M., Process and apparatus for the production of para-xylene, US 10,059,644, 2018.

18. Leflaive, P., Cocq, D.L.L., Hotier, G. and Wolff, L., Process and apparatus for simulated counter-current chromatographic separation using two adsorbers in parallel for optimized paraxylene production, US 9,731,220, 2017.
19. Porter, J.R., Bender, T.P. and Pilliod, D.L., Xylene separation process and apparatus, US 9,896,398, 2018.
20. Lee, J.-S. and Shin, N.-C., Method for separating aromatic compounds using simulated moving bed operation, US 8,013,202, 2011.
21. Miller, J.T., Williams, B.A., Doyle, R.A. and Zoia, G., Process for production of para-xylene incorporating pressure swing adsorption and simulated moving bed, US 6,573,418, 2003.
22. Dorsi, C.M., Agrawal, G., Weber, M.W., Pilliod, D.L. and Porter, J.R., Process for the recovering of paraxylene, US 10,300,404, 2019.
23. Agrawal, G., Weber, M.W., Salciccioli, M., Porter, J.R., Pilliod, D.L. and Bender, T.P., Paraxylene separation process, US 9,850,186, 2017.
24. Ou, J.D., Abichandani, J.S., Kawajiri, Y. and Guo, S., Xylene separation process, US 10,093,598, 2018.
25. Dreux, H., Leflaive, P. and Cocq, D.L.L., Method for the production of paraxylene, comprising two simulated moving bed separation and two isomerization units, one being in the gas phase, US 10,029,958, 2018.
26. ExxonMobil., Liquid phase xylenes isomerization (LPI). 2018 [cited 2019 July 1]; Available from: https://www.exxonmobilchemical.com/en/catalysts-and-technology-licensing/xylenes-production/liquid-phase-xylenes-isomerization.
27. Shen, Y., Fu, Q., Zhang, D. and Na, P., A systematic simulation and optimization of an industrial-scale p-xylene simulated moving bed process. *Separation and Purification Technology*, 2018. **191**: p. 48–60.
28. Sutanto, P.S., Lim, Y.-I. and Lee, J., Bed-line flushing and optimization in simulated moving-bed recovery of *para*-xylene. *Separation and Purification Technology*, 2012. **96**: p. 168–181.
29. Silva, M.S.P., Rodrigues, A.E. and Mota, J.P.B., Modeling and simulation of an industrial-scale parex process. *AIChE Journal*, 2015. **61**(4): p. 1345–1363.
30. Minceva, M. and Rodrigues, A.E., Influence of the transfer line dead volume on the performance of an industrial scale simulated moving bed for p-xylene separation. *Separation Science and Technology*, 2003. **38**(7): p. 1463–1497.
31. Ernst, G.A., Noe, J.L. and Mlinar, A.N., Process for a dual extract flush, US 10,124,273, 2018.
32. Silva, M.S.P., Mota, J.P.B. and Rodrigues, A.E., Adsorption equilibrium and kinetics of the Parex' feed and desorbent streams from batch experiments. *Chemical Engineering and Technology*, 2014. **37**(9): p. 1541–1551.
33. Silva, M.S.P., Rodrigues, A.E. and Mota, J.P.B., Effect of dead volumes on the performance of an industrial-scale simulated moving-bed Parex unit for p-xylene purification. *AIChE Journal*, 2016. **62**(1): p. 241–255.
34. Ben, L., Huiguo, W., Zhuo, Y., Dehua, W. and Wei, W., Commercial applications of paraxylene adsorbents RAX-2000A and RAX-3000. *China Petroleum Processing & Petrochemical Technology*, 2013. **15**(1): p. 10–13.
35. Yang, Y., Bai, P. and Guo, X., Separation of xylene isomers: A review of recent advances in materials. *Industrial & Engineering Chemistry Research*, 2017. **56**(50): p. 14725–14753.
36. Faruque Hasan, M.M., First, E.L. and Floudas, C.A., Discovery of novel zeolites and multi-zeolite processes for p-xylene separation using simulated moving bed (SMB) chromatography. *Chemical Engineering Science*, 2017. **159**: p. 3–17.

37. Moreira, M.A., Ferreira, A.F., Santos, J.C., Loureiro, J.M. and Rodrigues, A.E., Hybrid process for o-and p-xylene production in aromatics plants. *Chemical Engineering and Technology*, 2014. **37**(9): p. 1483–1492.

38. Moreira, M.A., Santos, M.P.S., Silva, C.G., Loureiro, J.M., Chang, J.-S., Serre, C., Ferreira, A.F.P. and Rodrigues, A.E., Adsorption equilibrium of xylene isomers and ethylbenzene on MIL-125(Ti)_NH2: The temperature influence on the para-selectivity. *Adsorption*, 2018. **24**(8): p. 715–724.

39. Ferreira, A.F.P., Adsorption equilibrium of xylene isomers and ethylbenzene on MIL-125(Ti)_NH2: The temperature influence on the para-selectivity, in *6th International Conference on Metal–Organic Frameworks & Open Framework Compounds*. 2018: Auckland, New Zeland.

40. Minceva, M., Gomes, P.S., Meshko, V. and Rodrigues, A.E., Simulated moving bed reactor for isomerization and separation of p-xylene. *Chemical Engineering Journal*, 2008. **140**(1–3): p. 305–323.

41. Bergeot, G., Leinekugel-Le-Cocq, D., Leflaive, P., Laroche, C., Muhr, L. and Bailly, M., Simulated moving bed reactor for paraxylene production. *Chemical Engineering Transactions*, 2009. **17**: p. 87–92.

42. Bergeot, G., Laroche, C., Leflaive, P., Leinekugel-Le-Cocq, D. and Wolff, L., Reactive simulated mobile bed for producing paraxylene, WO 2009/130402-A1, 2009.

43. Gonçalves, J.C. and Rodrigues, A.E., Simulated moving bed reactor for p-xylene production: Adsorbent and catalyst homogeneous mixture. *Chemical Engineering Journal*, 2014. **258**: p. 194–202.

44. Gonçalves, J.C. and Rodrigues, A.E., Simulated moving bed reactor for p-xylene production: Optimal particle size. *Canadian Journal of Chemical Engineering*, 2015. **93**(12): p. 2205–2213.

45. Gonçalves, J.C. and Rodrigues, A.E., Simulated moving bed reactor for p-xylene production: Dual-bed column. *Chemical Engineering and Processing: Process Intensification*, 2016. **104**: p. 75–83.

46. Gonçalves, J.C. and Rodrigues, A.E., Revamping an existing aromatics complex with simulated moving-bed reactor for p-xylene production. *Chemical Engineering & Technology*, 2015. **38**(12): p. 2340–2344.

47. Leinekugel-le-Cocq, D., Paraxylene production using temperature gradient simulated moving bed reactor, in *13th International Conference on the Fundamentals of Adsorption*. 2019: Cairns, Australia.

48. Shi, Q., Gonçalves, J.C., Ferreira, A.F.P. and Rodrigues, A.E., Simulated moving bed reactor for p-xylene production: Modeling, simulation, and optimization, in *4th North American Symposium on Chemical Reaction Engineering*. 2019: Houston, USA.

49. Johnson, J.A., Aromatic complexes, in *Handbook of Petroleum Refining Processes*, R.A. Meyers, Editor. 2003, McGraw-Hill: New York. p. 2.3–2.11.

50. Frey, S.J., Liquid industrial aromatics adsorbent separation, in *Zeolites in Industrial Separation and Catalysis*, S. Kulprathipanja, Editor. 2010, Wiley-VCH: Weinheim. p. 229–247.

51. Laroche, C., Leflaive, P., Baudot, A., Bouvier, L., Lutz, C. and Nicolas, S., Method for separating meta-xylene using a zeolitic adsorbent with a large external surface area, US 10,125,064, 2018.

52. Hotier, G., Leflaive, P. and Wolff, L., Process for producing high purity meta-xylene, comprising simulated moving bed adsorption and crystallization, US 7,928,276, 2011.

53. Turowicz, M., Enhance naphtha value and gasoline reformer performance using UOP's MaxEne™ process, in *1st IndianOil Petrochemical Conclave*. 2012, UOP LLC: Gurgaon, India.

54. Jun, C., Benxian, S. and Jichang, L., Optimal operation of simulated moving bed technology on utilization of naphtha resource. *Separation Science and Technology*, 2012. **48**(2): p. 246–253.

55. Sohn, S.W., Rice, L.H. and Kulprathipanja, S., Ethylene production by steam cracking of normal paraffins, US 8,283,511, 2012.

56. Do, T.X., Lim, Y.-i., Lee, J. and Lee, W., Techno-economic analysis of petrochemical complex retrofitted with simulated moving-bed for olefins and aromatics production. *Chemical Engineering Research and Design*, 2016. **106**: p. 222–241.

57. Ragil, K., Bailly, M., Jullian, S. and Clause, O., Process for chromatographic separation of a C5-C8 feed or an intermediate feed into three effluents, respectively rich in straight chain, mono-brnached and multi-branched paraffins, US 6,353,144, 2002.

58. Peralta, D., Chaplais, G., Simon-Masseron, A., Barthelet, K. and Pirngruber, G.D., Separation of C6 paraffins using zeolitic imidazolate frameworks: Comparison with zeolite 5A. *Industrial & Engineering Chemistry Research*, 2012. **51**(12): p. 4692–4702.

59. Jichang, L., Benxian, S. and Hui, S., Adsorption behaviour of normal paraffins in a fixed bed adsorber containing 5 Å molecular sieves. *Adsorption Science and Technology*, 2006. **24**(4): p. 311–320.

60. Rice, L.H., Isomerization process with adsorptive separation, US 7,514,590, 2009.

61. Sohn, S.W., UOP olex process for olefin recovery, in *Handbook of Petroleum Refining Processes*, R.A. Meyers, Editor. 2003, McGraw-Hill: New York. p. 10.79–10.81.

62. Sohn, S.W., Liquid industrial non-aromatics adsorptive separation, in *Zeolites in Industrial Separation and Catalysis*, S. Kulprathipanja, Editor. 2010, Wiley-VCH: Weinheim. p. 249–272.

63. Kulprathipanja, S., Aspects of mechanisms, processes, and requirements for zeolite separation, in *Zeolites in Industrial Separation and Catalysis*, S. Kulprathipanja, Editor. 2010, Wiley-VCH: Weinheim. p. 203–228.

64. Leflaive, P., Wolff, L., Cocq, D.L.L., Lamia, N., Rodrigues, A. and Grande, C., Process for separating propylene mixed with propane by adsorption in a simulated moving bed, US 10,000,430, 2018.

65. Campo, M.C., Baptista, M.C., Ribeiro, A.M., Ferreira, A., Santos, J.C., Lutz, C., Loureiro, J.M. and Rodrigues, A.E., Gas phase SMB for propane/propylene separation using enhanced 13X zeolite beads. *Adsorption*, 2014. **20**(1): p. 61–75.

66. Martins, V.F.D., Ribeiro, A.M., Plaza, M.G., Santos, J.C., Loureiro, J.M., Ferreira, A.F.P. and Rodrigues, A.E., Gas-phase simulated moving bed: Propane/propylene separation on 13X zeolite. *Journal of Chromatography A*, 2015. **1423**: p. 136–148.

67. Narin, G., Martins, V.F.D., Campo, M., Ribeiro, A.M., Ferreira, A., Santos, J.C., Schumann, K. and Rodrigues, A.E., Light olefins/paraffins separation with 13X zeolite binderless beads. *Separation and Purification Technology*, 2014. **133**: p. 452–475.

68. Martins, V.F.D., Ribeiro, A.M., Santos, J.C., Loureiro, J.M., Gleichmann, K., Ferreira, A. and Rodrigues, A.E., Development of gas-phase SMB technology for light olefin/paraffin separations. *AIChE Journal*, 2016. **62**(7): p. 2490–2500.

69. Martins, V.F.D., Ribeiro, A.M., Chang, J.-S., Loureiro, J.M., Ferreira, A. and Rodrigues, A.E., Towards polymer grade ethylene production with Cu-BTC: Gas-phase SMB versus PSA. *Adsorption*, 2018. **24**(2): p. 203–219.

70. Li, K., Beaver, M., Speronello, B. and García-Martínez, J., Surfactant-templated meso-structuring of zeolites: From discovery to commercialization, in *Mesoporous Zeolites Preparation, Characterization and Applications*, J. García-Martínez and K. Li, Editors. 2015, Wiley-VCH: Weinheim, Germany.

71. Aguado, S., Bergeret, G., Daniel, C. and Farrusseng, D., Absolute molecular sieve separation of ethylene/ethane mixtures with silver zeolite A. *Journal of the American Chemical Society*, 2012. **134**(36): p. 14635–14637.

72. Aguado, S. and Farrusseng, D., Separation of ethylene/ethane mixtures with silver zeolites, in *Small-Scale Gas to Liquid Fuel Synthesis*, N. Kanellopoulos, Editor. 2015, CRC Press: Boca Raton.

73. Lin, R.-B., Wu, H., Li, L., Tang, X.-L., Li, Z., Gao, J., Cui, H., Zhou, W. and Chen, B., Boosting ethane/ethylene separation within isoreticular ultramicroporous metal–organic frameworks. *Journal of the American Chemical Society*, 2018. **140**(40): p. 12940–12946.

74. Wu, Y., Chen, H., Liu, D., Qian, Y. and Xi, H., Adsorption and separation of ethane/ethylene on ZIFs with various topologies: Combining GCMC simulation with the ideal adsorbed solution theory (IAST). *Chemical Engineering Science*, 2015. **124**: p. 144–153.

75. Martins, V.F.D., Ribeiro, A.M., Kortunov, P., Ferreira, A. and Rodrigues, A.E., High purity ethane/ethylene separation by gas phase simulated moving bed using ZIF-8 adsorbent. *AIChE Journal*, 2019. **65**(8): p. e16619.

76. Sivakumar, S.V. and Rao, D.P., Adsorptive separation of gas mixtures: Mechanistic view, sharp separation and process intensification. *Chemical Engineering and Processing: Process Intensification*, 2012. **53**: p. 31–52.

77. Bieser, H.J. and De Rosset, A.J., Continuous countercurrent separation of saccharides with inorganic adsorbents. *Starch - Stärke*, 1977. **29**(11): p. 392–397.

78. Janakievski, F., Glagovskaia, O. and De Silva, K., 5 - Simulated moving bed chromatography in food processing, in *Innovative Food Processing Technologies*, K. Knoerzer, P. Juliano, and G. Smithers, Editors. 2016, Woodhead Publishing: Melbourne. p. 133–149.

79. Maruyama, R.T., Karnal, P., Sainio, T. and Rajendran, A., Design of bypass-simulated moving bed chromatography for reduced purity requirements. *Chemical Engineering Science*, 2019. **205**: p. 401–413.

80. Sreedhar, B. and Kawajiri, Y., Multi-column chromatographic process development using simulated moving bed superstructure and simultaneous optimization – Model correction framework. *Chemical Engineering Science*, 2014. **116**: p. 428–441.

81. David, L., Yun, J. and Nicoud, R.-M., Comparing multi-column chromatographic processes for purifying monosaccharides part I: A simplified approach. *Adsorption*, 2017. **23**(4): p. 577–591.

82. Yu, I.K.M., Ong, K.L., Tsang, D.C.W., Haque, M.A., Kwan, T.H., Chen, S.S., Uisan, K., Kulkarni, S. and Lin, C.S.K., Chemical transformation of food and beverage waste-derived fructose to hydroxymethylfurfural as a value-added product. *Catalysis Today*, 2018. **314**: p. 70–77.

83. Tangpromphan, P., Budman, H. and Jaree, A., A simplified strategy to reduce the desorbent consumption and equipment installed in a three-zone simulated moving bed process for the separation of glucose and fructose. *Chemical Engineering and Processing - Process Intensification*, 2018. **126**: p. 23–37.

84. Vignesh, S.V., Hariprasad, K., Athawale, P. and Bhartiya, S., An optimization-driven novel operation of simulated moving bed chromatographic separation. *IFAC-PapersOnLine*, 2016. **49**(7): p. 165–170.

85. Yao, C., Jing, K., Ling, X., Lu, Y. and Tang, S., Application of dodecahedron to describe the switching strategies of asynchronous simulated-moving-bed. *Computers & Chemical Engineering*, 2017. **96**: p. 69–74.

86. Sharma, G., Vignesh, S.V., Hariprasad, K. and Bhartiya, S., Control-relevant multiple linear modeling of simulated moving bed chromatography. *IFAC-PapersOnLine*, 2015. **48**(8): p. 477–482.

87. Yao, C., Tang, S., Lu, Y., Yao, H.-M. and Tade, M.O., Combination of space–time conservation element/solution element method and continuous prediction technique for accelerated simulation of simulated moving bed chromatography. *Chemical Engineering and Processing: Process Intensification*, 2015. **96**: p. 54–61.

88. Beste, Y.A., Lisso, M., Wozny, G. and Arlt, W., Optimization of simulated moving bed plants with low efficient stationary phases: Separation of fructose and glucose. *Journal of Chromatography A*, 2000. **868**(2): p. 169–188.

89. Vignesh, S.V., Hariprasad, K., Athawale, P., Siram, V. and Bhartiya, S., Optimal strategies for transitions in simulated moving bed chromatography. *Computers & Chemical Engineering*, 2016. **84**: p. 83–95.

90. Soepriatna, N., Wang, N.H.L. and Wankat, P.C., Standing wave design and optimization of nonlinear four-zone thermal simulated moving bed systems. *Industrial & Engineering Chemistry Research*, 2015. **54**(42): p. 10419–10433.

91. Heinonen, J., Sanlaville, Q., Niskakoski, H. and Sainio, T., Effect of separation material particle size on pressure drop and process efficiency in continuous chromatographic separation of glucose and fructose. *Separation and Purification Technology*, 2018. **193**: p. 317–326.

92. Borges da Silva, E.A., Ulson de Souza, A.A., de Souza, S.G.U. and Rodrigues, A.E., Analysis of the high-fructose syrup production using reactive SMB technology. *Chemical Engineering Journal*, 2006. **118**(3): p. 167–181.

93. Hashimoto, K., Adachi, S., Noujima, H. and Ueda, Y., A new process combining adsorption and enzyme reaction for producing higher-fructose syrup. *Biotechnology and Bioengineering*, 1983. **25**(10): p. 2371–2393.

94. Zhang, Y., Hidajat, K. and Ray, A.K., Modified reactive SMB for production of high concentrated fructose syrup by isomerization of glucose to fructose. *Biochemical Engineering Journal*, 2007. **35**(3): p. 341–351.

95. Toumi, A. and Engell, S., Optimization-based control of a reactive simulated moving bed process for glucose isomerization. *Chemical Engineering Science*, 2004. **59**(18): p. 3777–3792.

96. Sun, Z.-Y., Tang, Y.-Q., Iwanaga, T., Sho, T. and Kida, K., Production of fuel ethanol from bamboo by concentrated sulfuric acid hydrolysis followed by continuous ethanol fermentation. *Bioresource Technology*, 2011. **102**(23): p. 10929–10935.

97. Tan, L., Sun, Z., Zhang, W., Tang, Y., Morimura, S. and Kida, K., Production of biofuel ethanol from distilled grain waste eluted from Chinese spirit making process. *Bioprocess and Biosystems Engineering*, 2014. **37**(10): p. 2031–2038.

98. Sun, Z.-Y., Ura, T., Matsuura, H., Kida, K. and Jyo, A., Dowex 1X4 and Dowex 1X8 as substitute of Diaion MA03SS in simulated moving bed chromatographic separation of sulfuric acid and sugars in concentrated sulfuric acid hydrolysates of bamboo. *Separation and Purification Technology*, 2016. **166**: p. 92–101.

99. Mun, S., Optimization of production rate, productivity, and product concentration for a simulated moving bed process aimed atfucose separation using standing-wave-design and genetic algorithm. *Journal of Chromatography A*, 2018. **1575**: p. 113–121.

100. Lee, C.-G., Jo, C.Y., Song, Y.J. and Mun, S., Continuous-mode separation of fucose and 2,3-butanediol using a three-zone simulated moving bed process and its performance improvement by using partial extract-collection, partial extract-recycle, and partial desorbent-port closing. *Journal of Chromatography A*, 2018. **1579**: p. 49–59.

101. Hong, S.-B., Choi, J.-H., Chang, Y.K. and Mun, S., Production of high-purity fucose from the seaweed of *Undaria pinnatifida* through acid-hydrolysis and simulated-moving bed purification. *Separation and Purification Technology*, 2019. **213**: p. 133–141.

102. Choi, J.-H., Park, H., Park, C., Wang, N.-H.L. and Mun, S., Highly efficient recovery of xylobiose from xylooligosaccharides using a simulated moving bed method. *Journal of Chromatography A*, 2016. **1465**: p. 143–154.

103. Lee, C.-G., Choi, J.-H., Park, C., Wang, N.-H.L. and Mun, S., Standing wave design and optimization of a simulated moving bed chromatography for separation of xylobiose and xylose under the constraints on product concentration and pressure drop. *Journal of Chromatography A*, 2017. **1527**: p. 80–90.

104. Song, X., Tang, S., Jiang, L., Zhu, L. and Huang, H., Integrated biocatalytic process for trehalose production and separation from maltose. *Industrial & Engineering Chemistry Research*, 2016. **55**(40): p. 10566–10575.

105. Gomes, P.S., Minceva, M. and Rodrigues, A.E., Simulated moving bed technology: Old and new. *Adsorption*, 2006. **12**(5–6): p. 375–392.

106. Adam, P., Nicoud, R.N., Bailly, M. and Ludemann-Hombourger, O., Process and device for separation with variable-length, US 6136198 A, 2000.

107. Ludemann-Hombourger, O., Bailly, M. and Nicoud, R.M., Design of a simulated moving bed: Optimal particle size of the stationary phase. *Separation Science and Technology*, 2000. **35**(9): p. 1285–1305.

108. Ludemann-Hombourger, O., Nicoud, R.M. and Bailly, M., The "VARICOL" process: A new multicolumn continuous chromatographic process. *Separation Science and Technology*, 2000. **35**(12): p. 1829–1862.

109. Pais, L.S. and Rodrigues, A.E., Design of simulated moving bed and Varicol processes for preparative separations with a low number of columns. *Journal of Chromatography A*, 2003. **1006**(1): p. 33–44.

110. Yang, Y., Lu, K., Gong, R., Qu, D., Li, P., Yu, J. and Rodrigues, A.E., Separation of guaifenesin enantiomers by simulated moving bed process with four operation modes. *Adsorption*, 2019. **25**(6): p. 1–14.

111. Lin, X., Gong, R., Li, J., Li, P., Yu, J. and Rodrigues, A.E., Enantioseparation of racemic aminoglutethimide using asynchronous simulated moving bed chromatography. *Journal of Chromatography A*, 2016. **1467**: p. 347–355.

112. Yu, Y., Wood, K.R. and Liu, Y.A., Simulation and comparison of operational modes in simulated moving bed chromatography. *Industrial & Engineering Chemistry Research*, 2015. **54**(46): p. 11576–11591.

113. Perna, R., Cremasco, M. and Santana, C., Chromatographic separation of verapamil racemate using a Varicol continuous multicolumn process. *Brazilian Journal of Chemical Engineering*, 2015. **32**(4): p. 929–939.

114. Lourenço, T.C., Batista Jr, J.M., Furlan, M., He, Y., Nafie, L.A., Santana, C.C. and Cass, Q.B., Albendazole sulfoxide enantiomers: Preparative chiral separation and absolute stereochemistry. *Journal of Chromatography A*, 2012. **1230**: p. 61–65.

115. da Silva Jr., A.C., Salles Jr., A.G., Perna, R.F., Correia, C.R.D. and Santana, C.C., Chromatographic separation and purification of mitotane racemate in a Varicol multicolumn continuous process. *Chemical Engineering & Technology*, 2012. **35**(1): p. 83–90.

116. Zhang, Y., Hidajat, K. and Ray, A.K., Multi-objective optimization of simulated moving bed and Varicol processes for enantio-separation of racemic pindolol. *Separation and Purification Technology*, 2009. **65**(3): p. 311–321.

117. Ribeiro, A.E., Rodrigues, A.E. and Pais, L.S., Separation of nadolol stereoisomers by chiral liquid chromatography at analytical and preparative scales. *Chirality*, 2013. **25**(3): p. 197–205.

118. Arafah, R.S., Ribeiro, A.E., Rodrigues, A.E. and Pais, L.S., Separation of nadolol stereoisomers using chiralpak IA chiral stationary phase. *Chirality*, 2016. **28**(5): p. 399–408.

119. Lee, J.W. and Wankat, P.C., Design of pseudo-simulated moving bed process with multi-objective optimization for the separation of a ternary mixture: Linear isotherms. *Journal of Chromatography A*, 2010. **1217**(20): p. 3418–3426.

120. Jermann, S., Meijssen, M. and Mazzotti, M., Three column intermittent simulated moving bed chromatography: 3. Cascade operation for center-cut separations. *Journal of Chromatography A*, 2015. **1378**: p. 37–49.

121. Palacios, J.G., Kramer, B., Kienle, A. and Kaspereit, M., Experimental validation of a new integrated simulated moving bed process for the production of single enantiomers. *Journal of Chromatography A*, 2011. **1218**(16): p. 2232–2239.

122. Fuereder, M., Femmer, C., Storti, G., Panke, S. and Bechtold, M., Integration of simu-lated moving bed chromatography and enzymatic racemization for the production of single enantiomers. *Chemical Engineering Science*, 2016. **152**: p. 649–662.

123. Chankvetadze, B., Chapter 3 - Liquid chromatographic separation of enantiomers, in *Liquid Chromatography (Second Edition)*, S. Fanali, P.R. Haddad, C.F. Poole, and M.-L. Riekkola, Editors. 2017, Elsevier: Tbilisi. p. 69–86.

124. Festel, G., Economic aspects of industrial biotechnology, in: *Advances in Biochemical Engineering/Biotechnology*, G. Festel, Editor. 2018, Springer: Berlin, Heidelberg. p. 1–22.

125. Shmorhun, M., BER-myriant succinic acid biorefinery. 2015, BER-Myriant Succinic Acid Biorefinery: Lake Providence.

126. Nam, H.-G., Park, C., Jo, S.-H., Suh, Y.-W. and Mun, S., Continuous separation of suc-cinic acid and lactic acid by using a three-zone simulated moving bed process packed with Amberchrom-CG300C. *Process Biochemistry*, 2012. **47**(12): p. 2418–2426.

127. Mun, S., Comparison of the relative merits of port-location rearrangement and par-tial-feeding as the strategy for improving the performances of a three-zone simulated moving chromatography for separation of succinic acid and lactic acid. *Journal of Chromatography A*, 2014. **1341**: p. 8–14.

128. Mun, S., Enhanced performance of a three-zone simulated moving bed chromatogra-phy for separation of succinic acid and lactic acid by simultaneous use of port-loca-tion rearrangement and partial-feeding. *Journal of Chromatography A*, 2014. **1350**: p. 72–82.

129. Choi, J.-H., Nam, H.-G. and Mun, S., Enhancement of yield and productivity in the 3-zone nonlinear SMB for succinic-acid separation under overloaded conditions. *Journal of Industrial and Engineering Chemistry*, 2018. **58**: p. 222–228.

130. Park, C., Nam, H.-G., Lee, K.B. and Mun, S., Optimal design and experimental valida-tion of a simulated moving bed chromatography for continuous recovery of formic acid in a model mixture of three organic acids from *Actinobacillus* bacteria fermentation. *Journal of Chromatography A*, 2014. **1365**: p. 106–114.

131. Rezkallah, A., Method for purification of glycerol, US 7667081 B2, 2010.

132. Coelho, L.C.D., Filho, N.M.L., Faria, R.P.V., Ribeiro, A.M. and Rodrigues, A.E., Selection of a stationary phase for the chromatographic separation of organic acids obtained from bioglycerol oxidation. *Adsorption*, 2017. **23**(5): p. 627–638.

133. Coelho, L.C.D., Filho, N.M.L., Faria, R.P.V., Ferreira, A.F.P., Ribeiro, A.M. and Rodrigues, A.E., Separation of tartronic and glyceric acids by simulated moving bed chromatography. *Journal of Chromatography A*, 2018. **1563**: p. 62–70.

134. Hilaly, A.K., Sandage, R.D. and Soper, J.G., Separation of a mixture of polyhydric alcohols, US 8177980 B2, 2012.

135. Schultz, M.A., Havill, A.M. and Oroskar, A.R., Fermentation and simulated moving bed process, US 8980596 B2, 2015.

136. Liang, M.-T., Lin, C.-H., Tsai, P.-Y., Wang, H.-P., Wan, H.-P. and Yang, T.-Y., The sepa-ration of butanediol and propanediol by simulated moving bed. *Journal of the Taiwan Institute of Chemical Engineers*, 2016. **61**: p. 12–19.

137. Rodrigues, A.E., Pereira, C.S.M., Minceva, M., Pais, L.S., Ribeiro, A.M., Ribeiro, A., Silva, M., Graça, N. and Santos, J.C., *Simulated Moving Bed Technology: Principles, Design and Process Applications*. 2015, Elsevier Science: Oxford, UK and Waltham, MA.

138. Egidio, R.A., Yi-jiang, W., Miguel, M.L. and Faria, R.P.V., *Sorption Enhanced Reaction Processes*. Vol. 1. 2017, World Scientific: London.

139. Mitra Ray, N. and Ray, A.K., Modelling, simulation, and experimental study of a sim-ulated moving bed reactor for the synthesis of biodiesel. *The Canadian Journal of Chemical Engineering*, 2016. **94**(5): p. 913–923.

140. Faria, R.P.V., Graça, N.S. and Rodrigues, A.E., Chapter 7: Green fuels and fuel additives production in simulated moving bed reactors, in *Intensification of Biobased Processes*, A. Górak and A. Stankiewicz, Editors. 2018, The Royal Society of Chemistry: London. p. 145–165.

141. Silva, V.M.T.M. and Rodrigues, A.E., Novel process for diethylacetal synthesis. *AIChE Journal*, 2005. **51**(10): p. 2752–2768.

142. Graça, N.S., Delgado, A.E., Constantino, D.S., Pereira, C.S. and Rodrigues, A.E., Synthesis of a renewable oxygenated diesel additive in an adsorptive reactor. *Energy Technology*, 2014. **2**(9–10): p. 839–850.

143. Rodrigues, A.E., Pereira, C.S.M. and Santos, J.C., Chromatographic reactors. *Chemical Engineering & Technology*, 2012. **35**(7): p. 1171–1183.

144. Pereira, C.S.M., Zabka, M., Silva, V.M.T.M. and Rodrigues, A.E., A novel process for the ethyl lactate synthesis in a simulated moving bed reactor (SMBR). *Chemical Engineering Science*, 2009. **64**(14): p. 3301–3310.

145. Walsh, G., Biopharmaceutical benchmarks 2018. *Nature Biotechnology*, 2018. **36**: p. 1136–1145.

146. Shaikh, S. and Jaiswal, P., Biopharmaceuticals market by type and application: Global opportunity analysis and industry forecast, 2018-2025. 2018, Allied Market Research: https://www.alliedmarketresearch.com/biopharmaceutical-market.

147. Girard, V., Hilbold, N.-J., Ng, C.K., Pegon, L., Chahim, W., Rousset, F. and Monchois, V., Large-scale monoclonal antibody purification by continuous chromatography, from process design to scale-up. *Journal of Biotechnology*, 2015. **213**: p. 65–73.

148. Jagschies, G., Lindskog, E., Lacki, K. and Galliher, P.M., *Biopharmaceutical Processing: Development, Design, and Implementation of Manufacturing Processes*. 2018, Elsevier: Amsterdam, Netherlands and Oxford, UK and Cambridge, MA.

149. Hashimoto, K., Adachi, S. and Shirai, Y., Continuous desalting of proteins with a simulated moving-bed adsorber. *Agricultural and Biological Chemistry*, 1988. **52**(9): p. 2161–2167.

150. Xie, Y., Mun, S., Kim, J. and Wang, N.-H.L., Standing wave design and experimental validation of a tandem simulated moving bed process for insulin purification. *Biotechnology Progress*, 2002. **18**(6): p. 1332–1344.

151. Jungbauer, A., Continuous downstream processing of biopharmaceuticals. *Trends in Biotechnology*, 2013. **31**(8): p. 479–492.

152. Steinebach, F., Müller-Späth, T. and Morbidelli, M., Continuous counter-current chromatography for capture and polishing steps in biopharmaceutical production. *Biotechnology Journal*, 2016. **11**(9): p. 1126–1141.

153. Wellhoefer, M., Sprinzl, W., Hahn, R. and Jungbauer, A., Continuous processing of recombinant proteins: Integration of inclusion body solubilization and refolding using simulated moving bed size exclusion chromatography with buffer recycling. *Journal of Chromatography A*, 2013. **1319**: p. 107–117.

154. Wellhoefer, M., Sprinzl, W., Hahn, R. and Jungbauer, A., Continuous processing of recombinant proteins: Integration of refolding and purification using simulated moving bed size-exclusion chromatography with buffer recycling. *Journal of Chromatography A*, 2014. **1337**: p. 48–56.

155. Saremirad, P., Wood, J.A., Zhang, Y. and Ray, A.K., Oxidative protein refolding on size exclusion chromatography: From batch single-column to multi-column counter-current continuous processing. *Chemical Engineering Science*, 2015. **138**: p. 375–384.

156. Li, P., Xiu, G. and Rodrigues, A.E., Proteins separation and purification by salt gradient ion-exchange SMB. *AIChE Journal*, 2007. **53**(9): p. 2419–2431.

157. Li, P., Yu, J., Xiu, G. and Rodrigues, A.E., Separation region and strategies for proteins separation by salt gradient ion-exchange SMB. *Separation Science and Technology*, 2008. **43**(1): p. 11–28.

158. Keβler, L.C., Gueorguieva, L., Rinas, U. and Seidel-Morgenstern, A., Step gradients in 3-zone simulated moving bed chromatography: Application to the purification of antibodies and bone morphogenetic protein-2. *Journal of Chromatography A*, 2007. **1176**(1–2): p. 69–78.

159. Wu, X., Arellano-Garcia, H., Hong, W. and Wozny, G., Improving the operating conditions of gradient ion-exchange simulated moving bed for protein separation. *Industrial & Engineering Chemistry Research*, 2013. **52**(15): p. 5407–5417.

160. Fischer, L.M., Wolff, M.W. and Reichl, U., Purification of cell culture-derived influenza A virus *via* continuous anion exchange chromatography on monoliths. *Vaccine*, 2018. **36**(22): p. 3153–3160.

161. Kröber, T., Wolff, M.W., Hundt, B., Seidel-Morgenstern, A. and Reichl, U., Continuous purification of influenza virus using simulated moving bed chromatography. *Journal of Chromatography A*, 2013. **1307**: p. 99–110.

162. Park, C., Nam, H.-G., Hwang, H.-J., Kim, J.-H. and Mun, S., Development of a three-zone simulated moving bed process based on partial-discard strategy for continuous separation of valine from isoleucine with high purity, high yield, and high product concentration. *Process Biochemistry*, 2014. **49**(2): p. 324–334.

163. Park, C., Nam, H.G., Kim, P.H. and Mun, S., Experimental evaluation of the effect of a modified port-location mode on the performance of a three-zone simulated moving-bed process for the separation of valine and isoleucine. *Journal of Separation Science*, 2014. **37**(11): p. 1215–1221.

164. Park, C., Nam, H.-G., Jo, S.-H., Wang, N.-H.L. and Mun, S., Continuous recovery of valine in a model mixture of amino acids and salt from *Corynebacterium* bacteria fermentation using a simulated moving bed chromatography. *Journal of Chromatography A*, 2016. **1435**: p. 39–53.

165. Harvey, D., Weeden, G. and Wang, N.-H.L., Speedy standing wave design and simulated moving bed splitting strategies for the separation of ternary mixtures with linear isotherms. *Journal of Chromatography A*, 2017. **1530**: p. 152–170.

166. Sainio, T., Unified design of chromatographic processes with timed events: Separation of ternary mixtures. *Chemical Engineering Science*, 2016. **152**: p. 547–567.

167. Cristancho, C.A.M. and Seidel-Morgenstern, A., Purification of single-chain antibody fragments exploiting pH-gradients in simulated moving bed chromatography. *Journal of Chromatography A*, 2016. **1434**: p. 29–38.

168. Chin, C.Y. and Wang, N.H.L., Simulated moving bed equipment designs. *Separation & Purification Reviews*, 2004. **33**(2): p. 77–155.

169. Faria, R.P.V. and Rodrigues, A.E., Instrumental aspects of simulated moving bed chromatography. *Journal of Chromatography A*, 2015. **1421**: p. 82–102.

170. Lutin, F., Bailly, M. and Bar, D., Process improvements with innovative technologies in the starch and sugar industries. *Desalination*, 2002. **148**(1): p. 121–124.

171. Katsuo, S. and Mazzotti, M., Intermittent simulated moving bed chromatography: 1. Design criteria and cyclic steady-state. *Journal of Chromatography A*, 2010. **1217**(8): p. 1354–1361.

172. Katsuo, S. and Mazzotti, M., Intermittent simulated moving bed chromatography: 2. Separation of Tröger's base enantiomers. *Journal of Chromatography A*, 2010. **1217**(18): p. 3067–3075.

173. Kearney, M.M. and Hieb, K.L., Time variable simulated moving bed process, US 5102553 A, 1992.

174. Kloppenburg, E. and Gilles, E.D., A new concept for operating simulated moving-bed processes. *Chemical Engineering & Technology*, 1999. **22**(10): p. 813–817.

175. Zhang, Z., Mazzotti, M. and Morbidelli, M., PowerFeed operation of simulated moving bed units: Changing flow-rates during the switching interval. *Journal of Chromatography A*, 2003. **1006**(1–2): p. 87–99.

176. Zang, Y. and Wankat, P.C., SMB operation strategy-partial feed. *Industrial & Engineering Chemistry Research*, 2002. **41**(10): p. 2504–2511.

177. Bae, Y.-S. and Lee, C.-H., Partial-discard strategy for obtaining high purity products using simulated moving bed chromatography. *Journal of Chromatography A*, 2006. **1122**(1–2): p. 161–173.

178. Sá Gomes, P. and Rodrigues, A.E., Outlet streams swing (OSS) and multifeed operation of simulated moving beds. *Separation Science and Technology*, 2007. **42**(2): p. 223–252.

179. Mun, S., Partial port-closing strategy for obtaining high throughput or high purities in a four-zone simulated moving bed chromatography for binary separation. *Journal of Chromatography A*, 2010. **1217**(42): p. 6522–6530.

180. Schramm, H., Kaspereit, M., Kienle, A. and Seidel-Morgenstern, A., Simulated moving bed process with cyclic modulation of the feed concentration. *Journal of Chromatography A*, 2003. **1006**(1–2): p. 77–86.

181. Schramm, H., Kienle, A., Kaspereit, M. and Seidel-Morgenstern, A., Improved operation of simulated moving bed processes through cyclic modulation of feed flow and feed concentration. *Chemical Engineering Science*, 2003. **58**(23–24): p. 5217–5227.

182. Bailly, M., Nicoud, R.M., Adam, P. and Ludemann-Hombourger, O., Method and device for chromatography comprising a concentration step, EP 1558355 B1, 2006.

183. Paredes, G., Rhee, H.-K. and Mazzotti, M., Design of simulated-moving-bed chromatography with enriched extract operation (EE-SMB): Langmuir isotherms. *Industrial & Engineering Chemistry Research*, 2006. **45**(18): p. 6289–6301.

184. Clavier, J.Y., Nicoud, R.M. and Perrut, M., A new efficient fractionation process: The simulated moving bed with supercritical eluent, in *High Pressure Chemical Engineering* Edited by Ph. Rudolf von Rohr. 1996, Elsevier: Zürich, Switzerland.

185. Antos, D. and Seidel-Morgenstern, A., Application of gradients in the simulated moving bed process. *Chemical Engineering Science*, 2001. **56**(23): p. 6667–6682.

186. Abel, S., Mazzotti, M. and Morbidelli, M., Solvent gradient operation of simulated moving beds: I. Linear isotherms. *Journal of Chromatography A*, 2002. **944**(1–2): p. 23–39.

187. Abel, S., Mazzotti, M. and Morbidelli, M., Solvent gradient operation of simulated moving beds: 2. Langmuir isotherms. *Journal of Chromatography A*, 2004. **1026**(1): p. 47–55.

188. Migliorini, C., Wendlinger, M., Mazzotti, M. and Morbidelli, M., Temperature gradient operation of a simulated moving bed unit. *Industrial & Engineering Chemistry Research*, 2001. **40**(12): p. 2606–2617.

189. Jin, W. and Wankat, P.C., Thermal operation of four-zone simulated moving beds. *Industrial & Engineering Chemistry Research*, 2007. **46**(22): p. 7208–7220.

190. Zang, Y. and Wankat, P.C., Three-zone simulated moving bed with partial feed and selective withdrawal. *Industrial & Engineering Chemistry Research*, 2002. **41**(21): p. 5283–5289.

191. Liang, M.-T. and Liang, R.-C., Fractionation of polyethylene glycol particles by simulated moving bed with size-exclusion chromatography. *Journal of Chromatography A*, 2012. **1229**: p. 107–112.

192. Nicolaos, A., Muhr, L., Gotteland, P., Nicoud, R.-M. and Bailly, M., Application of equilibrium theory to ternary moving bed configurations (four+ four, five+ four, eight and nine zones): I. Linear case. *Journal of Chromatography A*, 2001. **908**(1): p. 71–86.

193. Nicolaos, A., Muhr, L., Gotteland, P., Nicoud, R.-M. and Bailly, M., Application of the equilibrium theory to ternary moving bed configurations (4+ 4, 5+ 4, 8 and 9 zones): II. Langmuir case. *Journal of Chromatography A*, 2001. **908**(1): p. 87–109.

194. Wooley, R., Ma, Z. and Wang, N.-H., A nine-zone simulating moving bed for the recovery of glucose and xylose from biomass hydrolyzate. *Industrial & Engineering Chemistry Research*, 1998. **37**(9): p. 3699–3709.

195. Beste, Y.A. and Arlt, W., Side-stream simulated moving-bed chromatography for multi-component separation. *Chemical Engineering & Technology*, 2002. **25**(10): p. 956–962.
196. Xie, Y., Chin, C.Y., Phelps, D.S.C., Lee, C.-H., Lee, K.B., Mun, S. and Wang, N.-H.L., A five-zone simulated moving bed for the isolation of six sugars from biomass hydroly-zate. *Industrial & Engineering Chemistry Research*, 2005. **44**(26): p. 9904–9920.
197. Wankat, P.C., Simulated moving bed cascades for ternary separations. *Industrial & Engineering Chemistry Research*, 2001. **40**(26): p. 6185–6193.
198. Wankat, P.C., Systems and processes for performing separations using a simulated moving bed apparatus, US 6740243 B2, 2004.
199. Mun, S., Effect of subdividing the adsorbent bed in a five-zone simulated moving bed chromatography for ternary separation. *Journal of Liquid Chromatography & Related Technologies*, 2008. **31**(9): p. 1231–1257.
200. Sá Gomes, P. and Rodrigues, A.E., Simulated moving bed chromatography: From con-cept to proof-of-concept. *Chemical Engineering & Technology*, 2012. **35**(1): p. 17–34.
201. Tanimura, M., Tamura, M. and Teshima, T., Chromatographic separation method, JPH0746097B2, 1995.
202. Jermann, S., Katsuo, S. and Mazzotti, M., Intermittent simulated moving bed pro-cesses for chromatographic three-fraction separation. *Organic Process Research & Development*, 2012. **16**(2): p. 311–322.
203. Song, M., Cui, L., Kuang, H., Zhou, J., Yang, P., Zhuang, W., Chen, Y., Liu, D., Zhu, C., Chen, X., Ying, H. and Wu, J., Model-based design of an intermittent simulated moving bed process for recovering lactic acid from ternary mixture. *Journal of Chromatography A*, 2018. **1562**: p. 47–58.
204. Chung, J.-W., Kim, K.-M., Yoon, T.-U., Kim, S.-I., Jung, T.-S., Han, S.-S. and Bae, Y.-S., Power partial-discard strategy to obtain improved performance for simulated moving bed chromatography. *Journal of Chromatography A*, 2017. **1529**: p. 72–80.
205. Kim, K.-M., Song, J.-Y. and Lee, C.-H., Combined operation of outlet streams swing with partial-feed in a simulated moving bed. *Korean Journal of Chemical Engineering*, 2016. **33**(3): p. 1059–1069.
206. Song, J.-Y., Kim, K.-M. and Lee, C.-H., High-performance strategy of a simulated mov-ing bed chromatography by simultaneous control of product and feed streams under max-imum allowable pressure drop. *Journal of Chromatography A*, 2016. **1471**: p. 102–117.
207. Keßler, L.C. and Seidel-Morgenstern, A., Improving performance of simulated mov-ing bed chromatography by fractionation and feed-back of outlet streams. *Journal of Chromatography A*, 2008. **1207**(1): p. 55–71.
208. Li, S., Kawajiri, Y., Raisch, J. and Seidel-Morgenstern, A., Optimization of simulated moving bed chromatography with fractionation and feedback: Part I. Fractionation of one outlet. *Journal of Chromatography A*, 2010. **1217**(33): p. 5337–5348.
209. Li, S., Kawajiri, Y., Raisch, J. and Seidel-Morgenstern, A., Optimization of simulated moving bed chromatography with fractionation and feedback: Part II. Fractionation of both outlets. *Journal of Chromatography A*, 2010. **1217**(33): p. 5349–5357.
210. Kim, K.-M., Lee, H.-H. and Lee, C.-H., Improved performance of a simulated mov-ing bed process by a recycling method in the partial-discard strategy. *Industrial & Engineering Chemistry Research*, 2012. **51**(29): p. 9835–9849.
211. Kim, K.-M. and Lee, C.-H., Backfill-simulated moving bed operation for improving the separation performance of simulated moving bed chromatography. *Journal of Chromatography A*, 2013. **1311**: p. 79–89.
212. Lee, H.H., Kim, K.M. and Lee, C.H., Improved performance of simulated moving bed process using column-modified feed. *AIChE Journal*, 2011. **57**(8): p. 2036–2053.
213. Song, J.-Y., Oh, D. and Lee, C.-H., Effects of a malfunctional column on conven-tional and FeedCol-simulated moving bed chromatography performance. *Journal of Chromatography A*, 2015. **1403**: p. 104–117.

214. Nicoud, R.M., *Chromatographic Processes*. 2015, Cambridge University Press: Cambridge, UK.
215. Hotier, G., Nicoud, R.M. and Perrut, M., Procede et dispositif de fractionnement d'un melange en lit mobile simule en presence d'un gaz comprime, d'un fluide supercritique ou d'un liquide subcritique, EP0592646 A1, 1994.
216. Gottschlich, N. and Kasche, V., Purification of monoclonal antibodies by simulated moving-bed chromatography. *Journal of Chromatography A*, 1997. **765**(2): p. 201–206.
217. Guiochon, G. and Tarafder, A., Fundamental challenges and opportunities for preparative supercritical fluid chromatography. *Journal of Chromatography A*, 2011. **1218**(8): p. 1037–1114.
218. Johannsen, M. and Brunner, G., Supercritical fluid chromatographic separation on preparative scale and in continuous mode. *The Journal of Supercritical Fluids*, 2018. **134**: p. 61–70.
219. Liang, M.-T., Liang, R.-C., Huang, L.-R., Hsu, P.-H., Wu, Y.-H. and Yen, H.-E., Separation of sesamin and sesamolin by a supercritical fluid-simulated moving bed. *American Journal of Analytical Chemistry*, 2012. **3**(12): p. 931.
220. Liang, M.-T., Liang, R.-C., Huang, L.-R., Liang, K.-Y., Chien, Y.-L. and Liao, J.-Y., Supercritical fluids as the desorbent for simulated moving bed—Application to the concentration of triterpenoids from Taiwanofugus *camphorata*. *Journal of the Taiwan Institute of Chemical Engineers*, 2014. **45**(4): p. 1225–1232.
221. Liang, M.-T., Liang, R.-C., Yu, S.-Q. and Yan, R.-A., Separation of resveratrol and emodin by supercritical fluid-simulated moving bed chromatography. *Journal of Chromatography and Separation Techniques*, 2013. **4**(3): p. 1000175.
222. Lin, C.-H., Lin, H.-W., Wu, J.-Y., Houng, J.-Y., Wan, H.-P., Yang, T.-Y. and Liang, M.-T., Extraction of lignans from the seed of *Schisandra chinensis* by supercritical fluid extraction and subsequent separation by supercritical fluid simulated moving bed. *The Journal of Supercritical Fluids*, 2015. **98**: p. 17–24.
223. Soepriatna, N., Wang, N.L. and Wankat, P.C., Standing wave design of a four-zone thermal SMB fractionator and concentrator (4-zone TSMB-FC) for linear systems. *Adsorption*, 2014. **20**(1): p. 37–52.
224. Lee, J.W. and Wankat, P.C., Thermal simulated moving bed concentrator. *Chemical Engineering Journal*, 2011. **166**(2): p. 511–522.
225. Soepriatna, N., Wang, N.H.L. and Wankat, P.C., Standing wave design of 2-zone thermal simulated moving bed concentrator (TSMBC). *Industrial & Engineering Chemistry Research*, 2015. **54**(50): p. 12646–12663.
226. Barker, P. and Deeble, R., Sequential chromatographic equipment for the separation of a wide range of organic mixtures. *Chromatographia*, 1975. **8**(2): p. 67–79.
227. Barker, P.E., Irlam, G.A. and Abusabah, E.K.E., Continuous chromatographic separation of glucose-fructose mixtures using anion-exchange resins. *Chromatographia*, 1984. **18**(10): p. 567–574.
228. Shen, B., Chen, M., Jiang, H., Zhao, Y. and Wei, F., Modeling study on a three-zone simulated moving bed without zone I. *Separation Science and Technology*, 2011. **46**(5): p. 695–701.
229. Nowak, J., Antos, D. and Seidel-Morgenstern, A., Theoretical study of using simulated moving bed chromatography to separate intermediately eluting target compounds. *Journal of Chromatography A*, 2012. **1253**: p. 58–70.
230. da Silva, F.V.S. and Seidel-Morgenstern, A., Evaluation of center-cut separations applying simulated moving bed chromatography with 8 zones. *Journal of Chromatography A*, 2016. **1456**: p. 123–136.
231. Kiwala, D., Mendrella, J., Antos, D. and Seidel-Morgenstern, A., Center-cut separation of intermediately adsorbing target component by 8-zone simulated moving bed chromatography with internal recycle. *Journal of Chromatography A*, 2016. **1453**: p. 19–33.

232. Heinonen, J., Laatikainen, M. and Sainio, T., Chromatographic fractionation of a ternary mixture with an SMB cascade process: The effect of ion exchange resin cross-linkage on separation efficiency. *Separation and Purification Technology*, 2018. **206**: p. 286–296.

233. Agrawal, G. and Kawajiri, Y., Comparison of various ternary simulated moving bed separation schemes by multi-objective optimization. *Journal of Chromatography A*, 2012. **1238**: p. 105–113.

234. Jermann, S. and Mazzotti, M., Three column intermittent simulated moving bed chromatography: 1. Process description and comparative assessment. *Journal of Chromatography A*, 2014. **1361**: p. 125–138.

235. Jermann, S., Alberti, A. and Mazzotti, M., Three-column intermittent simulated moving bed chromatography: 2. Experimental implementation for the separation of Tröger's Base. *Journal of Chromatography A*, 2014. **1364**: p. 107–116.

236. Koo, Y.M., Lee, J.W. and Kim, J.-I., Chromatographic method of using three-zone simulated moving bed process comprising partial recycle, KR100926150 B1, 2009.

237. Kim, K.-M., Song, J.-Y. and Lee, C.-H., Three-port operation in three-zone simulated moving bed chromatography. *Journal of Chromatography A*, 2014. **1340**: p. 79–89.

238. Mun, S., Enhanced separation performance of a five-zone simulated moving bed process by using partial collection strategy based on alternate opening and closing of a product port. *Industrial & Engineering Chemistry Research*, 2010. **49**(19): p. 9258–9270.

239. Subramanian, G., *Continuous Biomanufacturing: Innovative Technologies and Methods*. 2017, John Wiley & Sons: Weinheim.

240. Theoleyre, M.-A., Baudouin, S., Merrien, A., Valery, E., Ludemann-Hombourger, O., Laurent, D. and Holzer, M., Multicolumn sequential separation process, US 8182696B2, 2012.

241. Holzer, M., Osuna-Sanchez, H. and David, L., Multicolumn chromatography. *BioProcess International*, 2008. **6**(8): p. 74–84.

242. Godawat, R., Brower, K., Jain, S., Konstantinov, K., Riske, F. and Warikoo, V., Periodic counter-current chromatography–design and operational considerations for integrated and continuous purification of proteins. *Biotechnology Journal*, 2012. **7**(12): p. 1496–1508.

243. Morbidelli, M. and Müller-Späth, T., Multicolumn counter current gradient chromatography for the purification of biopharmaceuticals, in *Continuous Processing in Pharmaceutical Manufacturing*, G. Subramanian, Editor. 2015, John Wiley & Sons: Weinheim. p. 227–254.

244. Ng, C.K., Rousset, F., Valery, E., Bracewell, D.G. and Sorensen, E., Design of high productivity sequential multi-column chromatography for antibody capture. *Food and Bioproducts Processing*, 2014. **92**(2): p. 233–241.

245. Pfister, D., David, L., Holzer, M. and Nicoud, R.-M., Designing affinity chromatographic processes for the capture of antibodies. Part I: A simplified approach. *Journal of Chromatography A*, 2017. **1494**: p. 27–39.

246. Bangtsson, P., Estrada, E., Lacki, K. and Skoglar, H., Method in a chromatography system, US 20120091063 A1, 2012.

247. Mahajan, E., George, A. and Wolk, B., Improving affinity chromatography resin efficiency using semi-continuous chromatography. *Journal of Chromatography A*, 2012. **1227**: p. 154–162.

248. Pollock, J., Bolton, G., Coffman, J., Ho, S.V., Bracewell, D.G. and Farid, S.S., Optimising the design and operation of semi-continuous affinity chromatography for clinical and commercial manufacture. *Journal of Chromatography A*, 2013. **1284**: p. 17–27.

249. Warikoo, V., Godawat, R., Brower, K., Jain, S., Cummings, D., Simons, E., Johnson, T., Walther, J., Yu, M. and Wright, B., Integrated continuous production of recombinant therapeutic proteins. *Biotechnology and Bioengineering*, 2012. **109**(12): p. 3018–3029.

250. Chmielowski, R.A., Mathiasson, L., Blom, H., Go, D., Ehring, H., Khan, H., Li, H., Cutler, C., Lacki, K. and Tugcu, N., Definition and dynamic control of a continuous chromatography process independent of cell culture titer and impurities. *Journal of Chromatography A*, 2017. **1526**: p. 58–69.

251. Aumann, L. and Morbidelli, M., Method and device for chromatographic purification, EP1716900, 2006.

252. Aumann, L. and Morbidelli, M., A continuous multicolumn countercurrent solvent gradient purification (MCSGP) process. *Biotechnology and Bioengineering*, 2007. **98**(5): p. 1043–1055.

253. Ströhlein, G., Aumann, L., Mazzotti, M. and Morbidelli, M., A continuous, countercurrent multi-column chromatographic process incorporating modifier gradients for ternary separations. *Journal of Chromatography A*, 2006. **1126**(1): p. 338–346.

254. Aumann, L. and Morbidelli, M., A semicontinuous 3-column countercurrent solvent gradient purification (MCSGP) process. *Biotechnology and Bioengineering*, 2008. **99**(3): p. 728–733.

255. Krättli, M., Steinebach, F. and Morbidelli, M., Online control of the twin-column countercurrent solvent gradient process for biochromatography. *Journal of Chromatography A*, 2013. **1293**: p. 51–59.

256. Steinebach, F., Krättli, M., Storti, G. and Morbidelli, M., Equilibrium theory based design space for the multicolumn countercurrent solvent gradient purification process. *Industrial & Engineering Chemistry Research*, 2017. **56**(45): p. 13482–13489.

257. Vogg, S., Ulmer, N., Souquet, J., Broly, H. and Morbidelli, M., Experimental evaluation of the impact of intrinsic process parameters on the performance of a continuous chromatographic polishing unit (MCSGP). *Biotechnology Journal*, 2019: p. 1800732.

258. Steinebach, F., Ulmer, N., Decker, L., Aumann, L. and Morbidelli, M., Experimental design of a twin-column countercurrent gradient purification process. *Journal of Chromatography A*, 2017. **1492**: p. 19–26.

259. Müller-Späth, T., Aumann, L., Ströhlein, G., Kornmann, H., Valax, P., Delegrange, L., Charbaut, E., Baer, G., Lamproye, A. and Jöhnck, M., Two step capture and purification of IgG2 using multicolumn countercurrent solvent gradient purification (MCSGP). *Biotechnology and Bioengineering*, 2010. **107**(6): p. 974–984.

260. Müller-Späth, T., Ulmer, N., Aumann, L., Ströhlein, G., Bavand, M., Hendriks, L.J., de Kruif, J., Throsby, M. and Bakker, A., Purifying common light-chain bispecific antibodies. *BioProcess International*, 2013. **11**(5): p. 36–45.

261. Broughton, D.B. and Gerhold, C.G., Distributing valve, US 3192954 A, 1965.

262. Nicoud, R.M., The simulated moving bed: A powerful chromatographic process. *LC-GC International*, 1992. **5**(5): p. 4347.

263. Hotier, G. and Nicoud, R.M., Separation process by simulated moving bed chromatography with correction for dead volume by desynchronisation of periods, EP0688589 B1, 2000.

264. Priegnitz, J.W., Small scale simulated moving bed separation apparatus and process, US 5470464 A, 1995.

265. Negawa, M. and Shoji, F., Simulated moving bed separation system, US 5456825 A, 1995.

266. Gomes, P.S., Zabkova, M., Zabka, M., Minceva, M. and Rodrigues, A.E., Separation of chiral mixtures in real SMB units: The FlexSMB-LSRE®. *AIChE Journal*, 2010. **56**(1): p. 125–142.

267. Wang, N.H.L. and Chin, C.Y., Versatile simulated moving bed systems, US 7141172 B2, 2006.

268. Ahlgren, B.K., Snyder, C.B. and Fawaz, I., Fluid-directing multiport rotary valve, US 6431202 B1, 2002.

269. Hotier, G., Rotary valve, US 6537451 B1, 2003.

270. Oroskar, A.R., Rotary valve apparatus for simulated moving bed separations, US 8349175 B1, 2013.
271. Rossiter, G.J. and Riley, R.J., Fluid-solid contacting apparatus, US 5676826 A, 1997.
272. Ketola, S., Fluid treatment device, US 20120152385 A1, 2012.
273. Bisschops, M.A.T., Pennings, J.A.M. and Tijsterman, J.A., Device for chromatographic separations, US 8828234 B2, 2014.
274. Wilke, A.P., Mierendorf, R.C., Grabski, A.C. and Cheng, C., Valve block assembly, US 8196603 B2, 2012.
275. Moran, M.G. and Fulton, S.P., Disposable chromatography valves and system, US 8920645 B2, 2014.

Index

Printed in the United States
by Baker & Taylor Publisher Services